Lecture Notes in Computer Science 13617

More information about this series at https://link.springer.com/bookseries/558

Diego Arroyuelo · Barbara Poblete (Eds.)

String Processing
and Information Retrieval

29th International Symposium, SPIRE 2022
Concepción, Chile, November 8–10, 2022
Proceedings

Springer

Editors
Diego Arroyuelo (ID)
Universidad Técnica Federico Santa María
Valparaíso, Chile

Millennium Institute for Foundational
Research on Data
Santiago, Chile

Barbara Poblete
Universidad de Chile
Santiago, Chile

ISSN 0302-9743 ISSN 1611-3349 (electronic)
Lecture Notes in Computer Science
ISBN 978-3-031-20642-9 ISBN 978-3-031-20643-6 (eBook)
https://doi.org/10.1007/978-3-031-20643-6

This Springer imprint is published by the registered company Springer Nature Switzerland AG
The registered company address is: Gewerbestrasse 11, 6330 Cham, Switzerland

Preface

The 29th International Symposium on String Processing and Information Retrieval, SPIRE 2022, was held during November 8–10, 2022, in Concepción, Chile. SPIRE started in 1993 as the South American Workshop on String Processing, and therefore it was held in Latin America until 2000. Then, SPIRE moved to Europe, and from then on it has been held in Australia, Japan, the UK, Spain, Italy, Finland, Portugal, Israel, Brazil, Chile, Colombia, Mexico, Argentina, Bolivia, Peru, the USA, and France. In this edition, SPIRE was back in Chile, continuing the long and well-established tradition of encouraging high-quality research at the broad nexus of string processing, information retrieval, and computational biology. After two years running online (because of the COVID-19 pandemic), this year SPIRE returned to onsite mode (allowing also online attendants).

This volume contains the accepted papers presented in SPIRE 2022. There was a total of 43 submissions. We thank all authors who submitted their work for consideration to SPIRE 2022. Each submission received at least three single blind reviews and, after intensive discussion, the Program Committee decided to accept 23 papers. These were classified into seven tracks: string algorithms, string data structures, string compression, information retrieval, computational biology, space-efficient data structures, and pattern matching. Authors of accepted papers come from 14 countries across four continents (Asia, Europe, North America, and South America). We thank the authors for their valuable contributions and presentations at the conference. We also want to especially thank the Program Committee members and the external reviewers for their valuable work during the review and discussion phases. The SPIRE 2022 program also included two invited talks:

- "De Bruijn Graphs: Solving Biological Problems in Small Space", by Leena Salmela, and
- "LZ-End Parsing: Upper Bounds", by Dominik Kempa,

and the tutorial "Graph Databases" by Aidan Hogan and Domagoj Vrgoč. We thank them for accepting our invitation and for their enlightening presentations.

We are also grateful to the organizing committee, chaired by José Fuentes and Cecilia Hernández (Universidad de Concepción), whose excellent work allowed SPIRE 2022 to become a reality. Also, we want to thank the financial support of the Institute for Foundational Research on Data (IMFD), the Centre for Biotechnology and Bioengineering (CeBiB), the Vicerrectoría and the Facultad de Ingeniería of Universidad de Concepción, and R9 Ingeniería, which was crucial to fund the invited speakers, tutorial, streaming service, free student registration (to encourage onsite student participation, after two years of online activities), and the auditorium for the conference.

To complete the event, SPIRE 2022 had a Best Paper Award sponsored by Springer, which was announced at the conference.

November 2022   Diego Arroyuelo
                                                                Barbara Poblete

Organization

Program Committee Chairs

Diego Arroyuelo   Universidad Técnica Federico Santa María and Millennium Institute for Foundational Research on Data, Chile

Barbara Poblete   University of Chile, Chile, and Amazon, USA

Program Committee

Amihood Amir	Bar-Ilan University, Israel
Ricardo Baeza-Yates	Northeastern University, USA, Pompeu Fabra University, Spain, and University of Chile, Chile
Hideo Bannai	Tokyo Medical and Dental University, Japan
Altigran da Silva	Universidade Federal do Amazonas, Brazil
Antonio Fariña	University of A Coruña, Spain
Gabriele Fici	University of Palermo, Italy
Travis Gagie	Dalhousie University, Canada
Pawel Gawrychowski	University of Wroclaw, Poland
Marcos Goncalves	Federal University of Minas Gerais, Brazil
Inge Li Gørtz	Technical University of Denmark, Denmark
Meng He	Dalhousie University, Canada
Wing-Kai Hon	National Tsing Hua University, Taiwan
Shunsuke Inenaga	Kyushu University, Japan
Dominik Köppl	Tokyo Medical and Dental University, Japan
Thierry Lecroq	University of Rouen Normandy, France
Zsuzsanna Lipták	University of Verona, Italy
Felipe A. Louza	Universidade Federal de Uberlândia, Brazil
Giovanni Manzini	University of Pisa, Italy
Joao Meidanis	University of Campinas and Scylla Bioinformatics, Brazil
Alistair Moffat	University of Melbourne, Australia
Viviane P. Moreira	Universidade Federal do Rio Grande do Sul, Brazil
Gonzalo Navarro	University of Chile, Chile
Nadia Pisanti	University of Pisa, Italy
Solon Pissis	Centrum Wiskunde & Informatica, The Netherlands
Nicola Prezza	Ca' Foscari University of Venice, Italy
Simon Puglisi	University of Helsinki, Finland
Rajeev Raman	University of Leicester, UK
Kunihiko Sadakane	The University of Tokyo, Japan

Srinivasa Rao Satti	Norwegian University of Science and Technology, Norway
Marinella Sciortino	University of Palermo, Italy
Diego Seco	University of A Coruña, Spain
Sharma V. Thankachan	University of Central Florida, USA
Rossano Venturini	University of Pisa, Italy
Nivio Ziviani	Federal University of Minas Gerais, Brazil

Steering Committee

Ricardo Baeza-Yates	Northeastern University, USA, Pompeu Fabra University, Spain, and University of Chile, Chile
Christina Boucher	University of Florida, USA
Nieves R. Brisaboa	University of A Coruña, Spain
Thierry Lecroq	University of Rouen Normandy, France
Simon Puglisi	University of Helsinki, Finland
Berthier Ribeiro-Neto	Google Inc. and Federal University of Minas Gerais, Brazil
Sharma Thankachan	University of Central Florida, USA
Hélène Touzet	CNRS, France
Nivio Ziviani	Federal University of Minas Gerais, Brazil

Organizing Committee

José Fuentes-Sepúlveda	Universidad de Concepción and Millennium Institute for Foundational Research on Data, Chile
Cecilia Hernández	Universidad de Concepción, Chile

Additional Reviewers

Paniz Abedin	Daniel Gibney
Fabiano Belém	Sara Giuliani
Luciana Bencke	Adrián Gómez-Brandón
Giulia Bernardini	Keisuke Goto
Itai Boneh	Veronica Guerrini
Davide Cenzato	Tomohiro I
Dustin Cobas	Michael Itzhaki
Guillermo De Bernardo	Varunkumar Jayapaul
Daniel Xavier De Sousa	Seungbum Jo
Jonas Ellert	Serikzhan Kazi
Massimo Equi	Eitan Kondratovsky
Celso França	William Kuszmaul
José Fuentes-Sepúlveda	Francesco Masillo
Younan Gao	Takuya Mieno
Samah Ghazawi	Yuto Nakashima

Abstracts of Invited Talks

De Bruijn Graphs: Solving Biological Problems in Small Space

Leena Salmela (iD)

Department of Computer Science, University of Helsinki, Helsinki, Finland
leena.salmela@helsinki.fi

Abstract. De Bruijn graphs have become a standard data structure in analysing sequencing data due to its ability to represent the information in a sequencing read set in small space. They represent the sequencing reads by the k-mers, i.e., substrings of length k occurring in the reads. Classically, the edges of a de Bruijn graph are defined to be the k-mers and the nodes are the $k - 1$-length prefixes and suffixes of the k-mers. The construction of a de Bruijn graph starts by counting the k-mers occurring in the reads. Many good methods exist for extracting exact k-mers from read data and counting the number of their occurrences. However, sequencing read sets can contain a significant number of sequencing errors, which limits the usefulness of counting exact k-mers to short k-mers. Recently, we have developed methods for extracting longer k-mers from noisy data by using spaced seeds and strobemers.

De Bruijn graphs were originally introduced for solving the genome assembly problem, where the goal is to reconstruct the genome based on sequencing reads. In practice, genome assembly is solved with de Bruijn graphs by reporting unitigs, which are non-branching paths in the de Bruijn graphs. The choice of k is a crucial matter in de-Bruijn-graph-based genome assembly. A too small k will make the graph tangled, resulting in short unitigs, while a too large k will fragment the graph, again resulting in short unitigs. A variable-order de Bruijn graph, which represents de Bruijn graphs of all orders k in a single data structure, has been presented as a solution to the choice of k. However, it is not clear how the definition of unitigs can be extended to variable-order de Bruijn graphs.

In this talk, we present a robust definition of assembled sequences in variable-order de Bruijn graphs and an algorithm for enumerating them. Apart from genome assembly, de Bruijn graphs are used in many other problems such as sequencing error correction, reference free variant calling, indexing read sets, and so on. At the end of this talk, we will review some of these applications and their de-Bruijn-graph-based solutions.

Keywords: de Bruijn graph · k-mer · Genome assembly

Supported by Academy of Finland (grant 323233).

LZ-End Parsing: Upper Bounds and Algorithmic Techniques

Dominik Kempa

Stony Brook University,
Stony Brook, New York, USA
kempa@cs.stonybrook.edu

Abstract. Lempel–Ziv (LZ77) compression is the most commonly used lossless compression algorithm. The basic idea is to greedily break the input string into blocks (called "phrases"), every time forming as a phrase the longest prefix of the unprocessed part that has an earlier occurrence. In 2010, Kreft and Navarro introduced a variant of LZ77 called LZ-End, that additionally requires the previous occurrence of each phrase to end at the boundary of an already existing phrase. Due to its excellent practical performance as a compression algorithm and a compressed index, they conjectured that it achieves a compression that can be provably upper-bounded in terms of the LZ77 size. Despite the recent progress in understanding such relation for other compression algorithms (e.g., the run-length encoded Burrows–Wheeler transform), no such result is known for LZ-End. In this talk, we give an overview of the recent progress on the above problem. More precisely, we prove that for any string of length n, the number z_e of phrases in the LZ-End parsing satisfies $z_e = \mathcal{O}(z \log^2 n)$, where z is the number of phrases in the LZ77 parsing. This is the first non-trivial upper bound on the size of LZ-End parsing in terms of LZ77, and it puts LZ-End among the strongest dictionary compressors. Using our techniques, we also derive bounds for other variants of LZ-End and with respect to other compression measures. Our second contribution is a data structure that implements random access queries to the text in $\mathcal{O}(z_e)$ space and $\mathcal{O}(\text{poly} \log n)$ time. This is the first linear-size structure on LZ-End that efficiently implements such queries. All previous data structures either incur a logarithmic penalty in the space or have slow queries. We also show how to extend these techniques to support longest-common-extension (LCE) queries. This work was carried out in collaboration with Barna Saha and was presented at the 2022 ACM-SIAM Symposium on Discrete Algorithms (SODA 2022).

Keywords: LZ-End · LZ77 · Dictionary compression

This work was supported by NIH HG011392, NSF DBI-2029552, Simons Foundation Junior Faculty Fellows Grant, NSF CAREER Award 1652303, NSF 1909046, NSF HDR TRIPODS Grant 1934846, and an Alfred P. Sloan Fellowship.

Contents

String Algorithms

Subsequence Covers of Words

Panagiotis Charalampopoulos[1]([✉])[ID], Solon P. Pissis[2,3][ID],
Jakub Radoszewski[4][ID], Wojciech Rytter[4][ID], Tomasz Waleń[4][ID],
and Wiktor Zuba[2][ID]

[1] Birkbeck, University of London, London, UK
p.charalampopoulos@bbk.ac.uk
[2] CWI, Amsterdam, The Netherlands
{solon.pissis,wiktor.zuba}@cwi.nl
[3] Vrije Universiteit, Amsterdam, The Netherlands
[4] University of Warsaw, Warsaw, Poland
{jrad,rytter,walen}@mimuw.edu.pl

Abstract. We introduce subsequence covers (s-covers, in short), a new
type of covers of a word. A word C is an *s-cover* of a word S if the
occurrences of C in S as subsequences cover all the positions in S.

The s-covers seem to be computationally much harder than standard
covers of words (cf. Apostolico et al., *Inf. Process. Lett.* 1991), but, on
the other hand, much easier than the related shuffle powers (Warmuth
and Haussler, *J. Comput. Syst. Sci.* 1984).

We give a linear-time algorithm for testing if a candidate word C is
an s-cover of a word S over a polynomially-bounded integer alphabet.
We also give an algorithm for finding a shortest s-cover of a word S,
which in the case of a constant-sized alphabet, also runs in linear time.
Furthermore, we complement our algorithmic results with a lower and an
upper bound on the length of a longest word without non-trivial s-covers,
which are both exponential in the size of the alphabet.

Keywords: String algorithms · Combinatorics on words · Covers ·
Shuffle powers · Subsequence covers

1 Introduction

The problem of computing covers in a word is a classic one in string algorithms;
see [1,2,11] and also [5] for a recent survey. In its most basic type, we say that
a word C is a *cover* of another longer word S if every position of S lies within
some occurrence of C as a factor (subword) in S [1].

In this paper we introduce a new type of cover, in which instead of subwords
we take subsequences (scattered subwords). Such covers turn out to be related to
shuffle problems [4,12,13]. Formally the new type of cover is defined as follows:

J. Radoszewski and T. Waleń—Supported by the Polish National Science Center, grant
no. 2018/31/D/ST6/03991.

Definition 1. *A word C is a* **subsequence cover** *(s-cover, in short) of a word S if every position in S belongs to an occurrence of C as a subsequence in S. We also write $S \in C^\otimes$, where $C^\otimes$ is the set of words having C as an s-cover.*

We say that an s-cover C of a word S is *non-trivial* if $|C| < |S|$. A word S is called *s-primitive* if it has no non-trivial s-cover.

An example s-primitive word is the Zimin word S_k [10], that is, a word over alphabet $\{1, \ldots, k\}$ given by recurrences of the form

$$S_1 = 1, \quad S_i = S_{i-1} i S_{i-1} \text{ for } i > 1.$$

The word S_k has length $2^k - 1$.

Clearly, if a word C is a (standard) cover of a word S, then C is an s-cover of S. However the converse implication is false: ab is an s-cover of aab, but is not a standard cover. For another example of an s-cover, see the following example.

Example 1. Figure 1 shows that $C = abcab$ is an s-cover of $S = abcbacab$. In fact C is a shortest s-cover of S.

$$
\begin{array}{llllllll}
a & b & c & a & & b & & \\
a & & & b & c & a & b & \\
a & b & c & & & a & b & \\
a & b & c & b & a & c & a & b
\end{array}
$$

Fig. 1. An illustration of the fact that $C = abcab$ is an s-cover of $S = abcbacab$.

We now provide some basic definitions and notation. An *alphabet* is a finite nonempty set of elements called *letters*. A *word* S is a sequence of letters over some alphabet. For a word S, by $|S|$ we denote its *length*, by $S[i]$, for $i = 0, \ldots, |S| - 1$, we denote its ith letter, and by $Alph(S)$ we denote the set of letters in S, i.e., $\{S[0], \ldots, S[|S| - 1]\}$. The *empty word* is the word of length 0.

For any two words U and V, by $U \cdot V = UV$ we denote their concatenation. For a word $S = PUQ$, where P, U, and Q are words, U is called a *factor* of S; it is called a *prefix* (resp. *suffix*) if P (resp. Q) is the empty word. By $S[i \mathinner{.\,.} j]$ we denote a factor $S[i] \ldots S[j]$ of S; we omit i if $i = 0$ and j if $j = |S| - 1$.

A word V is a *k-power* of a word U, for integer $k \geq 0$, if V is a concatenation of k copies of U, in which case we denote it by U^k. It is called a *square* if $k = 2$.

Remark 1. If a word S contains a non-empty square factor U^2, then S has a non-trivial s-cover resulting by removing any of the two consecutive copies of U. Further, if a word S has a factor being a gapped repeat UVU (see [9]), such that $Alph(V) \subseteq Alph(U)$, then S has a non-trivial s-cover resulting by removing VU from the gapped repeat. Moreover, if C is an s-cover of S, then C is an s-cover of S concatenated with any concatenation of suffixes of C.

A different version of covers, where we require that position-subsequences are disjoint, is the *shuffle closure* problem. The shuffle closure of a word U, denoted by $U^\odot$, is the set of words resulting by interleaving many copies of U; see [13]. The words in $U^\odot$ are sometimes called *shuffle powers* of U.

The following problems are NP-hard for constant-sized alphabets:

(1) Given two words U and S, test if $S \in U^\odot$; see [13].
(2) Given a word S, check if there exists a word U such that $|U| = |S|/2$ and $S \in U^\odot$ (this was originally called the *shuffle square* problem); see [4]. An NP-hardness proof for a binary alphabet was recently given in [3].
(3) Given a word S, find a shortest word U such that $S \in U^\odot$; its hardness is trivially reduced from (2).

The following observation links s-covers and shuffle closures.

Observation 1. *Let S be a word of length n. Then*

$$S \in C^\otimes \Rightarrow \exists r_0, r_1 \ldots, r_{n-1} \in \mathbb{Z}_+ : S[0]^{r_0} S[1]^{r_1} \ldots S[n-1]^{r_{n-1}} \in C^\odot.$$

In this paper we show that problems similar to (1) and (3) for s-covers, when we replace $\odot$ by $\otimes$, are tractable: notably, the first one is solved in linear time for any polynomially-bounded integer alphabet; and the last one in linear time for any constant-sized alphabet.

Our Results and Paper Organization:

– In Sect. 2 we present a linear-time algorithm for checking if a word C is an s-cover of a word S, assuming that C and S are over a polynomially-bounded integer alphabet $\{0, \ldots, |S|^{\mathcal{O}(1)}\}$. We also discuss why an equally efficient algorithm for this problem without this assumption is unlikely.

– Let $\gamma(k)$ denote the length of a longest s-primitive word over an alphabet of size k. In Sect. 3 we present general bounds on this function as well as its particular values for small values of k.

– In Sect. 4 we show that computing a non-trivial s-cover is fixed-parameter tractable for parameter $k = |Alph(S)|$. In particular we obtain a linear-time algorithm for computing a shortest s-cover of a word over a constant-sized alphabet.

– Finally in Sect. 5 we explore properties of s-covers that are significantly different from properties of standard covers. In particular, we show that a word can have exponentially many different shortest s-covers, which implies that computing all shortest s-covers of a word (over a superconstant alphabet) requires exponential time.

2   Testing if a Word is an s-Cover

Consider words $C = C[0\mathinner{.\,.}m-1]$ and $S = S[0\mathinner{.\,.}n-1]$. We would like to check whether C is an s-cover of S.

Let sequences $FirstOcc = (p_1, p_2, \ldots, p_m)$ and $LastOcc = (q_1, q_2, \ldots, q_m)$ be the lexicographically first and last position-subsequences of S containing C, where $p_1 = 0$ and $q_m = n - 1$. If there are no such subsequences of positions then C is not an s-cover, so we assume they exist and are well defined.

For all $i \in \{0, \ldots, n-1\}$, we define

$$Right[i] = \min(\{j \ : \ q_j > i\} \cup \{m+1\}),$$
$$Pref[i] = \max(\{j \ : \ p_j \leq i \ \wedge \ S[p_j] = S[i]\} \cup \{0\}).$$

Intuitively, if position i is in any subsequence occurrence of C in S, then there is a subsequence occurrence of C in S that consists of the prefix of $FirstOcc$ of length $Pref[i]$ and an appropriate suffix of $LastOcc$. All we have to do is check, for all i, whether such a pair of prefix and suffix exists. See Fig. 2 for an illustration of the argument and Lemma 1 for a formal statement of the condition that needs to be satisfied.

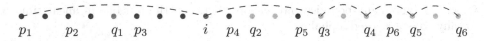

$$p_1 \qquad p_2 \qquad q_1 \ p_3 \qquad\qquad i \quad p_4 \ q_2 \qquad\quad p_5 \ q_3 \qquad\quad q_4 \ p_6 \ q_5 \qquad q_6$$

Fig. 2. Assume that for some words C and S the sequences $FirstOcc$ (red) and $LastOcc$ (green) are as in the figure. Further assume that $Pref[i] = 2$ (i.e., we have $S[i] = S[p_2] \neq S[p_3]$). As shown, we have $Right[i] = 2$. Thus, we have $Right[i] \leq Pref[i] + 1$ and consequently the position i is covered by an occurrence of C as a subsequence using positions $(p_1, i, q_3, q_4, q_5, q_6)$. (Color figure online)

Lemma 1. *Let us assume that $FirstOcc$ and $LastOcc$ are well defined. Then C is an s-cover of S if and only if for each position $0 \leq i \leq n - 1$ we have: $Pref[i] > 0$ and $Right[i] \leq Pref[i] + 1$.*

Proof. First, observe that if $Pref[i] = 0$ for any i, then C is not an s-cover of S. This follows from the greedy computation of $FirstOcc$, which implies that the prefix of C that precedes the first occurrence of $S[i]$ in C does not have a subsequence occurrence in $S[0\mathinner{.\,.}i-1]$; else, i would be in $FirstOcc$, a contradiction.

We henceforth assume that $Pref[i] > 0$ for every i and show that, in this case, C is an s-cover of S if and only if $Right[i] \leq Pref[i] + 1$ for all $i \in \{0, \ldots, n-1\}$.

($\Leftarrow$) Assume that $Right[i] \leq Pref[i] + 1$. In this case position i can be covered by a subsequence occupying positions $p_1, \ldots, p_{j-1}, i, q_{j+1}, \ldots, q_m$, for $j = Pref[i]$. As $S[p_j] = S[i]$ this subsequence is equal to C, and as $p_j \leq i$ and $q_{j+1} > i$ ($j+1 \geq Right[i]$) those positions form an increasing sequence (that is, we obtain a valid subsequence).

($\Rightarrow$) On the other hand assume that for some j there exists an increasing sequence

$$r_1, r_2, \ldots, r_{j-1}, i, r_{j+1}, \ldots, r_m,$$

such that $S[r_1]S[r_2]\ldots S[r_{j-1}]S[i]S[r_{j+1}]\ldots S[r_m] = C$.

By induction for $k = 1, \ldots, j$, $r_k \geq p_k$ (including $r_j = i$) and for $k = m, \ldots, j + 1$, $r_k \leq q_k$. But this means that $Pref[i] \geq j$ and $Right[i] \leq j + 1$. Hence $Right[i] \leq Pref[i] + 1$. This completes the proof. $\qquad\square$

The sequence $FirstOcc$ can be computed with a simple left-to-right pass over S and C; the computation of $LastOcc$ is symmetric. The table $Right$ can be computed via a right-to-left pass. The table $Pref[i]$ is computed on-line using an additional table $PRED$ indexed by the letters of the alphabet. The algorithm is formalized in the following pseudocode.

Algorithm 1: $TEST(C, S)$

Input: word $C = C[0 .. m-1]$ and word $S = S[0 .. n-1]$
Output: $true$ if and only if C is an s-cover of S

compute $FirstOcc = (p_1, \ldots, p_m)$ and $LastOcc = (q_1, \ldots, q_m)$

▷ compute $Right$
  $k := m + 1$
  for $i := n - 1$ **down to** 0 **do**
    $Right[i] := k$
    if $k > 1$ **and** $i = q_{k-1}$ **then** $k := k - 1$

▷ compute $Pref$
  $PRED[c] := 0 \; \forall c \in \Sigma$
  $k := 1$
  for $i := 0$ **to** $n - 1$ **do**
    if $i = p_k$ **then**
      $PRED[S[i]] := k$
      if $k < m$ **then** $k := k + 1$
    $Pref[i] := PRED[S[i]]$

return $\forall_{i=0,\ldots,n-1} \; (Pref[i] > 0 \text{ and } Right[i] \leq Pref[i] + 1)$

The correctness of the algorithm follows from Lemma 1 (inspect also Fig. 2). Note that, under the assumption of a polynomially-bounded integer alphabet, the table $PRED$ can be initialized and updated deterministically in linear total time by first sorting the letters of S. We thus arrive at the following result.

Theorem 1. *Given words C and S over an integer alphabet $\{0, \ldots, |S|^{\mathcal{O}(1)}\}$, we can check if C is an s-cover of S in $\mathcal{O}(|S|)$ time.*

In the standard setting (cf. [2]), one can check if a word C is a cover of a word S—what is more, find the shortest cover of S—in linear time for any (non-necessarily integer) alphabet. We show below that the existence of such an algorithm for testing a candidate s-cover is rather unlikely.

Let us introduce a slightly more general version of the s-cover testing problem in which, if C is an s-cover of S, we are to say, for each position i in S, which position j of C is actually used to cover $S[i]$; if there is more than one such position j, any one of them can be output. Let us call this problem the *witness s-cover testing* problem. In particular, our algorithm solves the witness s-cover testing problem with the answers stored in the *Pref* array. Actually it is hard to imagine an algorithm that solves the s-cover testing problem and not the witness version of it. We next give a comparison-based lower bound for the latter.

Theorem 2. *The witness s-cover testing problem for a word S of length n requires $\Omega(n \log n)$ time in the comparison model.*

Proof. Let us consider a word C of length m that is composed of m distinct letters and a family of words of the form $S = CTC$, where T is a word of length m such that $Alph(T) \subseteq Alph(C)$. Then C is an s-cover of each such word S. Each choice of word T implies a different output to the witness s-cover testing problem on C and S. There are m^m different outputs, so a decision tree for this problem must have depth $\Omega(\log m^m) = \Omega(m \log m) = \Omega(n \log n)$. $\square$

Let us further notice that even if C turns out not to be an s-cover of S, our algorithm actually computes the positions of S that can be covered using occurrences of C (they are exactly the positions i for which $Pref[i] > 0$ and $Right[i] \leq Pref[i]+1$). Hence our algorithm may be useful to find partial variants of s-covers, defined analogously as for the standard covers [6–8].

3   Maximal Lengths of s-Primitive Words

Let us recall that $\gamma(k)$ denotes the length of a longest s-primitive word over an alphabet of size k. It is obvious that $\gamma(2) = 3$; the longest s-primitive binary words are aba and bab. The case of ternary words is already more complicated; we study it in Sect. 3.1. General bounds on the function $\gamma(k)$ are shown in Sect. 3.2. A discussion on computing $\gamma(k)$ for small $k > 3$ is presented in Sect. 3.3. In particular, we were not able to compute the exact value of $\gamma(5)$.

3.1   Ternary Alphabet

Fact 1 $\gamma(3) = 8$.

Proof. The word $S = abcabacb$ is of length 8 and it is s-primitive, hence $\gamma(3) \geq 8$.

We still have to show that each 3-ary word of length 9 is not s-primitive (there are 19683 ternary words). The number of words to consider is substantially reduced by observing that relevant words are square-free and do not contain the structure specified in the following claim.

Claim. If a word S over a ternary alphabet contains a factor of the form $abXbc$ for some (maybe empty) word X and different letters a, b, c, then it is not s-primitive.

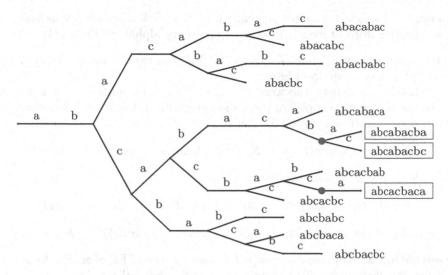

Fig. 3. A trie of all ternary square-free words starting with ab, truncated at words that are not s-primitive (in leaves). Only one word in a leaf (*abcabacba*) does not contain the structure specified in the proof of Fact 1, but it still has a non-trivial s-cover *abcba*. The trie has depth 9 (the leaves with words of length 9 are shown in frames and the internal nodes of depth 8 corresponding to s-primitive words are drawn as green circles), so $\gamma(3) = 8$. (Color figure online)

Proof. The factor $abXbc$ has abc as its s-cover, and thus it is not s-primitive. Consequently, the whole word S is not s-primitive.   □

Figure 3 shows a trie of all ternary square-free words starting with ab, truncated at words that are not s-primitive (in leaves). The words in all leaves but one contain the structure from the claim, and for the remaining word, a non-trivial s-cover can be easily given. The trie shows that words of length 9 over a ternary alphabet are not s-primitive.   □

3.2   General Alphabet

Definition 2. *For a word S over alphabet $Alph(S)$ of size k, let first(S) (resp. last(S)) denote the length-k word containing all the letters of $Alph(S)$ in the order of their first (resp. last) occurrence in S.*

Example 2. first($\underline{aba}d\underline{bc}d$) = abdc, last($ab\underline{adbcd}$) = abcd.

Lemma 2. *Let C be an s-cover of S. Then first(C) = first(S) and last(C) = last(S).*

Proof. Assume that letter a appears before letter b in first(S), but after letter b in first(C). Then $Pref[i] = 0$ (see Sect. 2 for the definition) for $i = \min\{j : S[j] = a\}$. This proves that first(C) = first(S); a proof that last(C) = last(S) follows by symmetry.   □

Example 3. Using a computer one can check that $S = abacadbabdcabcbadac$ is an s-primitive word of length 19 over a quaternary alphabet. Thus $\gamma(4) \geq 19$.

For a word X we define X_- (resp. X^-) as the word obtained from X by deleting the first (resp. last) letter.

By $shrink(S)$ we denote the word obtained from S by merging any non-zero number of consecutive copies of the same letter into just one copy. For example $shrink(abbacccbdd) = abacbd$. We define

$$\mathsf{FaLaFeL}(S) = shrink(F \cdot L^- \cdot F_- \cdot L), \text{ where } F = first(S), L = last(S).$$

Example 4. For $S = ababbacbcaabb$ we have

$$F = first(S) = abc, \ L = last(S) = cab, \ F_- = bc, \ L^- = ca, \text{ and}$$

$$shrink(FL) = abcab, \quad \mathsf{FaLaFeL}(S) = shrink(abc\,ca\,bc\,cab) = abc\,a\,bc\,ab.$$

Observation 2. *The word $shrink(FL)$ is an s-cover of $\mathsf{FaLaFeL}(S)$. However, it is possible that $shrink(FL)$ is an s-cover of S, while $\mathsf{FaLaFeL}(S)$ is not (as in the example).*

Lemma 3. *If the word $\mathsf{FaLaFeL}(S)$ is a subsequence of S, then $shrink(FL)$ is an s-cover of S.*

Proof. We need to show that each position i of S is covered by an occurrence of $shrink(FL)$ as a subsequence.

There exists a position j in S such that $shrink(FL^-)$ is a subsequence of $S[..j]$ and $shrink(F_-L)$ is a subsequence of $S[j..]$. We can assume that $i \leq j$; the other case is symmetric.

Let p be the index such that $F[p] = S[i]$. It suffices to argue that:

(1) $F[..p-1]$ is a subsequence of $S[..i-1]$; and

(2) $shrink(FL)[p+1..]$ is a subsequence of $S[i+1..]$.

Point (1) follows by the definition of $F = first(S)$.

As for point (2), if $i < j$, then $S[i+1..]$ has a subsequence $shrink(F_-L)$ by the definition of j and $shrink(FL)[p+1..]$ is a suffix of $shrink(F_-L)$.

If $p > 0$, then $S[i..]$ has a subsequence $shrink(F_-L)$ and so $S[i+1..]$ has a subsequence $shrink(FL)[2..]$.

Finally if $i = j$ and $p = 0$, then $S[i+1..]$ has a subsequence $shrink(F_-L)$ because $F_-[0] \neq F[0] = S[i]$. $\qquad\square$

We will apply the following lemma for $Z = \mathsf{FaLaFeL}(S)$.

Lemma 4. *Let S, Z be words and x be a positive integer such that $|Alph(S)| = k$, $|S| = 2kx + 1$ and $|Z| \leq 4k - 2$. We assume that each factor of S of length $x + 1$ contains all k letters, and the length-$(x+1)$ prefix/suffix of S contains, as a subsequence, the length-k prefix/suffix of Z, respectively. If $shrink(Z) = Z$, then S contains Z as a subsequence.*

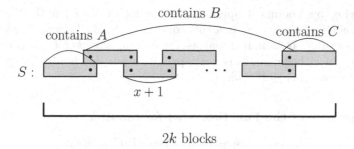

Fig. 4. Illustration of the proof of Lemma 4. Let $Z = ABC$ where $|A| = |C| = k$. Each block represents a factor of length $x + 1$ containing all letters and starting at a given position. The blocks overlap by one letter. We have $|S| = 2kx + 1$.

Proof. Let us cover S with $2k$ blocks, each of length $x + 1$, with overlaps of one position between consecutive blocks; see Sect. 4.

Let $Z = ABC$ where $|A| = |C| = k$. By the assumption of the lemma, the first and the last block in S contain A and C as a subsequence, respectively. Let us choose some $2k$ positions in S that form these occurrences. Each of the remaining $2k - 2$ blocks contains a copy of each of the letters in $Alph(S)$; in particular, we can choose the letters from the word B in them. No two consecutive letters in B are the same, so we will not choose the same position twice.   □

Observation 3. *If a letter a occurs in a word $S = S'aS''$ only once, then every s-cover of S has a form $C'aC''$, where C', C'' are s-covers of S', S'', respectively.*

Theorem 3. *For $k \geq 4$ we have*

$$5 \cdot 2^{k-2} - 1 \leq \gamma(k) \leq 2^{k-1} k!.$$

Proof. We separately prove the lower and upper bounds.

Lower Bound. We can take the sequence of words $S_4 = abacadbabdcabcbadac$, and for $k > 4$:

$$S_k = S_{k-1}a_k S_{k-1}, \text{ where } a_k \text{ is a new letter.}$$

We have $|S_k| = 5 \cdot 2^{k-2} - 1$, and S_k has no non-trivial s-cover, due to Observation 3 and Example 3. Hence $\gamma(k) \geq 5 \cdot 2^{k-2} - 1$.

Upper Bound. We will show that

$$\gamma(k) \leq 2k \cdot \gamma(k - 1). \tag{1}$$

Let us assume that $|Alph(S)| = k$ and $|S| > 2k \cdot \gamma(k-1) + 1$. Let $x = \gamma(k-1)$. If any factor U of S of length $x + 1$ does not contain all k letters, then U is not s-primitive by the definition of γ. If $S = PUQ$ where U has a non-trivial s-cover C, then PCQ is a non-trivial s-cover of S and, consequently, S is not s-primitive.

Otherwise, by Lemma 4 applied for a prefix of S of length $2kx + 1$ and $Z = \mathsf{FaLaFeL}(S)$, $\mathsf{FaLaFeL}(S)$ is a subsequence of S. By Lemma 3, $shrink(FL)$ is an s-cover of S. It is non-trivial as for $k \geq 3$, $shrink(FL)$ is shorter than $\mathsf{FaLaFeL}(S)$. In either case, S is not s-primitive and (1) holds. Using a simple induction we get $\gamma(k) \leq 2^{k-1}k!$.   □

3.3   Behaviour of the Function $\gamma(k)$ for Small k

The values of γ for small k are as follows (see also Table 1):

- $\gamma(1) = 1$ – trivial;
- $\gamma(2) = 3$ – using square-free words;
- $\gamma(3) = 8$ – due to Fact 1 and Fig. 3;
- $\gamma(4) = 19$ – through computer experiments[1];
- $39 \leq \gamma(5) \leq 190$ – due to Inequality (1) and $\gamma(4) = 19$.

Table 1. The values of γ for small alphabet-size k.

k	$\gamma(k)$	Examples of s-primitive words
1	1	a
2	3	aba
3	8	$abcabacb$
4	19	$abacadbabdcabcbadac$
		$abcdabacadbdcbabdac$
5	≥ 39	$abacadbabdcabcbadaceabacadbabdcabcbadac$

Remark 2. There are $2 \cdot 3! = 12$ s-primitive words of length $\gamma(3) = 8$ over ternary alphabet (cf. Sect. 3.1 and Fig. 3, for each pair of distinct letters there are two s-primitive words starting with these letters). This accounts for less than 0.2% among all 3^8 ternary words of length 8. For a 4-letter alphabet, our program shows that the relative number of s-primitive words of length $\gamma(4) = 19$ is very small. There are exactly 2496 such words, out of 4^{19}, which gives a fraction less than 10^{-8}. This suggests that s-primitive 5-ary words of length $\gamma(5)$ are extremely sparse and finding an s-primitive word over a 5-letter alphabet of length $\gamma(5)$, if $\gamma(5) > 39$, could be a challenging task.

4   Computing s-Covers

The following observation is a common property of s-covers and standard covers.

---

[1] The optimized C++ code used for the experiments can be found at https://www.mimuw.edu.pl/~jrad/code.cpp. The program reads k and computes $\gamma(k)$; it finishes within 1 min for $k \leq 4$.

Observation 4. *If C is an s-cover of S and C' is an s-cover of C, then C' is an s-cover of S.*

Theorem 4. *Let S be a length-n word over an integer alphabet of size $k = n^{\mathcal{O}(1)}$.*

(a) A shortest s-cover of S can be computed in $\mathcal{O}(n \cdot \min(2^n, k^{\gamma(k)}))$ time.
(b) One can check if S is s-primitive and, if not, return a non-trivial s-cover of S in $\mathcal{O}(n + 2^{\gamma(k)}\gamma(k))$ time.
(c) An s-cover of S of length at most $\gamma(k)$ can be computed in $\mathcal{O}(n2^{\gamma(k)}\gamma(k))$ time.

Proof. (a) By Theorem 3, there are $\mathcal{O}(k^{\gamma(k)})$ s-primitive k-ary words and, by Observations 4, the shortest s-cover of S must be one of them. On the other hand, there are 2^n subsequences of S. Hence, there are $\min(2^n, k^{\gamma(k)})$ candidates to be checked. With the aid of the algorithm from Lemma 1 we can check each candidate in $\mathcal{O}(n)$ time. This gives the desired complexity.

(b) If $n \leq \gamma(k)$, we can use the algorithm from (a) which works in $\mathcal{O}(2^{\gamma(k)}\gamma(k))$ time. Otherwise, we know by Theorem 3 that S is not s-primitive. We can find a non-trivial s-cover of S as follows. Let $S = S'S''$ where $|S'| = \gamma(k) + 1$. We can use the algorithm from (a) to compute a shortest s-cover C of S' in $\mathcal{O}(2^{\gamma(k)}\gamma(k))$ time. By Theorem 3, C is a non-trivial s-cover of S'. Then, we can output CS'' as a non-trivial s-cover of S. This takes $\mathcal{O}(n + 2^{\gamma(k)}\gamma(k))$ time.

(c) By Observations 4, any s-cover of an s-cover of S will be an s-cover of S. We can thus repeatedly apply the algorithm underlying (b); apart from outputting the computed non-trivial s-cover. As each application of this algorithm removes at least one letter of S, the number of steps is at most $n - \gamma(k)$. Each step takes $\mathcal{O}(2^{\gamma(k)}\gamma(k))$ time and hence the conclusion follows. □

Corollary 1. *A shortest s-cover of a word over a constant-sized alphabet can be computed in linear time.*

5 The Number of Distinct Shortest s-Covers

In the case of standard covers, if a word S has two covers C, C', then one of C, C' is a cover of the other. This property implies, in particular, that a word has exactly one shortest cover.

In this section we show that analogous properties do not hold for s-covers. There exist words S having two s-covers C, C' such that none of C, C' is an s-cover of the other; e.g. $S = abcabcabcb$, $C = abcb$ and $C' = abcacb$. Moreover, a word can have many different shortest s-covers, as shown in Theorem 5.

Theorem 5. *For every positive integer n there exists a word of length n over an alphabet of size $\mathcal{O}(\log n)$ that has at least $2^{\lfloor \frac{n+1}{16} \rfloor}$ different shortest s-covers.*

Proof. We start with an example of a word with two different shortest s-covers and then extend it recursively.

$$a \qquad b\ c\ a \qquad d \quad c\ b\ a$$
$$a\ b\ c\ a\ d \qquad c\ b \quad a$$
$$a\ b\ c\ a\ d\ b\ c\ a\ c\ b\ d\ a\ c\ b\ a$$

Fig. 5. $C_1 = abca\ d\ cba$ is a shortest s-cover of $S = abca\ d\ bcacb\ d\ acba$. S is a palindrome, hence $C_2 = abc\ d\ acba$ is also its s-cover.

Claim. The word $S = abca\ d\ bcacb\ d\ acba$ has two different s-covers of length 8, $C_1 = abca\ d\ cba$ and $C_2 = abc\ d\ acba$ (cf. Fig. 5). It does not have any shorter s-cover.

Proof. Any s-cover of this word must contain the letter d and before its first occurrence letters a, b, c (in that order) must appear. Symmetrically, after this letter, letters c, b, a must appear. The only word of length smaller than 8 which satisfies this property is $abc\ d\ cba$; however, this is not an s-cover of S (as it does not cover the middle letter a in S). $\square$

We now construct a sequence of words T_i such that $T_0 = S$ and $T_i = T_{i-1} a_i T_{i-1}$ for $i > 0$, where a_i is a new letter.

The word T_i has length $16 \cdot 2^i - 1 = 2^{i+4} - 1$. Let us consider an infinite word $T = \lim_{i \to \infty} T_i$ (this word is well defined as each T_i is a prefix of T_{i+1}).

We show by induction, using Observations 3, that $T[0 .. n - 1]$ has at least $2^{\lfloor \frac{n+1}{16} \rfloor}$ different shortest covers.

The base case for $n \le 15$ holds as every word has a shortest s-cover and for $n = 15$ we apply the previous claim as $T[.. n - 1] = S$. Assume that $n > 15$. Let i be a non-negative integer such that $2^{i+4} \le n < 2^{i+5}$. Then $T[0 .. n - 1] = T_{i+1}[.. n - 1] = T_i\ a_i\ T_i[.. n - 2^{i+4} - 1]$. By Observations 3, the number of shortest s-covers of $T[.. n - 1]$ is the number of shortest s-covers of T_i times the number of shortest s-covers of $T[.. n - 2^{i+4} - 1]$, that is, at least

$$2^{\frac{2^{i+4}}{16}} \cdot 2^{\lfloor \frac{n - 2^{i+4} + 1}{16} \rfloor} = 2^{\lfloor \frac{n+1}{16} \rfloor}, \text{ as desired.} \qquad \square$$

6 Final Remarks

There are several natural questions concerning the following problems:

1. Is a given word s-primitive?
2. What is its shortest s-cover?
3. What is the number of its different s-covers?
4. What is the exact value of $\gamma(5)$?
5. Let us define $\gamma'(1) = 1$, $\gamma'(k + 1) = 2\gamma'(k) + k$ for $k > 1$.
   We have $\gamma(k) = \gamma'(k)$ for $1 \le k < 5$. Is it always true?
6. Is there a really short, understandable and computer-avoiding proof of s-primitiveness of the word $a\,b\,a\,c\,a\,d\,b\,a\,b\,d\,c\,a\,b\,c\,b\,a\,d\,a\,c$?

We believe that the first three problems are NP-hard for general alphabets.

Acknowledgements. We thank Juliusz Straszyński for his help in conducting computer experiments.

References

1. Apostolico, A., Farach, M., Iliopoulos, C.S.: Optimal superprimitivity testing for strings. Inf. Process. Lett. **39**(1), 17–20 (1991). https://doi.org/10.1016/0020-0190(91)90056-N
2. Breslauer, D.: An on-line string superprimitivity test. Inf. Process. Lett. **44**(6), 345–347 (1992). https://doi.org/10.1016/0020-0190(92)90111-8
3. Bulteau, L., Vialette, S.: Recognizing binary shuffle squares is NP-hard. Theor. Comput. Sci. **806**, 116–132 (2020). https://doi.org/10.1016/j.tcs.2019.01.012
4. Buss, S., Soltys, M.: Unshuffling a square is NP-hard. J. Comput. Syst. Sci. **80**(4), 766–776 (2014). https://doi.org/10.1016/j.jcss.2013.11.002
5. Czajka, P., Radoszewski, J.: Experimental evaluation of algorithms for computing quasiperiods. Theor. Comput. Sci. **854**, 17–29 (2021). https://doi.org/10.1016/j.tcs.2020.11.033
6. Flouri, T., et al.: Enhanced string covering. Theor. Comput. Sci. **506**, 102–114 (2013). https://doi.org/10.1016/j.tcs.2013.08.013
7. Kociumaka, T., Pissis, S.P., Radoszewski, J., Rytter, W., Waleń, T.: Fast algorithm for partial covers in words. Algorithmica **73**(1), 217–233 (2014). https://doi.org/10.1007/s00453-014-9915-3
8. Kociumaka, T., Pissis, S.P., Radoszewski, J., Rytter, W., Walen, T.: Efficient algorithms for shortest partial seeds in words. Theor. Comput. Sci. **710**, 139–147 (2018). https://doi.org/10.1016/j.tcs.2016.11.035
9. Kolpakov, R., Podolskiy, M., Posypkin, M., Khrapov, N.: Searching of gapped repeats and subrepetitions in a word. In: Kulikov, A.S., Kuznetsov, S.O., Pevzner, P. (eds.) CPM 2014. LNCS, vol. 8486, pp. 212–221. Springer, Cham (2014). https://doi.org/10.1007/978-3-319-07566-2_22
10. Lothaire, M.: Algebraic Combinatorics on Words. Encyclopedia of Mathematics and its Applications, Cambridge University Press (2002). https://doi.org/10.1017/CBO9781107326019
11. Moore, D.W.G., Smyth, W.F.: A correction to "An optimal algorithm to compute all the covers of a string". Inf. Process. Lett. **54**(2), 101–103 (1995). https://doi.org/10.1016/0020-0190(94)00235-Q
12. Rizzi, R., Vialette, S.: On recognizing words that are squares for the shuffle product. In: Bulatov, A.A., Shur, A.M. (eds.) CSR 2013. LNCS, vol. 7913, pp. 235–245. Springer, Heidelberg (2013). https://doi.org/10.1007/978-3-642-38536-0_21
13. Warmuth, M.K., Haussler, D.: On the complexity of iterated shuffle. J. Comput. Syst. Sci. **28**(3), 345–358 (1984). https://doi.org/10.1016/0022-0000(84)90018-7

Maximal Closed Substrings

Golnaz Badkobeh[1] [ID], Alessandro De Luca[2] [ID], Gabriele Fici[3]([⊠]) [ID],
and Simon J. Puglisi[4] [ID]

[1] Department of Computing, Goldsmiths University of London, London, UK
`g.badkobeh@gold.ac.uk`
[2] DIETI, Università di Napoli Federico II, Naples, Italy
`alessandro.deluca@unina.it`
[3] Dipartimento di Matematica e Informatica, Università di Palermo, Palermo, Italy
`gabriele.fici@unipa.it`
[4] Department of Computer Science, University of Helisnki, Helsinki, Finland
`simon.puglisi@helsinki.fi`

Abstract. A string is closed if it has length 1 or has a nonempty border without internal occurrences. In this paper we introduce the definition of a *maximal closed substring* (MCS), which is an occurrence of a closed substring that cannot be extended to the left nor to the right into a longer closed substring. MCSs with exponent at least 2 are commonly called *runs*; those with exponent smaller than 2, instead, are particular cases of *maximal gapped repeats*. We show that a string of length n contains $\mathcal{O}(n^{1.5})$ MCSs. We also provide an output-sensitive algorithm that, given a string of length n over a constant-size alphabet, locates all m MCSs the string contains in $\mathcal{O}(n \log n + m)$ time.

Keywords: Closed word · Maximal closed substring · Run

1 Introduction

The distinction between open and closed strings was introduced by the third author in [8] in the context of Sturmian words.

A string is *closed* (or *periodic-like* [6]) if it has length 1 or it has a border that does not have internal occurrences (i.e., it occurs only as a prefix and as a suffix). Otherwise the string is *open*. For example, the strings *a*, *abaab* and *ababa* are closed, while *ab* and *ababaab* are open. In particular, every string whose exponent — the ratio between the length and the minimal period — is at least 2, is closed [1].

In this paper, we consider occurrences of closed substrings in a string with the property that the substring cannot be extended to the left nor to the right into another closed substring. These are called the *maximal closed substrings* (MCS)

Gabriele Fici is partly supported by MIUR project PRIN 2017 ADASCOML – 2017K7XPAN. Simon J. Puglisi is partly supported by the Academy of Finland, through grant 339070.

D. Arroyuelo and B. Poblete (Eds.): SPIRE 2022, LNCS 13617, pp. 16–23, 2022.
https://doi.org/10.1007/978-3-031-20643-6_2

of the string. For example, if $S = abaabab$, then the set of pairs of starting and ending positions of the MCSs of S is

$$\{(1,1),(1,3),(1,6),(2,2),(3,4),(4,8),(5,5),(6,6),(7,7),(8,8)\}$$

This notion encompasses that of a *run* (maximal repetition) which is a MCS with exponent 2 or larger. It has been conjectured by Kolpakov and Kucherov [12] and then finally proved, after a long series of papers, by Bannai et al. [2], that a string of length n contains less than n runs.

On the other hand, maximal closed substrings with exponent smaller than 2 are particular cases of *maximal gapped repeats* [11]. An α-gapped repeat ($\alpha \geq 1$) in a string S is a substring uvu of S such that $|uv| \leq \alpha|u|$. It is maximal if the two occurrences of u in it cannot be extended simultaneously with the same letter to the right nor to the left. Gawrychowski et al. [10] proved that there are words that have $\Theta(\alpha n)$ maximal α-gapped repeats.

In this paper, we address the following problems:

1. How many MCSs can a string of length n contain?
2. What is the running time of an algorithm that, given a string S of length n, returns all the occurrences of MCSs in S?

We show that:

1. A string of length n contains $\mathcal{O}(n^{1.5})$ MCSs.
2. There is an algorithm that, given a string of length n over a constant-size alphabet, locates all m MCSs the string contains in $\mathcal{O}(n \log n + m)$ time.

2   Preliminaries

Let $S = S[1..n] = S[1]S[2] \cdots S[n]$ be a string of n letters drawn from an alphabet Σ of constant size. The length n of a string S is denoted by $|S|$. The *empty string* has length 0. A *prefix* (resp. a *suffix*) of S is any string of the form $S[1..i]$ (resp. $S[i..n]$) for some $1 \leq i \leq n$. A *substring* of S is any string of the form $S[i..j]$ for some $1 \leq i \leq j \leq n$. It is also commonly assumed that the empty string is a prefix, a suffix and a substring of any string.

An integer $p \geq 1$ is a *period* of S if $S[i] = S[j]$ whenever $i \equiv j \pmod{p}$. For example, the periods of $S = aabaaba$ are 3, 6 and every $n \geq 7 = |S|$.

We recall the following classical result:

Lemma 1 (Periodicity Lemma (weak version) [9]). *If a string S has periods p and q such that $p + q \leq |S|$, then $\gcd(p,q)$ is also a period of S.*

Given a string S, we say that a string $\beta \neq S$ is a *border* of S if β is both a prefix and a suffix of S (we exclude the case $\beta = S$ but we do consider the case $|\beta| = 0$). Note that if β is a border of S, then $|S| - |\beta|$ is a period of S; conversely, if $p \leq |S|$ is a period of S, then S has a border of length $|S| - p$.

The following well-known property of borders holds:

Property 1. If a string has two borders β and β', with $|\beta| < |\beta'|$, then β is a border of β'.

The *border array* $B_S[1..n]$ of string $S = S[1..n]$ is the integer array where $B_S[i]$ is the length of the longest border of $S[1..i]$. When the string S is clear from the context, we will simply write B instead of B_S.

For any $1 \leq i \leq n$, let $B^1[i] = B[i]$ and $B^j[i] = B[B^{j-1}[i]]$ for $j \geq 2$. We set

$$B^+[i] = \{|\beta| \mid \beta \text{ is a border of } S[1..i]\}.$$

By Property 1, we have $B^+[i] = \{B^j[i] \mid j \geq 1\}$.

For example, in the string $S = aabaaaabaaba$, we have $B^+[6] = \{0,1,2\}$. Indeed, $B[6] = 2$, and $B^2[6] = B[2] = 1$, while $B^j[6] = 0$ for $j > 2$.

The *OC array* [5] $OC_S[1..n]$ of string S is a binary array where $OC_S[i] = 1$ if $S[1,i]$ is closed and $OC_S[i] = 0$ otherwise.

We also define the array P_S where $P_S[i]$ is the length of the longest repeated prefix of $S[1..i]$, that is, the longest prefix of $S[1..i]$ that has at least two occurrences in $S[1..i]$. Again, if S is clear from the context, we omit the subscripts.

Let S be a string of length n. Since for every $1 \leq i \leq n$, the longest repeated prefix v_i of $S[1..i]$ is the longest border of $S[1..j]$, where $j \leq i$ is the ending position of the second occurrence of v_i, we have that

$$P[i] = \max_{1 \leq j \leq i} B[j]. \tag{1}$$

Lemma 2 ([7]). *Let S be a string of length n. For every $1 \leq i \leq n$, one has*

$$P[i] = \sum_{j=1}^{i} OC[j] - 1, \tag{2}$$

that is, $P[i]$ is the rank *of 1's in $OC[1..i]$ minus one.*

Proof. For every repeated prefix v of S, the second occurrence of v in S determines a closed prefix of S; conversely, every closed prefix of S of length greater than 1 ends where the second occurrence of a repeated prefix of S ends. Indeed, the length of the longest repeated prefix increases precisely in those positions in which we have a closed prefix. That is, $P[i] = P[i-1] + OC[i]$, for any $1 < i \leq n$, which, together with $P[1] = 0 = OC[1] - 1$, yields (2). $\square$

As a consequence of (1) and (2), if two strings have the same border array, then they have the same OC array, but the converse is not true in general (take for example *aaba* and *aabb*).

The OC array of a string can be obtained from its P array by taking the differences of consecutive values, putting 1 in the first position (cf. [8]). Since the border array can be easily computed in linear time [13], it is possible to compute the OC array in linear time.

Example 1. The OC, B, and P arrays for $S = aabaaaabaaba$ are shown in the following table:

i	1	2	3	4	5	6	7	8	9	10	11	12
S	a	a	b	a	a	a	a	b	a	a	b	a
OC	1	1	0	0	1	0	0	1	1	1	0	0
B	0	1	0	1	2	2	2	3	4	5	3	4
P	0	1	1	1	2	2	2	3	4	5	5	5

3   A Bound on the Number of MCS

The goal of this section is to prove our bound $\mathcal{O}(n^{1.5})$ in the number of MCSs in a string of length n. This will be derived from a bound on the number of runs (in the sense of maximal blocks of identical symbols, not to be confused with maximal repetitions mentioned in the introduction) in the OC array.

In the next lemmas, we gather some structural results on the OC array.

Lemma 3 ([7, Remark 8]). *If* $OC[i] = 1$, *then* $B[i] = P[i]$, *and* $B[i-1] = P[i-1]$ *(provided* $i > 1$*).*

Lemma 4. *For all* i *and* k *such that* $OC[i+1..i+k+1] = 0^k 1$, *if* $P[i] \geq k$ *then* $P[i] - k \in B^+[i]$.

Proof. By Lemma 2 and Lemma 3, $P[i+k+1] = P[i] + 1$ is the length of the longest border of S at position $i+k+1$. The assertion is then a consequence of the following simple observation: Let u, v and x be strings; if ux is a border of vx, then u is a border of v. In fact, letting $v = S[1..i]$, and $x = S[i+1..i+k+1]$, as $B^+[i+k+1] > k$, the longest border of vx can be written as ux for some u of length $P[i] + 1 - k - 1 = P[i] - k$. □

Lemma 5. *For all* i *and* k *such that* $OC[i..i+k+1] = 10^k 1$, *if* $P[i] \geq k$ *then* $P[i] - k \in B^+[P[i]]$.

Proof. Immediate by Lemmas 3 and 4, as $B[i] = P[i]$ and $P[i] - k \in B^+[i]$. □

Lemma 6. *If* $OC[i..i+k_1+k_2+t+1] = 10^{k_1}1^t 0^{k_2}1$ *and* $k_1, k_2 > 0$, *then* $P[i] < k_1 + k_2$.

Proof. By contradiction. Assume $P[i] \geq k_1 + k_2$. Then by Lemma 5 we have $P[i] - k_1 \in B^+[P[i]]$, which implies that k_1 is a period of $S[1..P[i]]$. Similarly, k_2 is a period of $S[1..P[i]+t]$ and then of $S[1..P[i]+1]$ and $S[1..P[i]]$, since $P[i] \geq k_2$. By the Periodicity Lemma 1 we know that $K = \gcd(k_1, k_2)$ is also a period of $S[1..P[i]]$. Note that $k_1 - k_2$ is divisible by K.

Furthermore, $S[i+1] \neq S[i+1+k_1]$ because $OC[i+1]$ is not 1. By Lemma 4, we have $P[i] + 1 - k_1 \in B^+[i+1]$, which implies $S[i+1] = S[P[i]+1-k_1]$.

However, $S[i+1+k_1] = S[P[i]+1] = S[P[i]+1-k_2] = S[P[i]+1-k_2 - (k_1 - k_2)] = S[P[i]+1-k_1] = S[i+1]$, which is a contradiction. □

Theorem 1. *Let S be a string of length n. Then the number of runs in its* OC *array is $\mathcal{O}(\sqrt{n})$.*

Proof. Let $\mathsf{OC}_S = 1^{t_1}0^{k_1}\cdots 1^{t_m}0^{k_m}$, where $k_m \geq 0$ and all other exponents are positive. By Lemma 6, we have for $1 < i < m$,

$$k_{i-1} + k_i \geq \sum_{r=1}^{i-1} t_r \geq i - 1.$$

This implies

$$n = \sum_{i=1}^{m}(t_i + k_i) \geq m + \sum_{j=1}^{\lfloor \frac{m-1}{2} \rfloor}(k_{2j-1} + k_{2j}) \geq m + \sum_{j=1}^{\lfloor \frac{m-1}{2} \rfloor}(2j-1) = m + \left\lfloor \frac{m-1}{2} \right\rfloor^2$$

so that $n = \Omega(m^2)$ and then $m = \mathcal{O}(\sqrt{n})$. □

The bound in the previous proposition is tight. Indeed, there exists a binary string whose OC array is $\prod_{k>0} 10^k$. Actually, the string is uniquely determined by its OC array and can be defined by $u = a\prod_{k>0} \overline{u[k]}u[1..k] = abaaabbababababa \cdots$.

The following proposition is a direct consequence of the definition of MCS. Essentially, it says that we can check if $S[i..j]$ is a MCS by looking at the OC array of the suffixes starting at position i and $i-1$.

Proposition 1. *Let S be a string of length n. If $S[i..j]$ is a MCS, then $\mathsf{OC}_{S[i..n]}[j - i + 1] = 1$ and either $j - i + 1 = n$ or $\mathsf{OC}_{S[i..n]}[j - i + 2] = 0$. Moreover, either $i = 1$ or $\mathsf{OC}_{S[i-1..n]}[j - i + 2] = 0$.*

Example 2. Let $S = aabaaaabaaba$. The OC arrays of the first few suffixes of S are displayed below.

S	a	a	b	a	a	a	a	b	a	a	b	a
$\mathsf{OC}_{S[1..n]}$	1	1	0	0	1	0	0	1	1	1	0	0
$\mathsf{OC}_{S[2..n]}$		1	0	1	0	0	0	1	1	1	0	0
$\mathsf{OC}_{S[3..n]}$			1	0	0	0	0	1	1	1	0	0
$\mathsf{OC}_{S[4..n]}$				1	1	1	1	0	0	0	0	0
$\mathsf{OC}_{S[5..n]}$					1	1	1	0	0	0	0	0
$\mathsf{OC}_{S[6..n]}$						1	1	0	0	1	1	1

One can check for instance that $S[4..7]$ is a MCS, because the $4 = (7-4+1)$th entry of $\mathsf{OC}_{S[4..n]}$ is a 1 which does not have another 1 on its right nor on top of it (i.e., in the OC array of the previous suffix). Similarly, $S[6..12]$ is a MCS because the last entry of $\mathsf{OC}_{S[6..n]}$ is 1 with a 0 on top.

As a consequence of the previous proposition, the number of MCSs in S is bounded from above by the total number of runs of 1s in all the OC arrays of the suffixes of S.

From Theorem 1, we therefore have a bound of $\mathcal{O}(n\sqrt{n})$ on the number of MCSs in a string of length n.

4 An Algorithm for Locating All MCS

In the previous section, we saw that one can locate all MCSs of S by looking at the OC arrays of all suffixes of S. However, since the OC array of a string of length n requires $\Omega(n)$ time to be constructed, this yields an algorithm that needs $\Omega(n^2)$ time to locate all MCSs.

We now describe a more efficient algorithm for computing all the maximal closed substrings in a string S of length n. For simplicity of exposition we assume that S is on a binary alphabet $\{a, b\}$, however the algorithm is easily adapted for strings on any constant-sized alphabet. The running time is asymptotically bounded by $n \log n$ plus the total number of MCSs in S.

The inspiration for our approach is an algorithm for finding maximal pairs under gap constraints due to Brodal, Lyngsø, Pedersen, and Stoye [3]. The central data structure is the suffix tree of the input string, which we now define.

Definition 1 (Suffix tree). *The suffix tree $T(S)$ of the string S is the compressed trie of all suffixes of S. Each leaf in $T(S)$ represents a suffix $S[i..n]$ of S and is annotated with the index i. We refer to the set of indices stored at the leaves in the subtree rooted at node v as the leaf-list of v and denote it $LL(v)$. Each edge in $T(S)$ is labelled with a nonempty substring of S such that the path from the root to the leaf annotated with index i spells the suffix $S[i..n]$. We refer to the substring of S spelled by the path from the root to node v as the path-label of v and denote it $L(v)$.*

At a high level, our algorithm for finding MCSs processes the suffix tree (which is a binary tree, for binary strings) in a bottom-up traversal. At each node the leaf lists of the (two, for a binary string) children are intersected. For each element in the leaf list of the smaller child, the successor in the leaf list of the larger child is found. Note that because the element from the smaller child and its successor in the larger child come from different subtrees, they represent a pair of occurrences of substring $L(v)$ that are right-maximal. To ensure left maximality, we must take care to only output pairs that have different preceding characters. We explain how to achieve this below.

Essential to our algorithm are properties of AVL trees that allow their efficient merging, and the so-called "smaller-half trick" applicable to binary trees. These proprieties are captured in the following lemmas.

Lemma 7 (Brown and Tarjan [4]). *Two AVL trees of size at most n and m can be merged in time $\mathcal{O}(\log \binom{n+m}{n})$.*

Lemma 8 (Brodal et al. [3], Lemma 3.3). *Let T be an arbitrary binary tree with n leaves. The sum over all internal nodes v in T of terms that are $\mathcal{O}(\log\binom{n_1+n_2}{n_1})$, where n_1 and n_2 are the n_1 numbers of leaves in the subtrees rooted at the two children of v, is $\mathcal{O}(n\log n)$.*

As stated above, our algorithm traverses the suffix tree bottom up. At a generic step in the traversal, we are at an internal node v of the suffix tree. Let the two children of node v be v_ℓ and v_r (recall the tree is a binary suffix tree, so every internal node has two children). The leaf lists of each child of v are maintained in two AVL trees — note, there are *two AVL trees for each of the two children*, two for v_ℓ and two for v_r. For a given child, say v_r, one of the two AVL trees contains positions where $L(v_r)$ is preceded by an a symbol, and the other AVL tree contains positions where $L(v_r)$ is preceded by a b symbol in S. Call these the a-tree and b-tree, respectively.

Without loss of generality, let v_r be the smaller of v's children. We want to search for the successor and predecessor of each of the elements of v_r's a-tree amongst the elements v_ℓ's b-tree, and, similarly the elements of v_r's b-tree with the elements from v_ℓ's a-tree. Observe that the resulting pairs of elements represent a pair of occurrences of $L(v)$ that are both right and left maximal: they have different preceding characters and so will be left maximal, and they are siblings in the suffix tree and so will be right maximal. These are candidate MCSs. What remains is to discard pairs that are not consecutive occurrences of $L(v)$, to arrive at the MCSs. Discarding is easy if we process the elements of $LL(v_r)$ in order (which is in turn easy, because they are stored in two AVL trees). To see this, consider two consecutive candidates that have the same right border position (a successor found in $LL(v_\ell)$; discarding for left borders is similar). The first of these candidates can clearly be discarded because there is an occurrence of $L(v)$ (from $LL(v_r)$) in between the two borders, preventing the pair of occurrences from forming an MCS. Because we only compute a successor/predecessor for each of the elements of the smaller of v's children, by Lemma 8 the total time for all successor/predecessor searches will be $\mathcal{O}(n\log n)$ (discarding also takes time proportional to the smaller subtree, and so does not increase this complexity). After this, the a-tree and b-tree of the smaller child are merged with their counterparts from the larger child.

Thus, by Lemmas 7 and 8, the overall processing is bounded by $\mathcal{O}(n\log n)$ in addition to the number of MCSs that are found.

The above approach is easily generalized from strings on binary alphabets to those on any alphabet of constant size by replacing nodes of the suffix tree having degree $d > 2$ with binary trees of height $\log d$. This does not increase the height of the suffix tree asymptotically and so preserves the runtime stated above. It would be interesting to design algorithms for general alphabets, and we leave this as an open problem.

References

1. Badkobeh, G., Fici, G., Lipták, Z.: On the number of closed factors in a word. In: Dediu, A.-H., Formenti, E., Martín-Vide, C., Truthe, B. (eds.) LATA 2015. LNCS, vol. 8977, pp. 381–390. Springer, Cham (2015). https://doi.org/10.1007/978-3-319-15579-1_29
2. Bannai, H.I.T., Inenaga, S., Nakashima, Y., Takeda, M., Tsuruta, K.: The "runs" theorem. SIAM J. Comput. **46**(5), 1501–1514 (2017)
3. Brodal, G.S., Lyngsø, R.B., Pedersen, C.N.S., Stoye, J.: Finding maximal pairs with bounded gap. In: Crochemore, M., Paterson, M. (eds.) CPM 1999. LNCS, vol. 1645, pp. 134–149. Springer, Heidelberg (1999). https://doi.org/10.1007/3-540-48452-3_11
4. Brown, M.R., Tarjan, R.E.: A fast merging algorithm. J. ACM **26**(2), 211–226 (1979)
5. Bucci, M., De Luca, A., Fici, G.: Enumeration and structure of trapezoidal words. Theor. Comput. Sci. **468**, 12–22 (2013)
6. Carpi, A., de Luca, A.: Periodic-like words, periodicity, and boxes. Acta Informatica **37**(8), 597–618 (2001). https://doi.org/10.1007/PL00013314
7. De Luca, A., Fici, G., Zamboni, L.Q.: The sequence of open and closed prefixes of a Sturmian word. Adv. Appl. Math. **90**, 27–45 (2017)
8. Fici, G.: Open and Closed Words. Bull. Eur. Assoc. Theor. Comput. Sci. EATCS **123**, 140–149 (2017)
9. Fine, N.J., Wilf, H.S.: Uniqueness theorems for periodic functions. P. Am. Math. Soc. **16**(1), 109–114 (1965)
10. Gawrychowski, P., I, T., Inenaga, S., Köppl, D., Manea, F.: Tighter bounds and optimal algorithms for all maximal α-gapped repeats and palindromes. Theor. Comput. Syst. **62**(1), 162–191 (2017). https://doi.org/10.1007/s00224-017-9794-5
11. Kolpakov, R., Podolskiy, M., Posypkin, M., Khrapov, N.: Searching of gapped repeats and subrepetitions in a word. J. Discrete Algorithms **46–47**, 1–15 (2017). https://doi.org/10.1007/978-3-319-07566-2_22
12. Kolpakov, R.M., Kucherov, G.: Finding maximal repetitions in a word in linear time. In: 40th Annual Symposium on Foundations of Computer Science, FOCS 1999, 17–18 October 1999, New York, NY, USA, pp. 596–604. IEEE Computer Society (1999)
13. Morris, J.H., Pratt, V.R.: A linear pattern-matching algorithm. Technical Report 40, University of California, Berkeley (1970)

Online Algorithms for Finding Distinct Substrings with Length and Multiple Prefix and Suffix Conditions

Laurentius Leonard[1]([✉])[iD], Shunsuke Inenaga[2][iD], Hideo Bannai[3][iD], and Takuya Mieno[4][iD]

[1] Department of Information Science and Technology, Kyushu University, Fukuoka, Japan
laurentius.leonard.705@s.kyushu-u.ac.jp
[2] Department of Informatics, Kyushu University, Fukuoka, Japan
inenaga@inf.kyushu-u.ac.jp
[3] M&D Data Science Center, Tokyo Medical and Dental University, Tokyo, Japan
hdbn.dsc@tmd.ac.jp
[4] Department of Computer and Network Engineering, University of Electro-Communications, Chofu, Japan
tmieno@uec.ac.jp

Abstract. Let two static sequences of strings P and S, representing prefix and suffix conditions respectively, be given as input for preprocessing. For the query, let two positive integers k_1 and k_2 be given, as well as a string T given in an online manner, such that T_i represents the length-i prefix of T for $1 \leq i \leq |T|$. In this paper we are interested in computing the set ans_i of distinct substrings w of T_i such that $k_1 \leq |w| \leq k_2$, and w contains some $p \in P$ as a prefix and some $s \in S$ as a suffix. More specifically, the counting problem is to output $|ans_i|$, whereas the reporting problem is to output all elements of ans_i, for each iteration i. Let σ denote the alphabet size, and for a sequence of strings A, $\|A\| = \sum_{u \in A} |u|$. Then, we show that after $O((\|P\| + \|S\|) \log \sigma)$-time preprocessing, the solutions for the counting and reporting problems for each iteration up to i can be output in $O(|T_i| \log \sigma)$ and $O(|T_i| \log \sigma + |ans_i|)$ total time. The preprocessing time can be reduced to $O(\|P\| + \|S\|)$ for integer alphabets of size polynomial with regard to $\|P\| + \|S\|$. Our algorithms have possible applications to network traffic classification.

Keywords: Pattern matching · Counting algorithm · Suffix array · Suffix tree

1 Introduction

Pattern matching has long been a central topic in the field of string algorithms [4], leading to various applications, including DNA analysis in bioin-

© The Author(s), under exclusive license to Springer Nature Switzerland AG 2022
D. Arroyuelo and B. Poblete (Eds.): SPIRE 2022, LNCS 13617, pp. 24–37, 2022.
https://doi.org/10.1007/978-3-031-20643-6_3

formatics [7,13] as well as packet classification [3,6,16,21] and anti-spam email filtering in network security [17].

In this paper, we propose algorithms for counting and reporting distinct substrings of an online text T that have some $p \in P$ as a prefix and some $s \in S$ as a suffix, and whose length is within the interval $[k_1..k_2]$, where P and S are static sequences of strings given as input for preprocessing, and integers k_1, k_2 and the characters of T are given as query. A similar yet different problem where patterns are given in the form of a pair of a prefix and a suffix condition, i.e. $p\Sigma^*s$ patterns, rather than a pair of sequences where one is of prefixes and the other of suffixes, is well-studied as the *followed-by* problem [2,14] or the *Dictionary Recognition with One Gap (DROG)* [1,12,20] problem. The problem of this paper is also of importance with possible applications in network traffic classification: All the application signatures in [19], for example, can be expressed as an instance of our problem, as discussed in Sect. 4 and demonstrated in Appendix A. Note that while traffic classification via these signatures was shown to be highly accurate, there are still cases of false positives. When analyzing which patterns give false positives, we may be interested in which patterns match the signatures, in which case the distinct condition of our problem helps prevent wasting computation time on repeated occurrences of each pattern. Another possible application is in computing the distinct substrings of an online text T whose lengths are at least k and belong to a (k,r)-*TTSS* language [18]; the words of length $\geq k$ in a (k,r)-*TTSS* language defined by the 4-tuple $(I_k, F_k, T_{k,r}, g)$ are the words that have some element of I_k as a prefix, some element of F_k as a suffix, and includes, for each $t \in T_{k,r}$, at most $g(t)$ occurrences of t as substring. Here, the elements of I_k and F_k are strings of length $k-1$ and the elements of $T_{k,r}$ are strings of length between 1 to k inclusive, and g is a function that projects $T_{k,r} \rightarrow \{0,1,\cdots,r-1\}$. A direct application of our algorithms can check whether w fulfills the prefix and suffix condition, while the condition of restricted segments, i.e. the number of occurrences of $T_{k,r}$ can also be considered by implementing the following modification: for each iteration i, maintain the minimum start-index res of the suffix of T_i that meets the condition of restricted segments, and use it to exclude any suffixes longer than $T_i[res..i]$ from the solution.

Our proposed algorithms take $O((\|P\|+\|S\|)\log\sigma)$ preprocessing time, while processing T itself in an online manner and outputting the solutions up to iteration i takes $O(|T_i|\log\sigma)$ and $O(|T_i|\log\sigma+|ans_i|)$ cumulative time for the counting and reporting problems respectively, using $O(|T_i|+\|P\|+\|S\|)$ working space. Here, T_i denotes the length-i prefix of T, $\|P\|$ and $\|S\|$ denote the total length of strings in P and S respectively, σ is the alphabet size, $|ans_i|$ is the number of substrings reported for each T_i, and cumulative time refers to the total amount of running time up to iteration i, as opposed to the running time of only iteration i. In addition, the preprocessing time can be reduced to $O(\|P\|+\|S\|)$ in the case of integer alphabets of size polynomial in $\|P\|+\|S\|$.

Also note that, the problems addressed in this paper differ from those of [11], in which a different set of solution strings is output for each $p \in P$ where each solution must have that specific element of P as a prefix, unlike the problem in this paper where only a single set of solution strings is output, where its elements can have any $p \in P$ as a prefix. Also, in [11] there is only one suffix condition and no length condition, and the algorithm is offline w.r.t. T, unlike in this paper.

2   Preliminaries and Definitions

2.1   Strings

Let Σ be an alphabet of size σ. An element of the set Σ^* is a string. The length of a string w is denoted by $|w|$. The empty string is denoted by ε. That is, $|\varepsilon| = 0$. For a string $w = pts$, p, t, and s are called a *prefix*, *substring*, and *suffix* of w, respectively. A prefix p (resp. suffix s) of a string w is called a *proper prefix* (resp. *proper suffix*) of w if $|p| < |w|$ (resp. $|s| < |w|$). For a string w, $w[i]$ denotes the i-th symbol of w for $1 \leq i \leq |w|$, and $w[i..j]$ denotes the substring $w[i]w[i+1]\cdots w[j]$ for $1 \leq i \leq j \leq |w|$. For a sequence S of strings, let $\|S\| = \sum_{u \in S} |u|$.

2.2   Suffix Array and LCP Array

The suffix array [15] of a string w is a lexicographically sorted array of suffixes of w, where each suffix is represented by its start-index. The LCP array is an auxiliary array commonly used alongside the suffix array, that stores the length of the longest common prefix of each adjacent pair of suffixes in the suffix array. More specifically, if SA and LCP are the suffix array and LCP array of the same string w, for $x \in [2..|w|]$, $LCP[x]$ is the length of longest common prefix of the suffixes $w[SA[x]..|w|]$ and $w[SA[x-1]..|w|]$. In this paper, we will use suffix arrays for some strings, in which we denote by LCP the LCP array of the same string of the suffix array being discussed. It is well-known that the suffix array and LCP array of a string w can be built in $O(|w|)$ time for integer alphabets of polynomial size in $|w|$ [8–10], and in $O(|w| \log \sigma)$ time for general ordered alphabets [22].

2.3   The Problems

The problems considered in the paper are as follows.

Definition 1 (Online substring counting and reporting problem with distinctness, multiple prefixes, multiple suffixes and length range conditions).   *Given two sequences of strings $P = (p_1, \cdots, p_n)$ and $S = (s_1, \cdots, s_m)$, two integers k_1 and k_2, and a string T given in an online manner (i.e., $T_0 = \varepsilon$ and for each iteration $i = 1, \ldots, |T|$, the i-th character is appended to T_{i-1} to form T_i), let ans_i denote the set of distinct substrings of T_i that have some $p \in P$ as a prefix, some $s \in S$ as a suffix, and whose length falls within the interval $[k_1..k_2]$.*

The counting problem. *On each iteration i, output $|ans_i|$.*
The reporting problem. *On each iteration i, output $ans_i \setminus ans_{i-1}$.*

This paper excludes the empty string ε from the solutions.

3 Algorithm

3.1 Sketch of Algorithm

In this section, we describe the general idea of our algorithm. During each iteration i, we need either to compute the size of $ans_i \setminus ans_{i-1}$ to add it to the counting solution, or to report all its elements. All elements of $ans_i \setminus ans_{i-1}$ must be suffixes of T_i, and thus for the suffix condition, T_i itself must have some element of S as a suffix; otherwise clearly $ans_i \setminus ans_{i-1} = \emptyset$ and there is no need to output a solution for the current iteration. Thus, let us consider the case where T_i has at least one element of S as a suffix, and call the shortest of them s. To help keep track of which suffixes of T_i fulfill the prefix condition, let us maintain a linked list $pList$ that contains in increasing order, all distinct indices j such that there is an element of P that occurs in T_i with start-index j. Clearly, the elements of $pList$ represent a bijection to the suffixes of T_i that have an element of P as a prefix, which are candidates for elements of the solution set $ans_i \setminus ans_{i-1}$. Specifically, for all $j \in pList$, the string $u = T_i[j..i] \in ans_i \setminus ans_{i-1}$ iff u fulfills all the following conditions:

(a) u has s as a suffix.
(b) u does not occur in T_{i-1}.
(c) $|u| \geq k_1$.
(d) $|u| \leq k_2$.

Here, (a) u has s as a suffix iff $|u| \geq |s|$, and (b) u does not occur in T_{i-1} iff $|u| > |lrs_i|$, where lrs_i denotes the longest repeating suffix of T_i, i.e. the longest suffix of T_i that occurs at least twice in T_i. Thus, conditions (a) to (c) set a lower bound for the length of suffixes of T_i corresponding to elements of $pList$ that can be a solution while condition (d) sets an upper bound. If we visualize $pList$ horizontally as shown in Fig. 1, conditions (a) to (c) exclude some elements from the right while condition (d) excludes some element from the left.

Take the maximum among the number of elements excluded by conditions (a) to (c) and denote it by $excludeRight$, and denote the number of elements excluded by condition (d) by $excludeLeft$. Then, $|ans_i \setminus ans_{i-1}| = \max(0, |pList| - excludeLeft - excludeRight)$, giving us the solution for the counting problem. For the reporting problem, let us maintain $start$, a pointer to the smallest element in $pList$ not excluded by condition (d). Then report all elements traversed by starting at $start$ and moving to the right $|ans_i \setminus ans_{i-1}| - 1$ times.

Example 1. The example in Fig. 1 occurs when $T_i = $ coldcocoaold, $P = $ (cave, coco, cocoa, d, oao, old), $S = $ (aold, oaold), $k_1 = 3$, and $k_2 = 8$. Then, $pList = (2, 4, 5, 8, 10, 12)$. $10, 12$ correspond to old, d which are shorter than

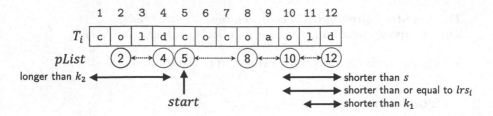

Fig. 1. A visualized example of *pList*.

$s =$ `aold`, and thus excluded by condition (a). $lrs_i =$ `old` and thus condition (b) also excludes 10 and 12, while condition (c) excludes only 12 which corresponds to `d`. Therefore, $excludeRight = \max(2, 2, 1) = 2$. Meanwhile, condition (d) excludes 2 and 4 which correspond to `oldcocoaold` and `dcocoaold` and so $excludeLeft = 2$. Thus, we have that $|ans_i \setminus ans_{i-1}| = \max(0, 6 - 2 - 2) = 2$. For the reporting solution, we have that *start* points to 5. Traversing 2 elements starting from 8 gives us 5 and 8, each corresponding to `cocoaold` and `oaold`, exactly the elements of $ans_i \setminus ans_{i-1}$.

3.2   Removing Redundant Elements

We say that $p_k \in P$ is *redundant* iff there exists $p_{k'}$ of P s.t. either p_k has $p_{k'}$ as a proper prefix, or $p_k = p_{k'} \wedge k > k'$. Similarly, $s_k \in S$ is *redundant* iff there exists $s_{k'}$ of S s.t. either s_k has $s_{k'}$ as a proper suffix, or $s_k = s_{k'} \wedge k > k'$.

It is not hard to see why they are called redundant; when multiple copies of the same string exist in P, keeping only one copy suffices, and when $p_k \in P$ has $p_{k'} \in P$ as a proper prefix, the strings that have p_k as a prefix is a subset of strings that have $p_{k'}$ as a prefix, and thus the solution remains the same even if we delete p_k from P. The same can be said for redundant elements of S.

As one part of the preprocessing, we rebuild P and S so that the redundant elements are deleted. First, we describe how to rebuild P. Let $P_{seq} = \$p_1\$p_2\$ \cdots \$p_n\$$, where $\$ \notin \Sigma$ and $\$ \prec c$ for all $c \in \Sigma$, and let SA be the suffix array of P_{seq}.

Then, each $SA[x]$ for $x \in [2..n + 1]$ corresponds to the start-index of $\$p$ in P_{seq}, for some $p \in P$. Starting from $x = 2$, output the corresponding p, namely the unique $p \in P$ s.t. $\$p\$$ occurs on index $SA[x]$. Then, increment x (at least once) until we have that $LCP[x] < |\$p|$, i.e. until we find an index that corresponds to $\$p'$ where $p' \in P$ does not have p as a prefix. Output the element of P corresponding to the new x, then again increment x in the same manner. Repeat this until $x > n + 1$, and we have that all non-redundant elements of P are output.

Other than the construction of the suffix array and LCP array, clearly this takes $O(\|P\|)$ time, and the same method can be used to compute non-redundant elements of S: Let $S^{-1} = (s_1^{-1}, \ldots, s_m^{-1})$ be the sequence of reversed elements of S, then apply the above algorithm to S^{-1} and reverse each string in the output to get the non-redundant elements of S. Thus, both P and S are rebuilt to exclude redundant elements in $O(\|P\| + \|S\|)$ time, in addition to the construction time of the suffix array and LCP array, which depends on the alphabet.

Example 2. Let $P = (\text{abc}, \text{ab}, \text{acc}, \text{ab}, \text{cab})$. Then, $P_{seq} = \text{\$abc\$ab\$acc\$ab\$cab\$}$ and we have the table as shown in Fig. 2. $x = 2$ corresponds to the occurrence of $\text{\$ab}$, which occurs on P_{seq} on index $SA[2] = 5$. Thus, ab is determined to be non-redundant, and we have that $|\text{\$ab}| = 3$, so increment x until we have that $LCP[x] < 3$. This skips over $x = 3, 4$, correctly determining their corresponding elements of P, namely the second ab and abc to be redundant. When $x = 5$, $LCP[x] = 2 < 3$ and so the corresponding $p = \text{acc}$ is output. Similarly, for $x = 6$, $LCP[x] = 1 < 4$ and thus cab is output. Afterwards, x is incremented beyond the interval $[2..n + 1]$ and thus the algorithm terminates and the non-redundant elements $(\text{ab}, \text{acc}, \text{cab})$ are output.

$$1234567890123456789$$
$$P_{seq} = \text{\$abc\$ab\$acc\$ab\$cab\$}$$

LCP	x	SA	suffix	p	|\$p|
-	1	19	$	-	-
1	2	5	abaccabcab$	ab	3
4	3	12	abcab$	ab	3
3	4	1	abcabaccabcab	abc	4
2	5	8	accabcab	acc	4
1	6	15	cab	cab	4
...	.	...	...	...	...

Fig. 2. The table for $P_{seq} = \text{\$abc\$ab\$acc\$ab\$cab}$

For the rest of the paper, we will assume that the above preprocessing is done and thus $P = (p_1, \ldots, p_n)$ and $S = (s_1, \ldots, s_m)$ from this point refer to the rebuilt sequences that have no redundant elements.

3.3   Detecting P and S Occurrences

As discussed, on each iteration i we need to detect whenever an element of S occurs as a suffix of T_i. Additionally, we will also need to detect when an element of P occurs as a suffix of T_i, in order to maintain $pList$. This can be done by building an Aho-Corasick automaton for P and S separately. Constructing both automata takes $O((\|P\| + \|S\|) \log \sigma)$ preprocessing time in general, and $O(\|P\| + \|S\|)$ time in the case of integer alphabets of size polynomial with regard to $\|P\| + \|S\|$ [5]. Running each automaton up to iteration i takes $O(|T_i| \log \sigma + occ)$ cumulative time, where occ is the number of occurrences detected. Here, the occurrences of elements of P in T_i must have distinct start-indices, as two occurrences with a shared start-index imply that one of them is redundant. Similarly, occurrences of elements of S must have distinct end-indices and thus $occ \in O(|T_i|)$ for both automata, and so the cumulative running time becomes $O(|T_i| \log \sigma)$.

3.4   Maintaining *pList*

To maintain *pList*, whenever some $p \in P$ is detected to occur as a suffix of T_i, its start-index j needs to be added to *pList* while maintaining the increasing order. Doing this naively would take $O(|pList|) = O(|T_i|)$ time for every insertion which gives quadratic time overall, so a more efficient scheme is necessary.

During any iteration i, the elements of P that occur as suffixes of T_i are detected in decreasing order of length, because we use Aho-Corasick automaton. Let p be such an element detected, and j be the start-index of its occurrence, i.e. $j = i - |p| + 1$. We need to add j into *pList* so that the increasing order of its elements are maintained.

To do that, we need to find the minimum j' among the current elements of *pList* such that $j' > j$. Here, j' being an element of *pList* implies that j' corresponds to an occurrence of $p' \in P$ starting at j' and ending at some $i' \leq i$. We can see that in fact $i' < i$, for if $i' = i$, j' was added to *pList* in the current iteration i before j, while $j' > j \wedge i' = i$ implies $|p'| < |p|$, contradicting the fact that the Aho-Corasick automaton detects the occurrences in decreasing order of length. Thus, $j' > j$ and $i' < i$, meaning the occurrence of p' falls completely within $p[2..|p| - 1]$.

Our scheme is then as follows: We precompute, for each $p_k \in P$, the minimum value y such that there is some $p_{k'} \in P$ that occurs in $p_k[2..|p_k| - 1]$ on start-index $y + 1$. If there is no such $p_{k'}$, then let $y = \infty$. Then, we will store the values on the array *successorOffset* that maps each $p_k \in P$ to its corresponding y.

Additionally, maintain also an array *pArray* such that *pArray*[j] points to the element of *pList* whose value is j if it exists, or *null* otherwise. Then, whenever some $p_k \in P$ occurs as a suffix of T_i with start-index j, we can just add j into *pList* exactly before the element pointed to by *pArray*[j + *successorOffset*[k]]. Clearly, once *successorOffset* is computed, adding each element of *pList* takes only constant time and thus maintaining *pList* and *pArray* takes cumulative $O(|T_i|)$ time, as the number of elements of *pList* for any given iteration i is bounded by $|T_i|$.

Computing successorOffset. Let $P_{concat} = p_1 \$ p_2 \$ \cdots \$ p_n \$$. Note that it differs from P_{seq} not only with regard to the positioning of \$, but also in that redundant elements of P are not included. For $k \in [1..n]$, let $PL[k]$ denote the start-index of p_k in P_{concat}, i.e. $PL[k] = 1 + \sum_{k' \in [1..k)} (|p_{k'}| + 1)$. For $j \in [1..|P_{concat}|]$, let $PI[j]$ be the index of the element of P covering index j in P_{concat}. That is, $PI[j] = k$

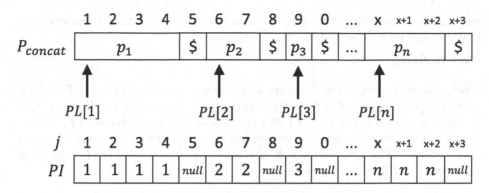

Fig. 3. Example of PL and PI.

where $j \in [PL[k]..PL[k] + |p_k| - 1]$ if such k exists, otherwise $PI[j] = null$. See Fig. 3 for an example.

Next, we construct the suffix array SAP of P_{concat}. We can then compute $successorOffset$. The general idea is that for each p_k, we find all its occurrences using the suffix array, and when the occurrence falls within some $p_{k'} \in P$, we update $successorOffset[k']$. A more detailed description is as follows:

- Initialize $successorOffset[k] = \infty$ for all $p_k \in P$.
- Using the LCP array, find the subinterval $[\ell..r]$ in SAP that corresponds to occurrences of p_k, i.e. each of $SAP[x]$ for all $x \in [\ell..r]$ corresponds to some start-index of occurrences of p_k in P_{concat}.
- For each such occurrence, whenever it falls within $p_{k'}[2..|p_{k'}| - 1]$ for some $p_{k'} \in P$, then assign to $successorOffset[k']$ the minimum value between itself and the offset distance $SAP[x] - PL[k']$. Formally, for all $x \in [\ell..r]$, if $P_{concat}[SAP[x] + |p|] \neq \$$, assign to $successorOffset[PI[SAP[x]]]$ the following value:

$$\min\left(successorOffset[PI[SAP[x]]], SAP[x] - PL[PI[SAP[x]]]\right) \qquad (1)$$

Computing P_{concat} and PI trivially takes $O(\|P\|)$ time. To compute the subinterval corresponding to occurrences of p_k, simply find the longest subinterval $[\ell..r]$ of SAP that includes the index $SAP^{-1}[PL[k]]$ and $LCP[x] \geq |p_k|$ for all $x \in [\ell + 1..r]$. This takes linear time w.r.t. to the subinterval length, which all adds up to the total number of occurrences of elements of P in P_{concat}. As no occurrence may share a start-index, it is bounded by $|P_{concat}| \in O(\|P\|)$. Thus, this preprocessing takes $O(\|P\|)$ time, in addition to the time required to compute SAP and LCP which depends on the alphabet.

3.5 Computing *excludeLeft*, *excludeRight*, and *start*

As discussed, the algorithm runs the Aho-Corasick automaton for S to check whether there is some $s \in S$ that occurs as a suffix of T_i, for each iteration i. In case such s exists, all of the computations below are performed, otherwise only maintaining *start* is necessary.

excludeRight

Exclusion by s. When $s \in S$ occurs as a suffix of T_i, the suffix $u = T_i[j..i]$ for each $j \in pList$ is excluded from the solution iff $j > i - |s| + 1$. That is, u fails to meet condition (a) of Sect. 3.1 iff j corresponds to an occurrence of $p \in P$ in $s[2..|s|]$. Thus, if we preprocess the number of occurrences of elements of P that occur within $s[2..|s|]$ for each $s \in S$, we can compute the number of elements of $pList$ excluded by s in constant time.

Example 3. In the case shown by Example 1, the elements 10 and 12 are excluded by s, which we can obtain from the fact that there are two occurrences of elements of P within $s[2..|s|] =$ old; one each of old and d.

A suffix array-based approach that is similar to what we used to compute *successorOffset* can be used here. Let

$$PS_{concat} = p_1\$ \cdots p_n\$s_1[2..|s_1|]\$ \cdots s_m[2..|s_m|],$$

and construct its suffix array $SAPS$. Define an array SI such that $SI[j] = k$ when j belongs to the part made up by $s_k[2..|s_k|]$ in PS_{concat}, similar to PI. Then, for all $p \in P$, compute the subinterval $[\ell..r]$ in $SAPS$ corresponding to suffixes of PS_{concat} that start with p, again using $SAPS^{-1}$ and LCP arrays. For all $x \in [\ell..r]$, then increment $sPCount[SI[SAPS[x]]]$ by one, where $sPCount$ is an array that maps each $s \in S$ to the number of occurrences of elements of P that occur within $s[2..|s|]$. Naturally, $sPCount$ initially maps all elements of S to zero before the counts are incremented.

The total of size of subintervals is bounded by the number of occurrences of elements of P in PS_{concat}, which is $O(|PS_{concat}|) = O(\|P\| + \|S\|)$. Thus, computing *successorOffset* takes $O(\|P\| + \|S\|)$ time, in addition to the construction time of $SAPS$ and LCP which depends on the alphabet. After the preprocessing, the number of elements of $pList$ excluded by s when s occurs as a suffix of T_i can be computed in constant time by simply referring to $sPCount[s]$.

Exclusion by lrs_i. It is known that Ukkonen's algorithm [22] maintains the locus of lrs_i during each iteration i where it builds the suffix tree of T_i, and thus we can compute its start-index $i - |lrs_i| + 1$ by simply running Ukkonen's algorithm. Then, clearly $j \in pList$ is excluded by condition (b) of Sect. 3.1 iff $j \geq i - |lrs_i| + 1$. Furthermore, the start-index is non-decreasing between iterations, i.e. $i - |lrs_i| + 1 \geq (i - 1) - |lrs_{i-1}| + 1$ for any iteration i. Thus, we can always maintain the count $|\{j \in pList \mid j \in [i - |lrs_i| + 1..i]\}|$ for each i as follows:

- During some iterations, $i - |lrs_i| + 1 > (i - 1) - |lrs_{i-1}| + 1$ which we will know from Ukkonen's algorithm. In that case, for each $j \in [(i-1) - |lrs_{i-1}| + 2..i - |lrs_i| + 1]$ such that $j \in pList$, increment the count by one.
- Whenever a new element j is added to $pList$ such that $j \in i - |lrs_i| + 1$, increment the count.

We can check whether $j \in pList$ in constant time for any j using $pArray$, and both the left end $i - |lrs_i| + 1$ and right end i of the interval only ever increases and is within 1 to i, so the above method takes cumulative $O(|T_i|)$ time, dominated by the runtime of Ukkonen's algorithm which is cumulative $O(|T_i| \log \sigma)$ time.

Exclusion by k_1. The same approach can be used to find the number of elements of $pList$ excluded by k_1: we maintain the number of $j \in pList$ such that $i - j + 1 < k_1 \Leftrightarrow j \in [i - k_1 + 2..i]$ for each iteration i. Since both endpoints of this interval can only increase and are always between 1 to i inclusive, maintaining the count can be done in cumulative $O(|T_i|)$ time.

excludeLeft and start

Maintaining excludeLeft. Similarly, *excludeLeft* is the number of $j \in pList$ such that $i - j + 1 > k_2 \Leftrightarrow j \in [1..i - k_2]$, and the same approach can be used to maintain this count in $O(|T_i|)$ time.

Maintaining start. We can easily maintain *start* so that it points to the minimum element of $pList$ of value at least $i + 1 - k_2$ in cumulative $O(|T_i|)$ time as follows:

- Initialize *start* to *null*.
- During each iteration i, if *start* is not *null* and $start < i + 1 - k_2$, then let *start* point to the next element in $pList$.
- Whenever a new element j is added to $pList$ such that $j \geq i + 1 - k_2$, if $start = null$ or $start > j$, let *start* point to j.

3.6   Summarizing the Algorithm

In the preprocessing, suffix arrays and LCP arrays are constructed and used to remove redundant elements of P and S, as well as compute *successorOffset*. Additionally, Aho-Corasick automata for P and S are built. For general ordered alphabets, this preprocessing takes $O((\|P\| + \|S\|) \log \sigma)$ time, with the construction time for Aho-Corasick automata and suffix arrays being the bottleneck. In the case of integer alphabets of size polynomial w.r.t. $\|P\| + \|S\|$, the construction times, and consequently the whole preprocessing time, can be reduced to $O(\|P\| + \|S\|)$ time.

For the query processing time up to any iteration i, *excludeLeft*, *excludeRight*, and *start* are computed in $O(|T_i| \log \sigma)$ cumulative time, giving us the solution for the counting problem. For the reporting problem, we traverse a total of $|ans_i|$ elements in $pList$, giving us $O(|T_i| \log \sigma + |ans_i|)$ cumulative time, assuming the solution strings are output in the form of index pairs.

Additionally, all the data structures used require only $O(|T_i| + \|P\| + \|S\|)$ total working space. Thus, we have the following results.

Theorem 1. *There are algorithms that solve the counting and solving problems from Definition 1 for general ordered alphabets, such that after $O((\|P\| + \|S\|) \log \sigma)$ preprocessing time, the solutions are output for each iteration up to i in $O(|T_i| \log \sigma)$ time for the counting problem, and $O(|T_i| \log \sigma + |ans_i|)$ time for the reporting problem, using $O(|T_i| + \|P\| + \|S\|)$ total working space.*

Corollary 1. *The preprocessing time in Theorem 1 can be reduced to $O(\|P\| + \|S\|)$ time in case of integer alphabets of size polynomial with regard to $\|P\| + \|S\|$.*

4   Applying the Algorithm for Traffic Classification

We show in Appendix A the input sets that match each of the application signatures described in [19].

Note that the signatures shown in [19] are generally characterized in the format of p followed by s, where $p \in P$ and $s \in S$ for sets or lists P and S. In general, this differs from our problem in that occurrences of p and s overlapping should not be counted as a match. For example, ab followed by bc means abc should not be counted as a match, while our algorithms do count this as a match. Nevertheless, such matches, which would be erroneous in the context of implementing the signatures, do not occur with the input sets listed in Appendix A, as the elements of P and S simply cannot overlap. For example, with Gnutella signatures each element of P ends with the character :, which no element of S contains, so no string $w \in \Sigma^*$ exists such that has some $p \in P$ as prefix, $s \in S$ as suffix, and p and s overlap (i.e. $|w| < |p| + |s|$). Note also that this problem also differs from the followed-by problem of [2,14], which does share the intolerance of such overlaps, but differs in that the inputs are given as pairs of p and s rather than pair of sets or lists P and S. Naively solving the pair-of-sets problem using algorithms for the pairs of p and s problem would take $|P| \times |S|$ queries, as we need one query for each pair of $p \in P$ and $s \in S$. This is clearly inefficient for large $|P|$ and $|S|$, and hence the necessity remains for our proposed algorithms.

5   Conclusion and Future Work

In this paper, we proposed online algorithms for counting and reporting all distinct substrings of an online text T that has some $p \in P$ as a prefix, some $s \in S$ as a suffix, and whose length is within the interval $[k_1..k_2]$, where P and S are static sequences of strings given as input for preprocessing, while positive integers k_1, k_2 and the characters of T are given as query. Our algorithms take $O((\|P\| + \|S\|) \log \sigma)$ preprocessing time for general ordered alphabets, which is reduced to $O(\|P\| + \|S\|)$ time for integer alphabets of size polynomial w.r.t. $\|P\| + \|S\|$. The computation up to the i-th character of T takes $O(|T_i| \log \sigma)$ cumulative time for the counting problem, and $O(|T_i| \log \sigma + |ans_i|)$ cumulative time for the reporting problem. Furthermore, we have shown that it has possible applications in traffic classification, by showing that all of the application signatures in [19] can be represented as input sets of our proposed problems.

A few problems remain to be considered as future work:

- As the discussion in Sect. 4 implies, solving the problem where the prefix and suffix strings are not allowed to overlap, i.e. substrings are in the form of $p\Sigma^k s$, where $k \in [k_1..k_2], p \in P, s \in S$, while retaining the distinctness condition as well as that the input sets be given as pairs of lists P and S, can

be useful in case we have a signature implemented with P, S such that there does exist a string w such that w has $p \in P$ as prefix, $s \in S$ as suffix and $|w| < |p| + |s|$, and we want to exclude such w from matches. Is it possible to devise an algorithm that solve this problem efficiently?
- In practice, how do the running times of our algorithms compare to the signature implementations used in [19]?

Acknowledgements. This work was supported by JSPS KAKENHI Grant Numbers JP20H04141 (HB) and JP22H03551 (SI), and by JST PRESTO Grant Number JPMJPR1922 (SI).

A   Appendix

Below, we show the input sets that match the each of the application signatures described in [19] (Table 1).

Table 1. Input sets corresponding to application signatures

Application	List of P elements	List of S elements	k_1	k_2
Gnutella	User — Agent :, UserAgent :, Server :	LimeWire, BearShare, Gnucleus, MorpheusOS, XoloX, MorpheusPE, gtkgnutella, Acquisition, Mutella — 0.4.1, MyNapster, Mutella0.4.1, Mutella — 0.4, Qtella, AquaLime, NapShare, Comeback, Go, PHEX, SwapNut, Mutella — 0.4.0, Shareaza, Mutella — 0.3.9b, Morpheus, FreeWire, Openext, Mutella — 0.3.3, Phex	1	∞
eDonkey	0xe3 (in hex)	(the packet length)	5-byte long	
DirectConnect	$MyNick, $Lock, $Key, $Direction, $GetListLen, $ListLen, $MaxedOut, $Error, $Send, $Get, $FileLength, $Canceled, $HubName, $ValidateNick, $ValidateDenide, $GetPass, $MyPass, $BadPass, $Version, $Hello, $LogedIn, $MyINFO, $GetINFO, $GetNickList, $NickList, $OpList, $To, $ConnectToMe, $MultiConnectToMe, $RevConnectToMe, $Search, $MultiSearch, $SR, $Kick, $OpForceMove, $ForceMove, $Quit	|	1	∞
BitTorrent	the 20-byte string where the first byte is 19 (0 × 13) and the next 19 bytes are the string 19BitTorrent protocol		20-byte long	
Kazaa	GET, HTTP	X — Kazaa	1	∞

References

1. Amir, A., Levy, A., Porat, E., Shalom, B.R.: Online recognition of dictionary with one gap. Inf. Comput. **275**, 104633 (2020)

2. Baeza-Yates, R.A., Gonnet, G.H.: Fast text searching for regular expressions or automaton searching on tries. J. ACM (JACM) **43**(6), 915–936 (1996)
3. Choi, Y.H., Jung, M.Y., Seo, S.W.: L+ 1-mwm: a fast pattern matching algorithm for high-speed packet filtering. In: IEEE INFOCOM 2008-The 27th Conference on Computer Communications, pp. 2288–2296. IEEE (2008)
4. Crochemore, M., Rytter, W.: Text algorithms. Maxime Crochemore (1994)
5. Dori, S., Landau, G.M.: Construction of Aho Corasick automaton in linear time for integer alphabets. In: Apostolico, A., Crochemore, M., Park, K. (eds.) Combinatorial Pattern Matching, pp. 168–177. Springer, Berlin Heidelberg, Berlin, Heidelberg (2005). https://doi.org/10.1007/11496656_15
6. Fuchino, T., Harada, T., Tanaka, K., Mikawa, K.: Acceleration of packet classification using adjacency list of rules. In: 2019 28th International Conference on Computer Communication and Networks (ICCCN) (2019). https://doi.org/10.1109/icccn.2019.8846923
7. Gusfield, D.: Algorithms on Strings, Trees, and Sequences - Computer Science and Computational Biology. Cambridge University Press (1997). https://doi.org/10.1017/cbo9780511574931
8. Kärkkäinen, J., Sanders, P., Burkhardt, S.: Linear work suffix array construction. J. ACM (JACM) **53**(6), 918–936 (2006)
9. Kasai, T., Lee, G., Arimura, H., Arikawa, S., Park, K.: Linear-time longest-common-prefix computation in suffix arrays and its applications. In: Amir, A. (ed.) CPM 2001. LNCS, vol. 2089, pp. 181–192. Springer, Heidelberg (2001). https://doi.org/10.1007/3-540-48194-X_17
10. Kim, D.K., Sim, J.S., Park, H., Park, K.: Constructing suffix arrays in linear time. J. Discrete Algorithms **3**(2), 126–142 (2005). https://doi.org/10.1016/j.jda.2004.08.019
11. Leonard, L., Tanaka, K.: Suffix tree-based linear algorithms for multiple prefixes, single suffix counting and listing problems (2022). https://doi.org/10.48550/ARXIV.2203.16908
12. Levy, A., Shalom, B.R.: Online parameterized dictionary matching with one gap. Theoret. Comput. Sci. **845**, 208–229 (2020). https://doi.org/10.1016/j.tcs.2020.09.016
13. Makinen, V., Belazzougui, D., Cunial, F., Tomescu, A.I.: Genome-Scale Algorithm Design. Cambridge University Press, Cambridge, England (May (2015))
14. Manber, U., Baeza-Yates, R.: An algorithm for string matching with a sequence of don't cares. Inf. Process. Lett. **37**(3), 133–136 (1991). https://doi.org/10.1016/0020-0190(91)90032-D
15. Manber, U., Myers, G.: Suffix arrays: a new method for on-line string searches. SIAM J. Comput. **22**(5), 935–948 (1993). https://doi.org/10.1137/0222058
16. Mikawa, K., Tanaka, K.: Run-based trie involving the structure of arbitrary bitmask rules. IEICE Trans. Inf. Syst. **E98.D**(6), 1206–1212 (2015). https://doi.org/10.1587/transinf.2013EDP7087
17. Pampapathi, R., Mirkin, B., Levene, M.: A suffix tree approach to anti-spam email filtering. Mach. Learn. **65**(1), 309–338 (2006). https://doi.org/10.1007/s10994-006-9505-y
18. Ruiz, J., España, S., García, P.: Locally threshold testable languages in strict sense: application to the inference problem. In: Honavar, V., Slutzki, G. (eds.) ICGI 1998. LNCS, vol. 1433, pp. 150–161. Springer, Heidelberg (1998). https://doi.org/10.1007/BFb0054072

19. Sen, S., Spatscheck, O., Wang, D.: Accurate, scalable in-network identification of p2p traffic using application signatures. In: Proceedings of the 13th International Conference on World Wide Web, pp. 512–521. WWW 2004, Association for Computing Machinery, New York, NY, USA (2004). https://doi.org/10.1145/988672.988742
20. Shalom, B.R.: Parameterized dictionary matching and recognition with one gap. Theoret. Comput. Sci. **854**, 1–16 (2021). https://doi.org/10.1016/j.tcs.2020.11.017
21. Tongaonkar, A.S.: Fast pattern-matching techniques for packet filtering. Ph.D. thesis, Stony Brook University (2004)
22. Ukkonen, E.: On-line construction of suffix trees. Algorithmica **14**(3), 249–260 (1995). https://doi.org/10.1007/BF01206331

The Complexity of the Co-occurrence Problem

Philip Bille[ID], Inge Li Gørtz[ID], and Tord Stordalen$^{(\boxtimes)}$[ID]

Technical University of Denmark, DTU Compute, Kgs. Lyngby, Denmark
{phbi,inge,tjost}@dtu.dk

Abstract. Let S be a string of length n over an alphabet Σ and let Q be a subset of Σ of size $q \geq 2$. The *co-occurrence problem* is to construct a compact data structure that supports the following query: given an integer w return the number of length-w substrings of S that contain each character of Q at least once. This is a natural string problem with applications to, e.g., data mining, natural language processing, and DNA analysis. The state of the art is an $O(\sqrt{nq})$ space data structure that—with some minor additions—supports queries in $O(\log \log n)$ time [CPM 2021].

Our contributions are as follows. Firstly, we analyze the problem in terms of a new, natural parameter d, giving a simple data structure that uses $O(d)$ space and supports queries in $O(\log \log n)$ time. The preprocessing algorithm does a single pass over S, runs in expected $O(n)$ time, and uses $O(d + q)$ space in addition to the input. Furthermore, we show that $O(d)$ space is optimal and that $O(\log \log n)$-time queries are optimal given optimal space. Secondly, we bound $d = O(\sqrt{nq})$, giving clean bounds in terms of n and q that match the state of the art. Furthermore, we prove that $\Omega(\sqrt{nq})$ bits of space is necessary in the worst case, meaning that the $O(\sqrt{nq})$ upper bound is tight to within polylogarithmic factors. All of our results are based on simple and intuitive combinatorial ideas that simplify the state of the art.

Keywords: Strings · Data structures · Lower bounds

1 Introduction

We consider the *co-occurrence problem* which is defined as follows. Let S be a string of length n over an alphabet Σ and let Q be a subset of Σ of size $q \geq 2$. For two integers i and j where $1 \leq i \leq j \leq n$, let $[i, j]$ denote the discrete interval $\{i, i+1, \ldots, j\}$, and let $S[i, j]$ denote the substring of S starting at $S[i]$ and ending at $S[j]$. The interval $[i, j]$ is a *co-occurrence* of Q in S if $S[i, j]$ contains each character in Q at least once. The goal is to preprocess S and Q into a data structure that supports the query

Philip Bille and Inge Li Gørtz are supported by Danish Research Council grant DFF-8021-002498.

D. Arroyuelo and B. Poblete (Eds.): SPIRE 2022, LNCS 13617, pp. 38–52, 2022.
https://doi.org/10.1007/978-3-031-20643-6_4

- $\mathrm{co}_{S,Q}(w)$: return the number of co-occurrences of Q in S that have length w, i.e., the number of length-w substrings of S that contain each character in Q at least once.

For example, let $\Sigma = \{\mathtt{A}, \mathtt{B}, \mathtt{C}, \mathtt{-}\}$, $Q = \{\mathtt{A}, \mathtt{B}, \mathtt{C}\}$ and

$$S = \underset{1}{\mathtt{-}}\ \underset{2}{\mathtt{-}}\ \underset{3}{\mathtt{-}}\ \underset{4}{\mathtt{-}}\ \underset{5}{\mathtt{B}}\ \underset{6}{\mathtt{C}}\ \underset{7}{\mathtt{-}}\ \underset{8}{\mathtt{A}}\ \underset{9}{\mathtt{C}}\ \underset{10}{\mathtt{C}}\ \underset{11}{\mathtt{B}}\ \underset{12}{\mathtt{-}}\ \underset{13}{\mathtt{-}}\ .$$

Then

- $\mathrm{co}_{S,Q}(3) = 0$, because no length-three substring contains all three characters A, B, and C.
- $\mathrm{co}_{S,Q}(4) = 2$, because both $[5, 8]$ and $[8, 11]$ are co-occurrences of Q.
- $\mathrm{co}_{S,Q}(8) = 6$, because all six of the length-eight substrings of S are co-occurrences of Q.

Note that only sublinear-space data structures are interesting. With linear space we can simply precompute the answer to $\mathrm{co}_{S,Q}(i)$ for each $i \in [0, n]$ and support queries in constant time.

This is a natural string problem with applications to, e.g., data mining, and a large amount of work has gone towards related problems such as finding frequent items in streams [6–8,10] and finding frequent sets of items in streams [1,3,4,9,11,12,16]. Furthermore, it is similar to certain string problems, such as episode matching [5] where the goal is to determine all the substrings of S that occur a certain number of times within a given distance from each other. Whereas previous work is mostly concerned with identifying frequent patterns either in the whole string or in a sliding window of fixed length, Sobel, Bertram, Ding, Nargesian and Gildea [14] introduced the problem of studying a given pattern across all window lengths (i.e., determining $\mathrm{co}_{S,Q}(i)$ for all i). They motivate the problem by listing potential applications such as training models for natural language processing (short and long co-occurrences of a set of words tend to represent respectively syntactic and semantic information), automatically organizing the memory of a computer program for good cache behaviour (variables that are used close to each other should be near each other in memory), and analyzing DNA sequences (co-occurrences of nucleotides in DNA provide insight into the evolution of viruses). See [14] for a more detailed discussion of these applications.

Our work is inspired by [14]. They do not consider fast, individual queries, but instead they give an $O(\sqrt{nq})$ space data structure from which they can determine $\mathrm{co}_{S,Q}(i)$ for each $i = 1, \ldots, n$ in $O(n)$ time. Supporting fast queries is a natural extension to their problem, and we note that their solution can be extended to support individual queries in $O(\log \log n)$ time using the techniques presented below.

A key component of our result is a solution to the following simplified problem. A co-occurrence $[i, j]$ is *left-minimal* if $[i + 1, j]$ is not a co-occurrence. The *left-minimal co-occurrence problem* is to preprocess S and Q into a data structure that supports the query

– $\mathsf{lmco}_{S,Q}(w)$: return the number of left-minimal co-occurrences of Q in S that have length w.

We first solve this more restricted problem, and then we solve the co-occurrence problem by a reduction to the left-minimal co-occurrence problem. To our knowledge this problem has not been studied before.

1.1   Our Results

Our two main contributions are as follows. Firstly, we give an upper bound that matches and simplifies the state of the art. Secondly, we provide lower bounds that show that our solution has optimal space, and that our query time is optimal for optimal-space data structures. As in previous work, all our results work on the word RAM model with logarithmic word size.

To do so we use the following parametrization. Let $\delta_{S,Q}$ be the difference encoding of the sequence $\mathsf{lmco}_{S,Q}(1), \ldots, \mathsf{lmco}_{S,Q}(n)$. That is, $\delta_{S,Q}(i) = \mathsf{lmco}_{S,Q}(i) - \mathsf{lmco}_{S,Q}(i-1)$ for each $i \in [2, n]$ (note that $\mathsf{lmco}(1) = 0$ since $|Q| \geq 2$). Let $Z_{S,Q} = \{i \in [2, n] \mid \delta(i) \neq 0\}$ and let $d_{S,Q} = |Z_{S,Q}|$. For the remainder of the paper we will omit the subscript on lmco, co, Z, and d whenever S and Q are clear from the context. Note that d is a parameter of the problem since it is determined exclusively by the input S and Q. We prove the following theorem.

Theorem 1. *Let S be a string of length n over an alphabet Σ, let Q be a subset of Σ of size $q \geq 2$, and let d be defined as above.*

(a) *There is an $O(d)$ space data structure that supports both $\mathsf{lmco}_{S,Q}$- and $\mathsf{co}_{S,Q}$-queries in $O(\log \log n)$ time. The preprocessing algorithm does a single pass over S, runs in expected $O(n)$ time and uses $O(d + q)$ space in addition to the input.*

(b) *Any data structure supporting either $\mathsf{lmco}_{S,Q}$- or $\mathsf{co}_{S,Q}$-queries needs $\Omega(d)$ space in the worst case, and any $d \log^{O(1)} d$ space data structure cannot support queries faster than $\Omega(\log \log n)$ time.*

(c) *The parameter d is bounded by $O(\sqrt{nq})$, and any data structure supporting either $\mathsf{lmco}_{S,Q}$- or $\mathsf{co}_{S,Q}$-queries needs $\Omega(\sqrt{nq})$ bits of space in the worst case.*

Theorem 1(a) and 1(b) together prove that our data structure has optimal space, and that with optimal space we cannot hope to support queries faster than $O(\log \log n)$ time. In comparison to the state of the art by Sobel et al. [14], Theorem 1(c) proves that we match their $O(\sqrt{nq})$ space and $O(\log \log n)$ time solution, and also that the $O(\sqrt{nq})$ space bound is tight to within polylogarithmic factors. All of our results are based on simple and intuitive combinatorial ideas that simplify the state of the art.

Given a set X of m integers from a universe U, the *static predecessor problem* is to represent X such that we can efficiently answer the query $\mathsf{predecessor}(x) = \max\{y \in X \mid y \leq x\}$. Tight bounds by Pătraşcu and Thorup [13] imply that

$O(\log \log |U|)$-time queries are optimal with $m \log^{O(1)} m$ space when $|U| = m^c$ for any constant $c > 1$. The lower bound on query time in Theorem 1(b) follows from the following theorem, which in turn follows from a reduction from the predecessor problem to the (left-minimal) co-occurrence problem.

Theorem 2. *Let $X \subseteq \{2, \ldots, u\}$ for some u and let $|X| = m$. Let n, q, and d be the parameters of the (left-minimal) co-occurrence problem as above. Given a data structure that supports lmco- or co-queries in $f_t(n, q, d)$ time using $f_s(n, q, d)$ space, we obtain a data structure that supports predecessor queries on X in $O(f_t(2u^2, 2, 8m))$ time using $O(f_s(2u^2, 2, 8m))$ space.*

In particular, if $f_s(n, q, d) = d \log^{O(1)} d$ then we obtain an $m \log^{O(1)} m$-space predecessor data structure on X. If also $u = m^c$, then it follows from the lower bound on predecessor queries that $f_t(2u^2, 2, 8m) = \Omega(\log \log u)$, which in turn implies that $f_t(n, q, d) = \Omega(\log \log n)$, proving the lower bound in Theorem 1(b).

The preprocessing algorithm and the proof of Theorem 2 can be found in Appendices A and B, respectively.

1.2 Techniques

The key technical insights that lead to our results stem mainly from the structure of δ.

To achieve the upper bound for lmco-queries we use the following very simple data structure. By definition, $\mathsf{lmco}(w) = \sum_{i=2}^{w} \delta(i)$. Furthermore, by the definition of Z it follows that for any $w \in [2, n]$ we have that $\mathsf{lmco}(w) = \mathsf{lmco}(w_p)$ where w_p is the predecessor of w in Z. Our data structure is a predecessor structure over the set of key-value pairs $\{(i, \mathsf{lmco}(i)) \mid i \in Z\}$ and answers lmco-queries with a single predecessor query. There are linear space predecessor structures that support queries in $O(\log \log |U|)$ time [15]. Here the universe U is $[2, n]$ so we match the $O(d)$ space and $O(\log \log n)$ time bound in Theorem 1(a).

Furthermore, we prove the $O(\sqrt{nq})$ upper bound on space by bounding $d = O(\sqrt{nq})$ using the following idea. In essence, each $\delta(z)$ for $z \in Z$ corresponds to some length-z *minimal co-occurrence*, which is a co-occurrence $[i, j]$ such that neither $[i + 1, j]$ nor $[i, j - 1]$ are co-occurrences (see below for the full details on δ). We bound the cumulative length of all the minimal co-occurrences to be $O(nq)$; then there are at most $d = O(\sqrt{nq})$ distinct lengths of minimal co-occurrences since $d = \omega(\sqrt{nq})$ implies that the cumulative length of the minimal co-occurrences is at least $1 + \ldots + d = \Omega(d^2) = \omega(nq)$.

To also support co-queries and complete the upper bound we give a straightforward reduction from the co-occurrence problem to the left-minimal co-occurrence problem. We show that by extending the above data structure to also store $\sum_{i=2}^{z} \mathsf{lmco}(i)$ for each $z \in Z$, we can support co-queries with the same bounds as for lmco-queries.

On the lower bounds side, we give all the lower bounds for the left-minimal co-occurrence problem and show that they extend to the co-occurrence problem. To prove the lower bounds we exploit that we can carefully design lmco-instances that result in a particular difference encoding δ by including minimal co-occurrences of certain lengths and spacing. Our lower bounds on space in

Theorem 1(b) and 1(c) are the results of encoding a given permutation or set in δ, respectively.

Finally, as mentioned above, we prove Theorem 2 (and, by extension, the lower bound on query time in Theorem 1(b)) by encoding a given instance of the static predecessor problem in an lmco-instance such that the predecessor of an element x equals $\mathsf{lmco}(x)$.

2 The Left-Minimal Co-occurrence Problem

2.1 Main Idea

Let $[i, j] \subseteq [1, n]$ be a co-occurrence of Q in S. Recall that then each character from Q occurs in $S[i, j]$ and that $[i, j]$ is left-minimal if $[i + 1, j]$ is not a co-occurrence. The goal is to preprocess S and Q to support the query $\mathsf{lmco}(w)$, which returns the number of left-minimal co-occurrences of length w.

We say that $[i, j]$ is a *minimal co-occurrence* if neither $[i + 1, j]$ nor $[i, j - 1]$ are co-occurrences and we denote the μ minimal co-occurrences of Q in S by $[\ell_1, r_1], \ldots, [\ell_\mu, r_\mu]$ where $r_1 < \ldots < r_\mu$. This ordering is unique since at most one minimal co-occurrence ends at a given index. To simplify the presentation we define $r_{\mu+1} = n + 1$. Note that also $\ell_1 < \ldots < \ell_\mu$ due to the following property.

Property 1. Let $[a, b]$ and $[a', b']$ be two minimal co-occurrences. Either both $a < a'$ and $b < b'$, or both $a' < a$ and $b' < b$.

Proof. If $a < a'$ and $b \geq b'$ then $[a, b]$ strictly contains another minimal co-occurrence $[a', b']$ and can therefore not be minimal itself. The other cases are analogous. $\square$

We now show that given all the minimal co-occurrences we can determine all the left-minimal co-occurrences.

Lemma 1. *Let $[\ell_1, r_1], \ldots, [\ell_\mu, r_\mu]$ be the minimal co-occurrences of Q in S where $r_1 < \ldots < r_\mu$ and let $r_{\mu+1} = n + 1$. Then*

(a) *there are no left-minimal co-occurrences that end before r_1, i.e., at an index $k < r_1$, and*

(b) *for each index k where $r_i \leq k < r_{i+1}$ for some i, the left-minimal co-occurrence ending at k starts at ℓ_i.*

Proof. See Fig. 1 for an illustration of the proof.

(a) If $[j, k]$ is a left-minimal co-occurrence where $k < r_1$, it must contain some minimal co-occurrence that ends before r_1—obtainable by shrinking $[j, k]$ maximally—leading to a contradiction.

(b) Let $r_i \leq k < r_{i+1}$. Then $[\ell_i, k]$ is a co-occurrence since it contains $[\ell_i, r_i]$. We show that it is left-minimal by showing that $[\ell_i + 1, k]$ is not a co-occurrence. If it were, it would contain a minimal co-occurrence $[s, t]$ where $\ell_i < s$ and $t < r_{i+1}$. By Property 1, $\ell_i < s$ implies that $r_i < t$. However, then $r_i < t < r_{i+1}$, leading to a contradiction. $\square$

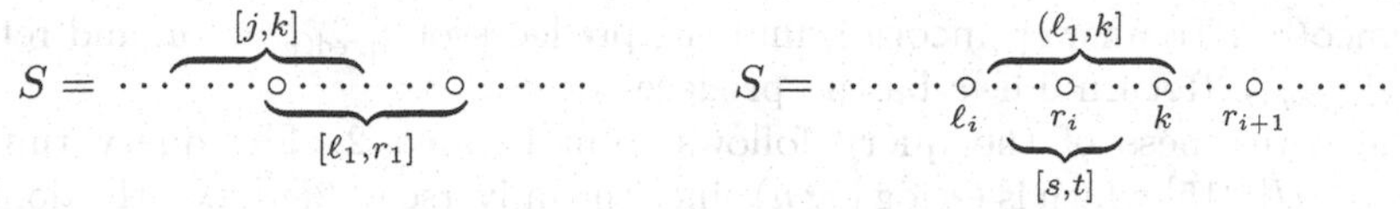

$$S = \cdots\cdots\overbrace{\cdots\underset{\ell_1}{\circ}\cdots\cdots\underset{}{\circ}\cdots\cdots}^{[j,k]} \qquad S = \cdots\cdots\overbrace{\underset{\ell_i}{\circ}\underset{r_i}{\cdots\circ\cdots}\underset{k}{\circ}\cdots\underset{r_{i+1}}{\circ}\cdots}^{(\ell_1,k]}$$

Fig. 1. *Left (Lemma 1(a)):* Any left-minimal co-occurrence $[j, k]$ must contain a minimal co-occurrence ending at or before k. If $k < r_1$ this contradicts that r_1 is the smallest endpoint of a minimal co-occurrence. *Right (Lemma 1(b)):* $[\ell_i, k]$ is a co-occurrence because it contains $[\ell_i, r_i]$. However, $[\ell_i + 1, k]$ is not a co-occurrence; if it were it would contain a minimal co-occurrence $[s, t]$ that ends between r_i and k, leading to a contradiction since $r_i < t < r_{i+1}$.

Let $\mathsf{len}(i, j) = j - i + 1$ denote the length of the interval $[i, j]$. Lemma 1 implies that each minimal co-occurrence $[\ell_i, r_i]$ gives rise to one additional left-minimal co-occurrence of length k for $k = \mathsf{len}(\ell_i, r_i), \ldots, \mathsf{len}(\ell_i, r_{i+1}) - 1$. Also, note that each left-minimal co-occurrence is determined by a minimal co-occurrence in this manner. Therefore, $\mathsf{lmco}(w)$ equals the number of minimal co-occurrences $[\ell_i, r_i]$ where $\mathsf{len}(\ell_i, r_i) \le w < \mathsf{len}(\ell_i, r_{i+1})$. Recall that $\delta(i) = \mathsf{lmco}(i) - \mathsf{lmco}(i - 1)$ for $i \in [2, n]$. Since $\mathsf{lmco}(1) = 0$ (because $|Q| \ge 2$) we have $\mathsf{lmco}(w) = \sum_{i=2}^{w} \delta(i)$. It follows that

$$\delta(w) = \sum_{i=1}^{\mu} \begin{cases} 1 & \text{if } \mathsf{len}(\ell_i, r_i) = w \\ -1 & \text{if } \mathsf{len}(\ell_i, r_{i+1}) = w \\ 0 & \text{otherwise} \end{cases}$$

since the contribution of each $[\ell_i, r_i]$ to the sum $\sum_{i=2}^{w} \delta(i)$ is one if $\mathsf{len}(\ell_i, r_i) \le w < \mathsf{len}(\ell_i, r_{i+1})$ and zero otherwise. We say that $[\ell_i, r_i]$ *contributes* plus one and minus one to $\delta(\mathsf{len}(\ell_i, r_i))$ and $\delta(\mathsf{len}(\ell_i, r_{i+1}))$, respectively.

However, note that only the non-zero $\delta(\cdot)$-entries affect the result of lmco-queries. Denote the elements of Z by $z_1 < z_2 < \ldots < z_d$ and define $\mathsf{pred}(w)$ such that $z_{\mathsf{pred}(w)}$ is the predecessor of w in Z, or $\mathsf{pred}(w) = 0$ if w has no predecessor. We get the following lemma.

Lemma 2. *For any $w \in [z_1, n]$ we have that*

$$\mathsf{lmco}(w) = \sum_{i=1}^{\mathsf{pred}(w)} \delta(z_i) = \mathsf{lmco}(z_{\mathsf{pred}(w)}).$$

For $w \in [0, z_1)$, w has no predecessor in Z and $\mathsf{lmco}(w) = 0$.

Proof. The proof follows from the fact that $\mathsf{lmco}(w) = \sum_{i=2}^{w} \delta(i)$ and $\delta(i) = 0$ for each $i \notin Z$. $\square$

2.2   Data Structure

The contents of the data structure are as follows. Store the linear space predecessor structure from [15] over the set Z, and for each key $z_i \in Z$ store the data $\mathsf{lmco}(z_i)$. To answer $\mathsf{lmco}(w)$, find the predecessor $z_{\mathsf{pred}(w)}$ of w and return $\mathsf{lmco}(z_{\mathsf{pred}(w)})$. Return 0 if w has no predecessor.

The correctness of the query follows from Lemma 2. The query time is $O(\log\log|U|)$ [15] which is $O(\log\log n)$ since the universe is $[2, n]$, i.e., the domain of δ. The predecessor structure uses $O(|Z|) = O(d)$ space, which we now show is $O(\sqrt{nq})$. We begin by bounding the cumulative length of the minimal co-occurrences.

Lemma 3. *Let* $[\ell_1, r_1], \ldots, [\ell_\mu, r_\mu]$ *be the minimal co-occurrences of* Q *in* S. *Then*

$$\sum_{i=1}^{\mu} \mathsf{len}(\ell_i, r_i) = O(nq).$$

Proof. We prove that for each $k \in [1, n]$ there are at most q minimal co-occurrences $[\ell_i, r_i]$ where $k \in [\ell_i, r_i]$; the statement in the lemma follows directly. Suppose that there are $q' > q$ minimal co-occurrences $[s_1, t_1], \ldots, [s_{q'}, t_{q'}]$ that contain k and let $t_1 < \ldots < t_{q'}$. By Property 1, and because each minimal occurrence contains k, we have

$$s_1 < \ldots < s_{q'} \leq k \leq t_1 < \ldots < t_{q'}$$

Furthermore, for each s_i we have that $S[s_i] = p$ for some $p \in Q$; otherwise $[s_i + 1, t_i]$ would be a co-occurrence and $[s_i, t_i]$ would not be minimal. Since $q' > q$ there is some $p \in Q$ that occurs twice as the first character, i.e., such that $S[s_i] = S[s_j] = p$ for some $i < j$. However, then $[s_i + 1, t_i]$ is a co-occurrence because it still contains $S[s_j] = p$, contradicting that $[s_i, t_i]$ is minimal. $\square$

By the definition of δ, we have that $\delta(k) \neq 0$ only if there is some minimal co-occurrence $[\ell_i, r_i]$ such that either $\mathsf{len}(\ell_i, r_i) = k$ or $\mathsf{len}(\ell_i, r_{i+1}) = k$. Using this fact in conjunction with Lemma 3 we bound the sum of the elements in Z.

$$\sum_{z \in Z} z = \sum_{k \text{ where } \delta(k) \neq 0} k \; \leq \sum_{i=1}^{\mu} \mathsf{len}(\ell_i, r_i) + \mathsf{len}(\ell_i, r_{i+1})$$

$$= \sum_{i=1}^{\mu} \mathsf{len}(\ell_i, r_i) + \Big(\mathsf{len}(\ell_i, r_i) + \mathsf{len}(r_i + 1, r_{i+1}) \Big)$$

$$= \sum_{i=1}^{\mu} 2 \cdot \mathsf{len}(\ell_i, r_i) + \sum_{i=1}^{\mu} \mathsf{len}(r_i + 1, r_{i+1})$$

$$= O(nq) + O(n)$$

Since the sum over Z is at most $O(nq)$ we must have $d = O(\sqrt{nq})$, because with $d = \omega(\sqrt{nq})$ distinct elements in Z we have

$$\sum_{z \in Z} z \geq 1 + 2 + \ldots + d = \Omega(d^2) = \omega(nq).$$

3   The Co-occurrence Problem

Recall that $co(w)$ is the number of co-occurrences of length w, as opposed to the number of left-minimal co-occurrences of length w. That is, $co(w)$ counts the number of co-occurrences among the intervals $[1, w], [2, w+1], \ldots, [n-w+1, n]$. We reduce the co-occurrence problem to the left-minimal co-occurrence problem as follows.

Lemma 4. *Let S be a string over an alphabet Σ, let $Q \subseteq \Sigma$ and let* lmco *be defined as above. Then*

$$co(w) = \left(\sum_{i=2}^{w} lmco(i) \right) - \max(w - r_1, 0).$$

Proof. For any index $k \geq w$, the length-w interval $[k - w + 1, k]$ ending at k is a co-occurrence if and only if the length of the left-minimal co-occurrence ending at index k is at most w. The sum $\sum_{i=2}^{w} lmco(i)$ counts the number of indices $j \in [1, n]$ such that the left-minimal co-occurrence ending at j has length at most w. However, this also includes the left-minimal co-occurrences that end at any index $j \in [r_1, w - 1]$. While all of these have length at most $w - 1$, none of the length-w intervals that end in the range $[r_1, w-1]$ correspond to substrings of S, so they are not co-occurrences. Therefore, the sum $\sum_{i=2}^{w} lmco(i)$ overestimates $co(w)$ by $w - r_1$ if $r_1 < w$ and by 0 otherwise. $\square$

We show how to represent the sequence $\sum_{i=2}^{2} lmco(i), \ldots, \sum_{i=2}^{n} lmco(i)$ compactly, in a similar way to what we did for lmco-queries. Recall that the elements of Z are denoted by $z_1 < \ldots < z_d$, that $z_{pred(x)}$ is the predecessor of x in Z, and that $pred(x) = 0$ if x has no predecessor. Then, for any $w \geq 2$ we get that

$$\sum_{i=2}^{w} lmco(i) = \sum_{i=2}^{w} \sum_{j=1}^{pred(i)} \delta(z_j)$$

$$= \sum_{k=1}^{pred(w)} \delta(z_k)(w - z_k + 1) \tag{1}$$

$$= \underbrace{(w + 1) \sum_{k=1}^{pred(w)} \delta(z_k) - \sum_{k=1}^{pred(w)} z_k \delta(z_k)}_{lmco(w)}$$

The first step follows by Lemma 2 and the second step follows because $\delta(z_k)$ occurs in $\sum_{j=1}^{pred(i)} \delta(z_j)$ for each of the $w - z_k + 1$ choices of $i \in [z_k, w]$.

To also support co-queries we extend our data structure from before as follows. For each z_k in the predecessor structure we store $\sum_{i=1}^{k} z_i \delta(z_i)$ in addition to $lmco(z_k)$. We also store r_1. Using Lemma 4 and Eq. 1 we can then answer co-queries with a single predecessor query and a constant amount of extra work, taking $O(\log \log n)$ time. The space remains $O(d) = O(\sqrt{nq})$. This completes the proof of Theorem 1(a), as well as the upper bound on space from Theorem 1(c).

4   Lower Bounds

In this section we show lower bounds on the space complexity of data structures that support lmco- or co-queries. In Sect. 4.1 we introduce a gadget that we use in Sect. 4.2 to prove that any data structure supporting lmco- or co-queries needs $\Omega(d)$ space (we use the same gadget in Appendix B to prove Theorem 2). In Sect. 4.3 we prove that any solution to the (left-minimal) co-occurrence problem requires $\Omega(\sqrt{nq})$ bits of space in the worst case.

All the lower bounds are proven by reduction to the left-minimal co-occurrence problem. However, they extend to data structures that support co-queries by the following argument. Store r_1 and any data structure that supports co on S and Q in time t per query. Then this data structure supports lmco-queries in $O(t)$ time, because by Lemma 4 we have that

$$co(w) - co(w-1)$$

$$= \left(\sum_{i=2}^{w} \mathsf{lmco}(i) - \max(w - r_1, 0) \right) - \left(\sum_{i=2}^{w-1} \mathsf{lmco}(i) - \max(w - 1 - r_1, 0) \right)$$

$$= \mathsf{lmco}(w) - \max(w - r_1, 0) + \max(w - 1 - r_1, 0)$$

4.1   The Increment Gadget

Let $Q = \{\mathtt{A}, \mathtt{B}\}$ and $U = \{2, \ldots, u\}$. For each $i \in U$ we define the *increment gadget*

$$G_i = \underbrace{\mathtt{A}\,\$\cdots\$\,\mathtt{B}}_{i}\ \underbrace{\$\cdots\$}_{u}$$

where $\$\cdots\$$ denotes a sequence of characters that are not in Q.

Lemma 5. *Let $Q = \{\mathtt{A}, \mathtt{B}\}$, $U = \{2, \ldots, u\}$, and let G_i be defined as above. Furthermore, for some $E = \{e_1, e_2, \ldots, e_m\} \subseteq U$ let S be the concatenation of $c_1 > 0$ copies of G_{e_1}, with $c_2 > 0$ copies of G_{e_2}, and so on. That is,*

$$S = \underbrace{G_{e_1} \cdots G_{e_1}}_{c_1}\ \cdots\cdots\cdots\ \underbrace{G_{e_m} \cdots G_{e_m}}_{c_m}$$

Then $\delta(e_i) = c_i$ for each $e_i \in E$ and $\delta(e) = 0$ for any $e \in U \setminus E$. Furthermore, $m \le d \le 8m$ and $n \le 2uC$ where $C = \sum_{i=1}^{m} c_i$ is the number of gadgets in S.

Proof. Firstly, $|G_j| = j + u \le 2u$ since $j \in U$, so the combined length of the C gadgets is at most $2uC$.

Now we prove that $\delta(e_i) = c_i$ for each $e_i \in E$ and $\delta(e) = 0$ for each $e \in U \setminus E$. Consider two gadgets G_j and G_k that occur next to each other in S.

$$\overbrace{\underbrace{\mathtt{A}\,\$\cdots\$\,\mathtt{B}}_{j}\ \underbrace{\$\cdots\$}_{u}}^{G_j}\ \overbrace{\underbrace{\mathtt{A}\,\$\cdots\$\,\mathtt{B}}_{k}\ \underbrace{\$\cdots\$}_{u}}^{G_k}$$

Three of the minimal co-occurrences in S occur in these two gadgets. Denote them by $[s_1, t_1], [s_2, t_2]$ and $[s_3, t_3]$.

$$\overbrace{\text{A \$}\cdots\text{\$ B}}^{[s_1,t_1]}\overbrace{\text{\$}\cdots\text{\$ A}}^{[s_2,t_2]}\overbrace{\text{\$}\cdots\text{\$ B}}^{[s_3,t_3]}\text{\$}\cdots\text{\$}$$

The two first minimal co-occurrences start in G_j. They contribute

- plus one to $\delta(x)$ for $x \in \{\text{len}(s_1, t_1), \text{len}(s_2, t_2)\} = \{j, u + 2\}$.
- minus one to $\delta(x)$ for $x \in \{\text{len}(s_1, t_2), \text{len}(s_2, t_3)\} = \{j + u + 1, k + u + 1\}$.

Hence, each occurrence of G_j contributes plus one to $\delta(j)$, and the remaining contributions are to $\delta(x)$ where $x \notin U$. The argument is similar also for the last gadget in S that has no other gadget following it. For each $e_i \in E$ there are c_i occurrences of G_{e_i} so $\delta(e_i) = c_i$. For each $e \in U \setminus E$ there are no occurrences of G_e so $\delta(e) = 0$.

Finally, note that each occurrence of $G_j G_k$ at different positions in S contributes to the same four $\delta(\cdot)$-entries. Therefore the number of distinct non-zero $\delta(\cdot)$-entries is linear in the number of distinct pairs (j, k) such that G_j and G_k occur next to each other in S. Here we have no more than $2m$ distinct paris since G_{e_i} is followed either by G_{e_i} or $G_{e_{i+1}}$. Each distinct pair contributes to at most four $\delta(\cdot)$-entries so $d \leq 8$ m. Finally each G_{e_i} contributes at least to $\delta(e_i)$ so $m \leq d$, concluding the proof.   □

4.2   Lower Bound on Space

We prove that any data structure supporting lmco-queries needs $\Omega(d)$ space in the worst case. Let U and Q be defined as in the increment gadget and let $P = p_2, \ldots, p_m$ be a sequence of length $m - 1$ where each $p_i \in U$ (the first element is named p_2 for simplicity). We let S be the concatenation of p_2 occurrences of G_2, with p_3 occurrences of G_3, and so on. That is,

$$S = \underbrace{G_2 \ldots G_2}_{p_2} \cdots\cdots\cdots \underbrace{G_m \ldots G_m}_{p_m}$$

Then any data structure supporting lmco on S and Q is a representation of P; by Lemma 5 we have that $\delta(i) = p_i$ for $i \in [2, m]$ and by definition we have $\delta(i) = \text{lmco}(i) - \text{lmco}(i - 1)$.

The sequence P can be any one of $(u - 1)^{m-1}$ distinct sequences, so any representation of P requires

$$\log((u - 1)^{m-1}) = (m - 1)\log(u - 1) = \Omega(m \log u)$$

bits—or $\Omega(m)$ words—in the worst case. By Lemma 5 this is $\Omega(d)$.

4.3 Lower Bound on Space in Terms of n and q

Here we prove that any data structure supporting lmco needs $\Omega(\sqrt{nq})$ bits of space in the worst case.

The main idea is as follows. Given an integer α and some $k \in \{2,\ldots,\alpha\}$, let V be the set of *even* integers from $\{k+1,\ldots,k\alpha\}$, and let T be some subset of V. We will construct an instance S and Q where

- the size of Q is $q = k$
- the length of S is $n = O(k\alpha^2)$
- for each $i \in V$ we have $\delta(i) = 1$ if and only if $i \in T$.

Then, as above, any data structure supporting lmco-queries on S and Q is a representation of T since $\delta(i) = \text{lmco}(i) - \text{lmco}(i-1)$. There are $2^{\Omega(k\alpha)}$ choices for T, so any representation of T requires

$$\log 2^{\Omega(k\alpha)} = \Omega(k\alpha) = \Omega(\sqrt{k^2\alpha^2}) = \Omega(\sqrt{nq})$$

bits in the worst case.

The reduction is as follows. Let $Q = \{c_1,\ldots,c_k\}$ and let \$ be a character not in Q. Assume for now that $|T|$ is a multiple of $k-1$ and partition T arbitrarily into $t = O(\alpha)$ sets $T_1,\ldots,T_t$, each of size $k-1$. Consider $T_j = \{e_1,\ldots,c_{k-1}\}$ where $e_1 < e_2 < \ldots < e_{k-1}$. We encode T_j in the gadget R_j where

- the length of R_j is $3k\alpha$.
- $R_j[1,k] = c_1 c_2 \ldots c_k$.
- $R_j[i + e_i] = c_i$ for each c_i except c_k. This is always possible since $i + e_i < (i+1) + e_{i+1}$.
- all other characters are \$.

See Fig. 2 for an illustration both of the layout of R_j and of the minimal co-occurrences contained within it. There are k minimal co-occurrences contained in R_j which we denote by $[s_1,t_1],\ldots,[s_k,t_k]$.

- The first one, $[s_1,t_1] = [1,k]$, starts and ends at the first occurrence of c_1 and c_k, respectively.
- For each $i \in [2,k]$, the minimal co-occurrence $[s_i,t_i]$ starts at the first occurrence of c_i and ends at the second occurrence of c_{i-1}, i.e., $[s_i,t_i] = [i, i-1+e_{i-1}]$.

Consider how these minimal co-occurrences contribute to δ. Each $[s_i,t_i]$ contributes plus one to $\delta(\text{len}(s_i,t_i))$. For the first co-occurrence, $\text{len}(s_1,t_1) = k$ (which is not a part of the universe V). Each of the other co-occurrences $[s_i,t_i]$ has length

$$\text{len}(s_i,t_i) = t_i - s_i + 1 = (i - 1 + e_{i-1}) - i + 1 = e_{i-1}$$

Therefore, the remaining minimal co-occurrences contribute plus one to of each the $\delta(\cdot)$-entries $e_1,\ldots,e_{k-1}$.

$$\overbrace{}^{\geq k\alpha}$$

$$\underset{1}{\mathsf{C}_1}\ \underset{2}{\mathsf{C}_2}\ \ldots\ \underset{k}{\mathsf{C}_k}\ \$\cdots\$\ \underset{1+e_1}{\mathsf{C}_1}\ \$\cdots\$\ \ldots\ldots\ \$\cdots\$\ \underset{k-1+e_{k-1}}{\mathsf{C}_{k-1}}\ \$\cdots\cdots\$$$

$$\overbrace{}^{[s_2,t_2]}$$

$$\underbrace{\mathsf{C}_1\ \mathsf{C}_2\ \ldots\ \mathsf{C}_k}_{[s_1,t_1]}\ \underbrace{\$\cdots\$\ \mathsf{C}_1\ \$\cdots\$\ \ldots\ldots\ \$\cdots\$\ \mathsf{C}_{k-1}\ \$\cdots\cdots\$}_{[s_k,t_k]}$$

Fig. 2. *Top:* Shows the layout of the gadget R_j, where $\$\cdots\$$ denotes a sequence of characters not in Q. The first k characters are $\mathsf{C}_1\ldots\mathsf{C}_k$. For $i \in [1, k-1]$ there is another occurrence of C_i at index $i + e_i$. Note that the second occurrence of C_1 occurs before the second occurrence of C_2, and so on. All other characters are $\$$. Since $|R_j| = 3k\alpha$ and $k-1+e_{k-1} \leq 2k\alpha$, R_j ends with at least $k\alpha$ characters that are not in Q. *Bottom:* Shows the k minimal co-occurrences in R_j denoted by $[s_1, t_1], \ldots, [s_k, t_k]$. Each of the $k-3$ minimal co-occurrences that are not depicted start at the first occurrence of some C_i and ends at the second occurrence of C_{i-1}.

Furthermore, each $[s_i, t_i]$ contributes negative one to $\delta(\mathsf{len}(s_i, t_{i+1}))$, where we define $t_{k+1} = |R_j|+1$. For $i < k$ we have that $\mathsf{len}(s_i, t_{i+1}) = 1 + \mathsf{len}(s_{i+1}, t_{i+1}) = 1 + e_i$ since $s_i + 1 = s_{i+1}$. Note that $1 + e_i$ is odd since e_i is even, and therefore not in V. For $i = k$, we get that $t_{i+1} = t_{k+1} = |R_j| + 1$. The last $k\alpha$ (at least) characters of R_j are not in Q, so $\mathsf{len}(s_k, |R_j| + 1) > k\alpha$ and therefore not in V.

Hence, R_j contributes plus one to $\delta(e_1), \ldots, \delta(e_{k-1})$ and does not contribute anything to $\delta(i)$ for any other $i \in V \setminus T_j$. To construct S, concatenate $R_1, \ldots, R_t$. Note that any minimal co-occurrence that crosses the boundary between two gadgets will only contribute to $\delta(i)$ for $i > k\alpha$ due to the trailing characters of each gadget that are not in Q. Since S consists of $t = O(\alpha)$ gadgets that each have length $O(k\alpha)$, we have $n = O(k\alpha^2)$ as stated above.

Finally, note that the assumption that $|T|$ is a multiple of $k - 1$ is not necessary. We ensure that the size is a multiple of $k - 1$ by adding at most $k - 2$ *even* integers from $\{k\alpha + 1, \ldots, 2k\alpha\}$ and adjusting the size of the gadgets accordingly. The reduction still works because we add even integers, the size of S is asymptotically unchanged, and any minimal co-occurrence due to the extra elements will have length greater than $k\alpha$ and will not contribute to any relevant δ-entries.

Acknowledgement. We would like to thank the anonymous reviewers for their comments, which improved the presentation of the paper.

A   Preprocessing

Finding Minimal Co-occurrences. To build the data structure, we need to find all the minimal co-occurrences in order to determine δ. For $j \geq r_1$, let $\mathsf{lm}(j)$ denote the length of the left-minimal co-occurrence ending at index j. By Lemma 1,

$\mathsf{lm}(r_i) = \mathsf{len}(\ell_i, r_i)$ for each $i \in [1, \mu]$. Furthermore, for $j \in [r_i + 1, r_{i+1} - 1]$ we have $\mathsf{lm}(j) = \mathsf{lm}(j - 1) + 1$ since both of the left-minimal co-occurrences ending at these two indices start at ℓ_i. However, $\mathsf{lm}(r_i) \leq \mathsf{lm}(r_i - 1)$ for each $i \in [2, \mu]$; the left-minimal co-occurrence ending at r_i starts at least one index further to the right than the left-minimal co-occurrence ending at $r_i - 1$ because $\ell_{i-1} < \ell_i$, so it cannot be strictly longer.

We determine $\ell_1, \ldots, \ell_\mu$ and $r_1, \ldots, r_\mu$ using the following algorithm. Traverse S and maintain $\mathsf{lm}(j)$ for the current index j. Whenever $\mathsf{lm}(j) \neq \mathsf{lm}(j-1)+1$ the interval $[j - \mathsf{lm}(j) + 1, \; j]$ is one of the minimal co-occurrences. Note that this algorithm finds the minimal co-occurrences in order by their rightmost endpoint. We maintain $\mathsf{lm}(j)$ as follows. For each character $p \in Q$ let $\mathsf{dist}(j, p)$ be the distance to the closest occurrence of p on the left of j. Then $\mathsf{lm}(j)$ is the maximum $\mathsf{dist}(j, \cdot)$-value. As in [14], we maintain the $\mathsf{dist}(j, \cdot)$-values in a linked list that is dynamically reordered according to the well-known *move-to-front* rule. The algorithm works as follows. Maintain a linked list over the elements in Q, ordered by increasing dist-values. Whenever you see some $p \in Q$, access its node in expected constant time through a dictionary and move it to the front of the list. The least recently seen $p \in Q$ (i.e., the p with the largest $\mathsf{dist}(j, \cdot)$-value) is found at the back of the list in constant time. The algorithm uses $O(q)$ space and expected constant time per character in S, thus it runs in expected $O(n)$ time.

Building the Data Structure We build the data structure as follows. Traverse S and maintain the two most recently seen minimal co-occurrences using the algorithm above. We maintain the non-zero $\delta(\cdot)$-values in a dictionary D that is implemented using chained hashing in conjunction with universal hashing [2]. When we find a new minimal co-occurrence $[\ell_{i+1}, r_{i+1}]$ we increment $D[\mathsf{len}(\ell_i, r_i)]$ and decrement $D[\mathsf{len}(\ell_i, r_{i+1})]$. Recall that $Z = \{z_1, \ldots, z_d\}$ where $z_j < z_{j+1}$ is defined such that $\delta(i) \neq 0$ if and only if $i \in Z$. After processing S the dictionary D encodes the set $\{(z_1, \delta(z_1)), \ldots, (z_d, \delta(z_d))\}$. Sort the set to obtain the array $E[j] = \delta(z_j)$. Compute the partial sum array over E, i.e. the array

$$F[j] = \sum_{i=1}^{j} E[i] = \sum_{i=1}^{j} \delta(z_i) = \mathsf{lmco}(z_j). \qquad \text{(we use 1-indexing)}$$

Build the predecessor data structure over Z and associate $\mathsf{lmco}(z_j)$ with each key z_j.

The algorithm for finding the minimal co-occurrences uses $O(q)$ space and the remaining data structures all use $O(d)$ space, for a total of $O(d + q)$ space. Finding the minimal co-occurrences and maintaining D takes $O(n)$ expected time, and so does building the predecessor structure from the sorted input.

Furthermore, we use the following sorting algorithm to sort the d entries in D with $O(d)$ extra space in expected $O(n)$ time. If $d < n/\log n$ we use merge sort which uses $O(d)$ extra space and runs in $O(d \log d) = O(n)$ time. If $d \geq n/\log n$ we use radix sort with base $\sqrt{n}$, which uses $O(\sqrt{n})$ extra space and $O(n)$ time. To elaborate, assume without loss of generality that $2k$ bits are necessary to represent n. We first distribute the elements into $2^k = O(\sqrt{n})$ buckets according to the most significant k bits of their binary representation, partially sorting the input. We then sort each bucket by distributing the elements in that bucket according to the *least* significant k bits of their binary representation, fully sorting the input. The algorithm runs in $O(n)$ time and uses $O(\sqrt{n}) = O(n/\log n) = O(d)$ extra space.

B   Lower Bound on Time

We now prove Theorem 2 by the following reduction from the predecessor problem. Let U, Q and G_i be as defined in Sect. 4.1 and let $X = \{x_1, x_2, \ldots, x_m\} \subseteq U$ where $x_1 < \ldots < x_m$. Define

$$S = \underbrace{G_{x_1} \cdots G_{x_1}}_{x_1} \underbrace{G_{x_2} \cdots G_{x_2}}_{x_2 - x_1} \cdots\cdots\cdots \underbrace{G_{x_m} \cdots G_{x_m}}_{x_m - x_{m-1}}$$

By Lemma 5 we have that $\delta(x_1) = x_1$, $\delta(x_i) = x_i - x_{i-1}$ for $i \in [2, m]$ and $\delta(i) = 0$ for $i \in U \setminus X$. Then, if the predecessor of some $x \in U$ is x_p, we have

$$\mathsf{lmco}(x) = \sum_{i=2}^{x} \delta(i) = x_1 + (x_2 - x_1) + \ldots + (x_p - x_{p-1}) = x_p$$

On the other hand, if $x < x_1$ then $\sum_{i=0}^{x} \delta(i) = 0$, unambiguously identifying that x has no predecessor.

Applying Lemma 5 again, we have $d \leq 8m$. Furthermore, there are $x_1 + (x_2 - x_1) + \ldots + (x_m - x_{m-1}) = x_m \leq u$ gadgets in total so $n \leq 2u^2$. Hence, given a data structure that supports lmco in $f_t(n, q, d)$ time using $f_s(n, q, d)$ space, we get a data structure supporting predecessor queries on X in $O(f_t(2u^2, 2, 8m))$ time and $O(f_s(2u^2, 2, 8m))$ space, proving Theorem 2.

References

1. Amagata, D., Hara, T.: Mining top-k co-occurrence patterns across multiple streams (extended abstract). In: Proceeding 34th ICDE, pp. 1747–1748 (2018). https://doi.org/10.1109/ICDE.2018.00231
2. Carter, L., Wegman, M.N.: Universal classes of hash functions. J. Comput. Syst. Sci. **18**(2), 143–154 (1979). https://doi.org/10.1016/0022-0000(79)90044-8
3. Chang, J.H., Lee, W.S.: Finding recently frequent item sets adaptively over online transactional data streams. Inf. Syst. **31**(8), 849–869 (2006). https://doi.org/10.1016/j.is.2005.04.001
4. Dallachiesa, M., Palpanas, T.: Identifying streaming frequent items in ad hoc time windows. Data Knowl. Eng. **87**, 66–90 (2013). https://doi.org/10.1016/j.datak.2013.05.007
5. Das, G., Fleischer, R., Gasieniec, L., Gunopulos, D., Kärkkäinen, J.: Episode matching. In: Proceeding 8th CPM, pp. 12–27 (1997). https://doi.org/10.1007/3-540-63220-4_46

6. Demaine, E.D., López-Ortiz, A., Munro, J.I.: Frequency estimation of internet packet streams with limited space. In: Proceeding 10th ESA, pp. 348–360 (2002). https://doi.org/10.1007/3-540-45749-6_33

7. Golab, L., DeHaan, D., Demaine, E.D., López-Ortiz, A., Munro, J.I.: Identifying frequent items in sliding windows over on-line packet streams. In: Proceeding 3rd ACM IMC, pp. 173–178 (2003). https://doi.org/10.1145/948205.948227

8. Karp, R.M., Shenker, S., Papadimitriou, C.H.: A simple algorithm for finding frequent elements in streams and bags. ACM Trans. Database Syst. **28**, 51–55 (2003). https://doi.org/10.1145/762471.762473

9. Li, H., Lee, S.: Mining frequent itemsets over data streams using efficient window sliding techniques. Expert Syst. Appl. **36**(2), 1466–1477 (2009). https://doi.org/10.1016/j.eswa.2007.11.061

10. Lim, Y., Choi, J., Kang, U.: Fast, accurate, and space-efficient tracking of time-weighted frequent items from data streams. In: Proceeding 23rd CIKM, pp. 1109–1118 (2014). https://doi.org/10.1145/2661829.2662006

11. Lin, C., Chiu, D., Wu, Y., Chen, A.L.P.: Mining frequent itemsets from data streams with a time-sensitive sliding window. In: Proceeding 5th SDM, pp. 68–79 (2005). https://doi.org/10.1137/1.9781611972757.7

12. Mozafari, B., Thakkar, H., Zaniolo, C.: Verifying and mining frequent patterns from large windows over data streams. In: Proceeding 24th ICDE, pp. 179–188 (2008). https://doi.org/10.1109/ICDE.2008.4497426

13. Patrascu, M., Thorup, M.: Randomization does not help searching predecessors. In: Proceedimg 18th SODA, pp. 555–564 (2007). https://dl.acm.org/citation.cfm?id=1283383.1283443

14. Sobel, J., Bertram, N., Ding, C., Nargesian, F., Gildea, D.: AWLCO: all-window length co-occurrence. In: Proceeding 32nd CPM, pp. 24:1–24:21. LIPIcs (2021). https://doi.org/10.4230/LIPIcs.CPM.2021.24

15. Willard, D.E.: Log-logarithmic worst-case range queries are possible in space $\Theta(N)$. Inf. Process. Lett. **17**(2), 81–84 (1983). https://doi.org/10.1016/0020-0190(83)90075-3

16. Yu, Z., Yu, X., Liu, Y., Li, W., Pei, J.: Mining frequent co-occurrence patterns across multiple data streams. In: Proceeding 1th EDBT, pp. 73–84 (2015). https://doi.org/10.5441/002/edbt.2015.08

String Data Structures

Reconstructing Parameterized Strings from Parameterized Suffix and LCP Arrays

Amihood Amir[1,2], Concettina Guerra[2], Eitan Kondratovsky[3],
Gad M. Landau[4,5], Shoshana Marcus[6(✉)], and Dina Sokol[7]

[1] Department of Computer Science, Bar-Ilan University, Ramat Gan, Israel
amir@esc.biu.ac.il
[2] College of Computing, Georgia Institute of Technology, 801 Atlantic Drive,
Atlanta, GA 30318, USA
[3] Cheriton School of Computer Science, Waterloo University, Waterloo, Canada
e2kondra@uwaterloo.ca
[4] Department of Computer Science, University of Haifa, Haifa 31905, Israel
landau@univ.haifa.ac.il
[5] NYU Tandon School of Engineering, New York University, Brooklyn, NY, USA
[6] Department of Mathematics and Computer Science, Kingsborough Community
College of the City University of New York, Brooklyn, NY, USA
shoshana.marcus@kbcc.cuny.edu
[7] Department of Computer and Information Science, Brooklyn College and The
Graduate Center, City University of New York, Brooklyn, NY, USA
sokol@sci.brooklyn.cuny.edu
https://u.cs.biu.ac.il/amir/, https://www.u.cs.biu.ac.il/kondrae/,
http://www.cs.haifa.ac.il/~landau/,
http://www.sci.brooklyn.cuny.edu/~sokol

Abstract. Reconstructing input from a data structure entails determining whether an instance of the data structure is in fact valid or not, and if valid, discovering the underlying data that it represents. In this paper we consider the parameterized suffix array (pSA) along with its corresponding parameterized longest-common-prefix (pLCP) array and solve the following problem. Given two arrays of numbers as input, A and P, does there exist a parameterized string S such that A is its pSA and P is its pLCP array? If the answer is positive, our algorithm produces a string S whose pSA is A and whose pLCP array is P. Although the naive approach would have to consider an exponential number of possibilities for such a string S, our algorithm's time complexity is only $O(n^2)$ for input arrays of size n.

A. Amir—Partially supported by Grant No. 2018141 from the United States-Israel Binational Science Foundation (BSF) and Israel Science Foundation Grant 1475-18.
C. Guerra—Partially supported by BSF Grant No. 2018141.
G. M. Landau—Partially supported by Grant No. 2018141 from the United States-Israel Binational Science Foundation (BSF) and Israel Science Foundation Grant 1475-18.
D. Sokol—Partially supported by BSF Grant No. 2018141.

D. Arroyuelo and B. Poblete (Eds.): SPIRE 2022, LNCS 13617, pp. 55–69, 2022.
https://doi.org/10.1007/978-3-031-20643-6_5

Keywords: Strings · Parameterized strings · Suffix array · Longest common prefix array

1 Introduction

Parameterized pattern matching, introduced by Baker [9,10], is a form of pattern matching that allows for interchange in the alphabet. More formally, a parameterized string (p-string) consists of characters from both a static alphabet Σ and a parameterized alphabet Π. Two p-strings of the same length are said to parameterized match (p-match) if one string can be transformed into the other by using a bijection on $\Sigma \cup \Pi$, with the restriction that the bijection must be the identity on the static characters of Σ. In other words, the bijection maps any $a \in \Sigma$ to a itself, while symbols in Π can be interchanged with a bijection. For example, let $\Pi = \{a, b, c\}$, $\Sigma = \{X, Y, Z\}$, $r =$ bcbXbcZ, $s =$ abaXabY and $t =$ bcbXbcY. We can say that p-strings s and t p-match each other, while r and s do not p-match. Parameterized pattern matching has many applications.

An optimal algorithm for *parameterized pattern matching* appeared in [4]. In this problem the pattern and the text are given as input and one seeks to report all *parameterized occurrences*. Approximate parameterized pattern matching was investigated in [7,9,17]. Idury and Schäffer [18] considered matching of multiple parameterized patterns.

Parameterized matching has proven useful in other contexts as well. An interesting problem is searching for color images (e.g. [3,8,20]). Assume, for example, that we are seeking a given icon in any possible color map. If the colors were fixed, then this is exact two-dimensional pattern matching [2]. However, if the color map is different, the exact matching algorithm would not find the pattern. Parameterized two-dimensional search is precisely what is needed. If, in addition, one is also willing to lose resolution, then a two dimensional function matching search should be used, where the renaming function is not necessarily a bijection [1].

Parameterized matching can be solved in linear time, when a constant-sized alphabet is considered [4]. Baker [9] showed that a parameterized suffix tree can be constructed in linear time for a text over a constant-sized alphabet. Lee et al. [19] showed that it can be constructed online in randomized linear time for unbounded alphabets, where online means that extensions by letters are supported, which then cost constant amortized time. Later, it was shown how to support extension of letters in worst case time per each extension [5]. This result was then improved [6].

Baker [9] introduced the *prev encoding* of a p-string which maintains each static character $\in \Sigma$ and maps each parameterized character $\in \Pi$ to a number, the distance to its previous occurrence in the p-string (or 0 if it is the first occurrence). Baker showed that two p-strings p-match iff their prev encodings are equivalent [14]. For example, the prev encodings of both $s =$ abaXabY and $t =$ bcbXbcY are 002X24Y. Thus, the parameterized matching problem amounts to efficiently comparing the prev encodings of p-strings.

In this paper we focus on the problem of reverse engineering a *parameterized* suffix array along with its corresponding *parameterized* longest common prefix (pLCP) array. We find a p-string whose p-suffix array and pLCP array are equal to the given arrays of integers, if there is a suitable p-string. We develop an algorithm that runs in $O(n^2)$ time for a p-string of size n. Reverse engineering of standard suffix arrays has already been worked out in linear time [11,13]. It is more challenging to derive a p-string from its p-suffix array than to derive a string from its suffix array since fundamental properties inherent in the relationship between strings and their substrings do not hold true for prev encodings of p-strings and prev encodings of their suffixes. We exploit properties of prev-encoded suffixes to derive an efficient algorithm to reconstruct a parameterized string from its parameterized suffix array and pLCP array, or determine that such a reconstruction is impossible.

This paper is organized as follows. We begin by formulating the problem and presenting key definitions in Sect. 2. Then we introduce an efficient algorithm in Sect. 3. All figures appear in the Appendix.

2 Preliminaries

Definition 1. *[14] The* prev *encoding of a p-string x of length n is the string* prev(x) *over the alphabet $\Sigma \cup \{0, ..., n-1\}$ and is defined as follows:*

$$prev(x)[i] = \begin{cases} x[i] & \text{if } x[i] \in \Sigma, \\ 0 & \text{if } x[i] \in \Pi \text{ and } x[i] \neq x[j] \text{ for any } 1 \leq j < i, \\ i - j & \text{if } x[i] \in \Pi, x[i] = x[j] \text{ and } x[i] \neq x[k] \text{ for any } j < k < i. \end{cases}$$

In this paper we call each numerical value in a prev-encoded string a *num-char*. The num-chars are by definition integers between 0 and $n-1$.

On every two p-strings T, S, we define an order based on the *prev* transformation such that $S < T$ if and only if $prev(S) < prev(T)$. Numbers are lexicographically smaller than the static letters, i.e., $n < \sigma$ for any $n \in \{0\} \cup \mathbb{N}, \sigma \in \Sigma$.

Definition 2. *[14] The* Parameterized Suffix Array (pSA) *of a p-string S of length n is an array $pSA[1 \ldots n]$ of integers such that $pSA[i] = j$ if and only if $prev(S[j \ldots n])$ is the ith lexicographically smallest string in $\{prev(S[i \ldots n]) | i = 1, \ldots, n\}$.*

Definition 3. *A p-suffix of a p-string S is the prev-encoding of the corresponding suffix. For $1 \leq i \leq n$, p-suffix $p_i = prev(S[i \ldots n])$.*

Definition 4. *The* Parameterized LCP (pLCP) Array *of a p-string S of length n is an array $pLCP[1 \ldots n-1]$ of integers such that $pLCP[i]$, for any $i \in \{1, \ldots, n-1\}$ is the longest common prefix between $prev(S[pSA[i] \ldots n])$ and $prev(S[pSA[i+1] \ldots n])$.*

Deguchi et al. [12] introduced the parameterized suffix array (pSA). Fujisato et. al. [14] developed the first linear time algorithm that directly computes the parameterized suffix and LCP arrays. For a string over static alphabet Σ and parameterized alphabet Π, the algorithm runs in $O(n\pi)$ time and $O(n)$ words of space, where π is the number of distinct symbols of Π in the string. (This is worst-case linear time when when there are a constant number of distinct parameterized symbols in the string.)

A suffix array lists the suffixes of a string in lexicographic sorted order. With standard strings, a substring is identical whether it occurs as a prefix, as a suffix, or as a stand-alone string. However, this equivalence does not hold for the prev encodings of p-strings. A substring of a prev encoding can be different than the prev encoding of the corresponding substring. The prev encoding is dependent on the context of a substring within a larger string. For example, consider the substring bba that occurs within abba. The long string abba has prev encoding 0013 and the substring bba has prev encoding 010. Even though bba is a substring of abba, 010 does not occur within 0013. Hence, algorithms for reconstructing standard strings from their suffix arrays do not readily extend to reconstructing parameterized strings from their p-suffix arrays. In this paper, the input pSA is augmented with its pLCP array which provides valuable information that our algorithm exploits.

It is interesting to note that the strings abba and abbc do not p-match, have different prev encodings, yet they share the same p-suffix array of $\{4, 3, 1, 2\}$ and pLCP array of $\{1, 2, 1\}$.

It is also interesting to observe that when $a < b$ and $c < d$, the strings abaab and dcddc have the same p-suffix arrays, yet different suffix arrays. However, the strings abaab and cdccd have both the same suffix arrays and p-suffix arrays.

The following lemma demonstrates that the static characters can be handled with no additional cost. Thus, in the rest of the paper we focus on p-strings that consist of only parameterized characters and no static characters.

Lemma 1. *Let Π, Σ be the parameterized and static alphabets, respectively. Assume A, P are the input under $\Pi \cup \Sigma$. Then, there exist A', P' such that A', P' are p-suffix array, and pLCP array under Π if and only if A, P are the p-suffix array, and pLCP array under $\Pi \cup \Sigma$.*

Proof. The proof will appear in the journal version.   □

3   Algorithm

Input: A pSA, A, of size n (a permutation of the numbers 1 through n) and a pLCP array, P, of size n.
Output: Does there exist a parameterized string S of size n such that A is its pSA and P is its pLCP array? If yes, construct such a parameterized string S.

The straightforward approach to solving this problem would be to attempt to reconstruct the p-string S directly from the pSA and pLCP arrays. Some characters would be trivial, such as those that are part of single character runs.

However, there are many decision points that will have a choice of characters, possibly the size of the alphabet. Once a decision is made, it is often necessary to backtrack later in the string, resulting in an algorithm that has exponential time complexity.

Our algorithm actually builds the prev-encoding of each suffix (i.e. each p-suffix) exploiting the data given in the pSA and pLCP arrays and then derives a p-string S from its list of p-suffixes. We are essentially reconstructing the underlying data structure which has size $O(n^2)$ and our algorithm accomplishes this in time and space $O(n^2)$. In some cases, there are inherent contradictions between the pSA and pLCP arrays and this always becomes apparent when attempting to build the p-suffixes. Thus, we conclude that the input is invalid when a state of contradiction is detected. In case we miss any contradictions along the way, we verify that the p-string we construct actually corresponds to the input pSA and pLCP array.

We discern in the Central Observation (Observation 1) that it is not necessary to reconstruct each p-suffix in its entirety in order to reconstruct S.

Observation 1 (Central Observation). *Any num-char in a p-suffix that is never compared to another num-char in a different p-suffix during the sort that generates a p-suffix array has no effect on the sorting of p-suffixes.*

We define two tables that the algorithm uses to reconstruct the p-suffixes. Both tables store the list of p-suffixes, however, one is left aligned, and one is right aligned, hence we call them the Left-Table and the Right-Table. The Left-Table imposes the constraints of the pSA and pLCP arrays and the Right-Table enforces consistency among the p-suffixes.

Left-Table: This table stores the truncated p-suffixes in the order of the given pSA. Formally, let $pSA[h] = i$, $1 \leq h \leq n$. Then row h of this table corresponds to a prefix of p_i. Since the p-suffixes are left-aligned, column c corresponds to location c in each p-suffix.

Right-Table: This table stores the truncated p-suffixes in reverse order of the original string S, that is, from p-suffix of length 1 to length n, with row i corresponding to the p-suffix beginning at $S[n - i + 1]$. The table is filled in a way that would result in the complete p-suffixes being right aligned, resulting in a lower right triangle in an $n \times n$ array with zeros on the diagonal. Column c in this table corresponds to position c in S, across all p-suffixes. Column c begins in row c, for $1 \leq c \leq n$.

Algorithm Outline

1. Fill in the Left-Table.
2. Resolve mismatch following pLCPs.
3. Fill in the Right-Table.
4. Reconstruct p-string.
5. Verify Output.

To reconstruct the p-suffixes, we begin by following the order of the pSA. The combination of the pSA and pLCP arrays guide us in identifying the prefixes of p-suffixes that are identical, along with the position of mismatch at which they diverge and the sorted order of the num-chars at the position of divergence. Then we put the prefixes of p-suffixes in the positional order in which they occur, corresponding to their positions of origin within the p-string. This way we can line up the corresponding positions in the different p-suffixes to ensure that we generate prev-encodings that are consistent with one another. It is sufficient for us to produce the prefixes of the p-suffixes that are used to establish the order of the p-suffix array. As we go along, we use both the sorting constraints and the rules of prev-encodings to resolve the unknown num-chars in the p-suffixes.

Our algorithm proceeds through all the steps sequentially unless it arrives at a state of contradiction. A contradiction can be expressed as either being forced to use two different num-chars at the same location in a p-suffix, or being forced to arrive at a p-suffix that is not a valid prev-encoding, or if the original pSA is not in fact the sorted order of the p-suffixes we construct, when adhering to the structure imposed by the p-suffix and pLCP arrays. If any of these scenarios occur, we conclude that it is not possible to reconstruct S.

We describe some of the key properties of the p-suffixes in Lemma 2 and properties of the pLCP array in Lemma 3 that are intuited by our algorithm. Then in the following subsections we detail the steps of the algorithm.

Lemma 2. *Let S be a p-string of length n over parameterized alphabet Π. The p-suffixes p_i, $1 \leq i \leq n$, of S have the following properties:*

1. *Every p-suffix begins with 0. Furthermore, every p-suffix begins with either 00 or 01 (except the shortest p-suffix which is simply 0).*
2. *Let z_i be the the number of zeros that occur in p-suffix p_i, $1 \leq i \leq n$. $z_i \leq |\Pi|$ and $z_i \geq z_{i+1}$. More specifically, either $z_i = z_{i+1}$ or $z_i = z_{i+1} + 1$*

Proof. The proof will appear in the journal version.   □

Lemma 3. *Let S be a p-string of length n over parameterized alphabet Π with a p-suffix array and pLCP array. The pLCP array has the following properties:*

1. *$pLCP[i] \geq 1$, for all $1 \leq i < n$. In addition, $pLCP[1] = 1$.*
2. *There exists at most one $1 < i < n$ such that $pLCP[i] = 1$. All p-suffixes p_j such that $pSA[h] = j$ and $h \leq i$ begin with 00 and all p-suffixes p_k such that $pSA[g] = k$ and $g > i$ begin with 01.*

Proof. The proof will appear in the journal version.   □

If the properties of Lemma 3 do not occur in the input, we conclude that the input is not reconstructable. Similarly, our algorithm halts if the input pSA is not a permutation of the numbers 1 through n.

3.1   Step 1: Left-Table

Our reconstruction process begins by following the order of the pSA to fill in the Left-Table. Lemma 4 and Corollary 1 demonstrate that it is sufficient for our algorithm to consider a specific prefix of each p-suffix p_i, $1 \leq i \leq n$. We can truncate each p-suffix after this point.

Lemma 4 (Unique substring). *Let α be a substring of the input string S that has a unique prev-encoding prev-α over all substrings of length $|\alpha|$. The parameterized characters in S following α are not related to those preceding α for the construction of the p-suffix array.*

Proof. The proof will appear in the journal version.   □

Corollary 1. *The pLCP array indicates the interesting num-chars in the p-suffixes. For each p-suffix, it is sufficient to reconstruct the positions up to and including the mismatch at the end of the pLCP with both of its neighbors in the p-suffix array.*

For $1 \leq i \leq n$, suppose $i = pSA[h]$, $j = pSA[h+1]$, and $k = pSA[h-1]$. Let $m_i = max\{pLCP(p_i, p_j), pLCP(p_k, p_i)\}$, i.e., the larger pLCP between p_i and its neighboring p-suffixes in the pSA. We work with the entire p_i if at least one of the pLCPs is as long as p_i, i.e., $m_i = n - i + 1$. Otherwise, when $m_i < n - i + 1$, we truncate p_i at the position of mismatch following m_i since the num-chars past that point are irrelevant to the reconstruction of S. From here on, when we refer to a p-suffix p_i, $1 \leq i \leq n$, we are referring to the truncated p-suffix of length $d_i = min(n - i + 1, m_i + 1)$ that we reconstruct in our algorithm. Lemma 4 and Corollary 1 show that this is sufficient.

We fill each row of the Left-Table with the first d_i num-chars in each p-suffix p_i. The pSA and pLCP arrays are a roadmap for reconstructing the p-suffixes since they express both the necessary similarity and dissimilarity that need to be incorporated in the p-suffixes. We use the actual num-chars when we know them and use upper-case letters as placeholders for num-chars that are as of yet unknown and need to be resolved as our algorithm proceeds. We know that each p-suffix begins with either 00 or 01, by Lemma 2. We know from Lemma 3 that in the sorted set of p-suffixes, the pLCP of 1 serves as a demarcation between the subset of p-suffixes beginning with 00 and the subset of p-suffixes beginning with 01. Thus, we go down the sorted list of p-suffixes in the Left-Table and fill in the first two num-chars; in the first subset we use 00 and in the second subset we use 01 (and the shortest p-suffix is simply a 0).

After filling in the first two num-chars of each p-suffix, we go down the Left-Table and copy num-chars based on the pLCPs between adjacent p-suffix array entries. We use *identical* placeholders for unknown num-chars that need to conform, occurring within a pLCP. In the case that a pLCP extends to the end of one of the p-suffixes it relates to, we do not need to identify the num-char at the position of mismatch in the longer p-suffix. On the other hand, when a pLCP does not extend to the end of either p-suffix in which it occurs, we compute the num-chars at the point of divergence in the next stage of the algorithm.

Consider the left side of Fig. 1 which depicts a set of input pSA and pLCP arrays alongside its Left-Table with the unknown placeholders representing mismatch following the pLCPs rendered in red. We have $pLCP(p_5, p_{11}) = 4$ and we know that both p_5 and p_{11} begin with 00. We record p_5 as $00ABC$ and p_{11} as $00ABD$. In the next step we will use Algorithm 1 to compute the values of C and D. On the other hand, even though $pLCP(p_{13}, p_3) = 4$, we record them both as $00PQ$ and do not incorporate an unknown mismatch num-char at the end of the pLCP, since p_{13} ends after these 4 num-chars.

3.2   Step 2: Reconstruct Point of Mismatch

In this section we present a deterministic method of computing each num-char at the point of mismatch following a pLCP. Algorithm 1 will identify each of these unknown num-chars unless there is an inconsistency in the pSA and pLCP arrays, in which case the algorithm will exit with failure.

Figure 1 shows a Left-Table before and after this step of the algorithm deploys Algorithm 1 to resolve the mismatches at the end of pLCPs to their values. The unknown placeholders that are resolved in this step appear in red. The figure depicts the placeholders in red on the left side and their corresponding values in red on the right side.

The pLCP array provides pLCPs of p-suffixes that are adjacent to one another in the pSA. With linear time preprocessing, we can use RMQ [15,16] in the pLCP array to obtain the pLCP between *any two* p-suffixes in $O(1)$ time. Then we go through each pair of p-suffixes p_i and p_j that are adjacent in the pSA, and we use Algorithm 1 to recover the mismatch following each pLCP, i.e., $p_i[lcp(p_i, p_j) + 1]$ and $p_j[lcp(p_i, p_j) + 1]$, when the pLCP does not span either complete p-suffix.

To recover the unknown num-char at the position of mismatch, the algorithm examines what happens to the pLCP and the sorted order as we eliminate one num-char at a time at the start of both p-suffixes. A way of deducing the unknown num-char is by removing the first num-char from both p-suffixes and seeing if this affects their sorted order. Generally, as we move along and remove the initial num-char from p-suffixes that share a pLCP, either the p-suffixes remain unchanged or they are both modified in the same way. However, when we remove a position that only one of them points to, that p-suffix has an additional 0. The first position at which this can occur is the position of mismatch at the end of the pLCPs.

Suppose the position of mismatch following the pLCP of p_i and p_j is a num-char $\Gamma > 0$, i.e., $p_i[lcp(p_i, p_j) + 1] = \Gamma$ or $p_j[lcp(p_i, p_j) + 1] = \Gamma$. After $lcp(p_i, p_j)$ iterations, the Γ will surely change to 0. We will uncover the missing value in the larger p-suffix before the smaller since a larger value will fall off more quickly, as it refers to an earlier position in the p-suffix.

As we move along and remove one position at a time from the beginning of the p-suffix, the pLCP either ends in the same place or grows (to one position further). If the pLCP continues to end at the same position, it is of interest if the sorted order changes. The inversion of the sorted order indicates that the larger value has become 0, and the smaller value remains non-zero. If the modified pLCP extends one position further, this indicates that the positions of mismatch have now both become 0.

In the algorithm, there are two cases. Either one of the values at the position of mismatch resolves to zero or they both resolve to non-zero values. In the former case, in which one of the values is 0, we discover the non-zero value once the pLCP grows, which indicates that the position of mismatch in both p-suffixes contains only zeros. In the latter case, in which both unknown num-chars are non-zero, the sorted order will change once we have lost the position that the larger value refers to. Then the smaller value is uncovered when we get to the state that both p-suffixes contain only zeros at the position of mismatch, resembling the first case. The sorted order will change only at the points that we discover a value and a non-zero num-char changes to 0 at the position of interest.

As we iterate through shrinking p-suffixes, the algorithm halts if it recognizes a contradiction in the input pSA and/or pLCP arrays. If the pLCP shrinks to end at an earlier position than it had previously, the algorithm halts since it has arrived at an impossible state.

Algorithm 1. Resolve unknown characters representing mismatch at end of pLCP

Input: i, j such that p_i and p_j are adjacent suffixes in the pSA, i.e. $j = pSA[h]$, $i = pSA[h+1]$, where $1 \leq h < n$, and $lcp(p_i, p_j) < n-i+1$ and $lcp(p_i, p_j) < n-j+1$.
Output: Recover $p_i[lcp(p_i, p_j) + 1]$ and $p_j[lcp(p_i, p_j) + 1]$.
$\ell \leftarrow lcp(p_i, p_j)$
$p_j[\ell + 1] \leftarrow 0$
for $k \leftarrow 1$ **to** $lcp(p_i, p_j)$ **do**   ▷ go through all lengths possible
$\quad \ell' \leftarrow lcp(p_{i+k}, p_{j+k})$
$\quad r' \leftarrow order(p_{i+k}, p_{j+k})$
$\quad r'' \leftarrow order(p_{i+k-1}, p_{j+k-1})$

$\quad$**if** $\ell - k < \ell'$ **then**
$\qquad\qquad$▷ The larger suffix with respect to the order r'' must have $\ell - k + 1$ at
position $\ell + 1$.
$\qquad$**if** $r'' = -1$ **then**   ▷ $p_{i+k-1} > p_{j+k-1}$
$\qquad\qquad p_i[\ell + 1] \leftarrow \ell - k + 1$
$\qquad$**else**   ▷ $p_{i+k-1} < p_{j+k-1}$
$\qquad\qquad p_j[\ell + 1] \leftarrow \ell - k + 1$
$\qquad$**end if**
$\qquad$Break
$\quad$**else if** $\ell - k = \ell'$ **then**
$\qquad$**if** $r' \neq r''$ **then**
$\qquad\qquad$▷ Larger suffix with respect to orig order r'' has $\ell - k + 1$ at position $\ell + 1$.
$\qquad\qquad p_i[\ell + 1] \leftarrow \ell - k + 1$
$\qquad$**end if**
$\quad$**else**
$\qquad$Halt   ▷ Contradiction
$\quad$**end if**
end for

We refer to Fig. 1 for several examples. First we look at an example that does not need to go through all possible values of k. We look at the computation of Q and R, which begins by considering $\ell = pLCP(p_3, p_9) = 3$. We initialize $p_3[4] = 0$. Then, when $k = 1$ we consider $\ell' = pLCP(p_4, p_{10}) = 5$. Since $\ell - k = 2 < \ell' = 5$, the algorithm terminates after the first iteration and sets $p_9[4] = 3$. Thus, we obtain $Q = 0$ and $R = 3$.

Now we go through an example in which one mismatch num-char is 0 and the other is non-zero. We look at the computation of C and D, which begins by setting $\ell = pLCP(p_5, p_{11}) = 4$. We initialize $p_5[5] = 0$. When $k = 1$, $\ell' = pLCP(p_6, p_{12}) = 3$. $\ell - k = 3 = \ell'$ and the order remains the same so we move on. When $k = 2$, $\ell' = pLCP(p_7, p_{13}) = 2$. $\ell - k = 2 = \ell'$ and the order remains the same so we move on. When $k = 3$, $\ell' = pLCP(p_8, p_{14}) = 1$ and the order remains the same so we move on to the last possible iteration. When $k = 4$, $\ell' = pLCP(p_9, p_{15}) = 2$. $\ell - k = 0 < \ell' = 2$ so we set $p_{11}[5] = 1$. Thus, we obtain $C = 0$ and $D = 1$.

Now we look at an example in which both mismatch num-chars are non-zero. We consider the computation of D and E, which begins by setting $\ell = pLCP(p_{11}, p_6) = 4$. When $k = 1$, $\ell' = pLCP(p_{12}, p_7) = 3$. $\ell - k = 3 = \ell'$ and the order remains the same so we move on. When $k = 2$, $\ell' = pLCP(p_{13}, p_8) = 2$. $\ell - k = 2 = \ell'$ and the order has switched, so we set $p_6[5] = 3$. We move on to $k = 3$, in which $\ell' = pLCP(p_{14}, p_9) = 1$. $\ell - k = 1 = \ell'$ so we move on. When $k = 4$, $\ell' = pLCP(p_{15}, p_{10}) = 1$. $\ell - k = 0 < \ell' = 1$ so we set $p_{11}[5] = 1$. Thus, we obtain $D = 1$ and $E = 3$.

Lemma 5 (Algorithm 1 Correctness). *For any i, j such that p_i and p_j are adjacent suffixes in the pSA, and both p_i and p_j end at least one position after the end of their pLCP, Algorithm 1 recovers the characters $p_i[lcp(p_i, p_j) + 1]$ and $p_j[lcp(p_i, p_j) + 1]$.*

Proof. The proof will appear in the journal version   □

Lemma 6 (Algorithm 1 Time). *For specific $i, j \in [1, n]$, Algorithm 1 runs in $O(n)$ time.*

Proof. The proof will appear in the journal version.   □

At most, we run Algorithm 1 for all $n - 1$ pairs of integers i, j such that p-suffixes p_i and p_j are adjacent in the pSA. This brings the total time for running Algorithm 1 to $O(n^2)$.

For each num-char Δ that has been resolved with Algorithm 1, we update any other occurrences of Δ in the other p-suffixes to enforce consistency among the LCPs.

3.3   Step 3: Right-Table

Once the Left-Table is complete, and we have resolved the unknown placeholders that occur where the pLCPs diverge, we shift our focus to the the Right-Table to enforce positional consistency among the p-suffixes. We copy the p-suffixes of the Left-Table into their corresponding positions in the Right-Table, maintaining all the unknowns as placeholders. Then we resolve the remaining unknowns to num-chars that they can possibly represent, so that we can reconstruct an underlying p-string, or determine that there is a contradiction that prevents the reconstruction from succeeding.

We demonstrate in Lemma 7 that each column of the Right Table follows a predictable structure.

Lemma 7. *(Property of Right-Table Columns) Consider column $1 \leq c \leq n$ in the Right-Table. Column c begins with a 0 in row c. If column c contains a num-char $\mu > 0$, then all entries on rows $c \leq r < c + \mu$ are 0 and all entries on rows $c + \mu \leq r \leq n$ are μ.*

Proof. The proof will appear in the journal version.   □

The following implicit constraints form a system of rules we must follow to correctly resolve the remaining unknown placeholders in the p-suffixes. If we need to violate any of the constraints, we are in a state of contradiction and halt.

1. **Consistency:** Ensure we consistently replace all occurrences of an unknown num-char, to enforce the pLCPs.
2. **Column:** Make sure each column of the Right-Table is valid, i.e., each column adheres to the Property of Right-Table Columns (Lemma 7).
3. **Row:** Make sure each row of the Right-Table forms a valid prev-encoded p-suffix. In other words, each p-suffix must consist of several linked lists that all begin with the num-char 0.

We have already filled in the first two num-chars in each p-suffix. Based on the Property of Right-Table Columns (Lemma 7), we can determine which unknowns resolve to 1. Any 1's in the Right-Table will appear in columns that begin with a 0 followed by a 1 and we can propagate these 1's down the columns in which they appear. In a similar way we can complete the columns of the Right-Table that contain a non-zero value computed by Algorithm 1.

Now we can arbitrarily assign values to the remaining unknowns so long as they match other copies of themselves, are consistent with other p-suffixes that span the same position, and each p-suffix is a valid prev-encoding. To satisfy the implicit constraints while keeping the algorithm simple, we will set all remaining unknowns to the num-char 0.

If we have succeeded this far in the reconstruction process, we can answer in the affirmative that *yes* there is a p-string S that corresponds to the input pSA A and pLCP array P.

The left side of Fig. 2 shows how we set up the Right-Table corresponding to the Left-Table of Fig. 1. First we copy each p-suffix from the Left-Table into its appropriate row in the Right-Table. Then we resolve the remaining unknown placeholders that can be determined from context within their columns using the Property of Right-Table Columns. In this way, S is set to 0 since 0 occurs below it (in both instances) and L is set to 1 since 1 appears above it (in both instances). Then we remain with only two unknown placeholders, T and U. Since their values cannot be determined from their context within their columns, we will set them to 0. The updated Right-Table is portrayed on the right side of Fig. 2.

3.4 Step 4 Reconstruct P-String

It is straightforward to reconstruct a p-string from left to right following the p-suffixes in the Right-Table. $S[i]$ corresponds to column i in the Right-Table, $1 \leq i \leq n$. We begin by choosing any character for $S[1]$ since column 1 contains a single 0. Then we use the following rule to insert a character for each subsequent $S[i]$, $2 \leq i \leq n$. If column i contains a num-char μ greater than 0, we set $S[i]$ to the character in $S[i - \mu]$. Otherwise, a column of only zeros indicates a new character for the p-suffixes that span this column in the Right-Table. In this case, we can choose any character that has not appeared yet in the lowest suffix in this column. That is, if column i spans p-suffixes $k \ldots k + a$, we set $S[i]$ to be any character that does not occur in the reconstructed $S[k \ldots i - 1]$. The right side of Fig. 2 shows a complete Right-Table along with two possible p-strings that can be reconstructed for it.

3.5 Step 5 Verify Output

The last step of our algorithm is to construct the pSA and pLCP arrays that correspond to the p-string we construct and verify that they are the same as the input. If these pSA and pLCP arrays do not match, we conclude that the input is not reconstructable. There can only be discrepancies between these pSA and pLCP arrays in the case that there are inherent contradictions in the input that prevent us from correctly constructing a p-string for the input.

3.6 Proofs of Correctness and Efficiency

Lemma 8. *If the given pSA and pLCP arrays are reconstructable then we can construct a Right-Table.*

Proof. The proof will appear in the journal version. □

Lemma 9. *Our algorithm correctly decides whether the input pSA and pLCP arrays are reconstructable.*

Proof. The proof will appear in the journal version. □

Lemma 10. *In $O(n)$ time we can reconstruct a possible p-string S of length n or determine that it is impossible to construct a p-string for the given pSA and pLCP arrays.*

Proof. The proof will appear in the journal version. □

We implemented the algorithm on all possible pSA and pLCP arrays of length up to 12 and verified that our algorithm gave the correct output each time. We have ideas on how to speed up the algorithm to $O(n \log n)$ time and also how to modify the algorithm to output the p-string with minimal alphabet size. The details will appear in the journal version of this paper.

A Appendix

pSA	pLCP	Left-Table						Left-Table					
16	1	0						0					
15	2	0	0					0	0				
5	4	0	0	A	B	C		0	0	0	0	0	
11	4	0	0	A	B	D		0	0	0	0	1	
6	3	0	0	A	B	E		0	0	0	0	3	
12	4	0	0	A	F	G		0	0	0	1	0	
8	4	0	0	A	F	H		0	0	0	1	3	
2	3	0	0	A	F	J		0	0	0	1	4	
7	5	0	0	A	K	L	M	0	0	0	3	L	3
1	2	0	0	A	K	L	N	0	0	0	3	L	4
13	4	0	0	P	Q			0	0	1	0		
3	3	0	0	P	Q			0	0	1	0		
9	1	0	0	P	R			0	0	1	3		
14	3	0	1	S				0	1	S			
4	5	0	1	S	T	U	V	0	1	S	T	U	0
10		0	1	S	T	U	X	0	1	S	T	U	1

Fig. 1. (left) The input p-suffix and pLCP arrays. (center) The Left-Table is constructed with unknown placeholders representing mismatch at the end of the pLCPs depicted in red. (right) The Left-Table after using Algorithm 1 to resolve the unknowns in red and updating other occurrences of the same unknowns, to maintain consistency. (Color figure online)

Right-Table

1	2	3	4	5	6	7	8	9	10	11	12	13	14	15	16	
														0	16	16
													0	0	15	15
												0	1	S	14	14
											0	0	1	0	13	13
										0	0	0	1	0	12	12
									0	0	0	0	1		11	11
								0	1	S	T	U	1		10	10
							0	0	1	3					9	9
						0	0	0	1	3					8	8
					0	0	0	3	L	3					7	7
				0	0	0	0	0	3						6	6
			0	0	0	0	0								5	5
		0	1	S	T	U	0								4	4
	0	0	1	0											3	3
0	0	0	1												2	2
0	0	0	3	L	4										1	1

Right-Table

1	2	3	4	5	6	7	8	9	10	11	12	13	14	15	16	
														0	16	16
													0	0	15	15
												0	1	0	14	14
											0	0	1	0	13	13
										0	0	0	1	0	12	12
									0	1	0	0	0	1		11	11
								0	1	0	0	0	1			10	10
							0	0	1	3						9	9
						0	0	0	1	3						8	8
					0	0	0	3	1	3						7	7
				0	0	0	0	0	3							6	6
			0	0	0	0	0									5	5
		0	1	0	0	0	0									4	4
	0	0	1	0												3	3
0	0	0	1	4												2	2
0	0	0	3	1	4											1	1

a	b	c	a	a	b	c	x	y	c	c	y	a	b	b	c
a	b	c	a	a	b	x	y	z	x	x	z	c	a	a	b

Fig. 2. (left) We initially fill the Right-Table by copying p-suffixes from the Left-Table into their places in the Right-Table. The unknown placeholders we are able to resolve from context within their columns are shown in yellow. The remaining unknown placeholders are colored orange. (right) The completed Right-Table and two different possible p-strings corresponding to it. (Color figure online)

References

1. Amir, A., Aumann, A., Lewenstein, M., Porat, E.: Function matching. SIAM J. Comput. **35**(5), 1007–1022 (2006)
2. Amir, A., Benson, G., Farach, M.: An alphabet independent approach to two dimensional pattern matching. SIAM J. Comp. **23**(2), 313–323 (1994)
3. Amir, A., Church, K.W., Dar, E.: Separable attributes: a technique for solving the submatrices character count problem. In: Proceedings 13th ACM-SIAM Symposium on Discrete Algorithms (SODA), pp. 400–401 (2002)
4. Amir, A., Farach, M., Muthukrishnan, S.: Alphabet dependence in parameterized matching. Inf. Process. Lett. **49**(3), 111–115 (1994). https://doi.org/10.1016/0020-0190(94)90086-8
5. Amir, A., Kondratovsky, E.: Sufficient conditions for efficient indexing under different matchings. In: Pisanti, N., Pissis, S.P. (eds.) 30th Annual Symposium on Combinatorial Pattern Matching, CPM 2019, 18–20 June 2019, Pisa, Italy. LIPIcs, vol. 128, pp. 6:1–6:12. Schloss Dagstuhl - Leibniz-Zentrum für Informatik (2019). https://doi.org/10.4230/LIPIcs.CPM.2019.6
6. Amir, A., Kondratovsky, E.: Towards a real time algorithm for parameterized longest common prefix computation. Theor. Comput. Sci. **852**, 132–137 (2021). https://doi.org/10.1016/j.tcs.2020.11.023
7. Apostolico, A., Erdös, P.L., Lewenstein, M.: Parameterized matching with mismatches. J. Discrete Algorithms **5**(1), 135–140 (2007). https://doi.org/10.1016/j.jda.2006.03.014
8. Babu, G., Mehtre, B., Kankanhalli, M.: Color indexing for efficient image retrieval. Multimedia Tools Appl. **1**(4), 327–348 (1995)
9. Baker, B.S.: Parameterized pattern matching: algorithms and applications. J. Comput. Syst. Sci. **52**(1), 28–42 (1996). https://doi.org/10.1006/jcss.1996.0003
10. Baker, B.S.: Parameterized duplication in strings: algorithms and an application to software maintenance. SIAM J. Comput. **26**(5), 1343–1362 (1997). https://doi.org/10.1137/S0097539793246707
11. Bannai, H., Inenaga, S., Shinohara, A., Takeda, M.: Inferring strings from graphs and arrays. In: Rovan, B., Vojtáš, P. (eds.) MFCS 2003. LNCS, vol. 2747, pp. 208–217. Springer, Heidelberg (2003). https://doi.org/10.1007/978-3-540-45138-9_15
12. Deguchi, S., Higashijima, F., Bannai, H., Inenaga, S., Takeda, M.: Parameterized suffix arrays for binary strings. In: Holub, J., Zdárek, J. (eds.) Proceedings of the Prague Stringology Conference 2008, Prague, Czech Republic, 1–3 September 2008, pp. 84–94. Prague Stringology Club, Department of Computer Science and Engineering, Faculty of Electrical Engineering, Czech Technical University in Prague (2008). http://www.stringology.org/event/2008/p08.html
13. Duval, J., Lefebvre, A.: Words over an ordered alphabet and suffix permutations. RAIRO Theor. Inform. Appl. **36**(3), 249–259 (2002). https://doi.org/10.1051/ita:2002012
14. Fujisato, N., Nakashima, Y., Inenaga, S., Bannai, H., Takeda, M.: Direct linear time construction of parameterized suffix and LCP arrays for constant alphabets. In: Brisaboa, N.R., Puglisi, S.J. (eds.) SPIRE 2019. LNCS, vol. 11811, pp. 382–391. Springer, Cham (2019). https://doi.org/10.1007/978-3-030-32686-9_27
15. Gabow, H.N., Bentley, J.L., Tarjan, R.E.: Scaling and related techniques for geometry problems. In: DeMillo, R.A. (ed.) Proceedings of the 16th Annual ACM Symposium on Theory of Computing, April 30 - May 2 1984, Washington, DC, USA, pp. 135–143. ACM (1984). https://doi.org/10.1145/800057.808675

16. Harel, D., Tarjan, R.E.: Fast algorithms for finding nearest common ancestors. SIAM J. Comput. **13**(2), 338–355 (1984). https://doi.org/10.1137/0213024
17. Hazay, C., Lewenstein, M., Sokol, D.: Approximate parameterized matching. In: Albers, S., Radzik, T. (eds.) ESA 2004. LNCS, vol. 3221, pp. 414–425. Springer, Heidelberg (2004). https://doi.org/10.1007/978-3-540-30140-0_38
18. Idury, R.M., Schäffer, A.A.: Multiple matching of parameterized patterns. In: Crochemore, M., Gusfield, D. (eds.) CPM 1994. LNCS, vol. 807, pp. 226–239. Springer, Heidelberg (1994). https://doi.org/10.1007/3-540-58094-8_20
19. Lee, T., Na, J.C., Park, K.: On-line construction of parameterized suffix trees. In: Karlgren, J., Tarhio, J., Hyyrö, H. (eds.) SPIRE 2009. LNCS, vol. 5721, pp. 31–38. Springer, Heidelberg (2009). https://doi.org/10.1007/978-3-642-03784-9_4
20. Swain, M., Ballard, D.: Color indexing. Int. J. Comput. Vision **7**(1), 11–32 (1991)

Computing the Parameterized Burrows–Wheeler Transform Online

Daiki Hashimoto[1], Diptarama Hendrian[1](✉)(iD), Dominik Köppl[2](iD),
Ryo Yoshinaka[1](iD), and Ayumi Shinohara[1](iD)

[1] Tohoku University, Sendai, Japan
daiki_hashimoto@shino.ecei.tohoku.ac.jp,
{diptarama,ryoshinaka,ayumis}@tohoku.ac.jp
[2] TMDU, Tokyo, Japan
koeppl.dsc@tmd.ac.jp
https://www.iss.is.tohoku.ac.jp, https://dkppl.de

Abstract. Parameterized strings are a generalization of strings in that
their characters are drawn from two different alphabets, where one is
considered to be the alphabet of static characters and the other to be
the alphabet of parameter characters. Two parameterized strings are a
parameterized match if there is a bijection over all characters such that
the bijection transforms one string to the other while keeping the static
characters (i.e., it behaves as the identity on the static alphabet). Ganguly et al. [SODA 2017] proposed the parameterized Burrows–Wheeler
transform (pBWT) as a variant of the Burrows–Wheeler transform for
space-efficient parameterized pattern matching. In this paper, we propose
an algorithm for computing the pBWT online by reading the characters
of a given input string one-by-one from right to left. Our algorithm works
in $O(|\Pi| \log n / \log \log n)$ amortized time for each input character, where
n and Π denote the size of the input string and the alphabet of the
parameter characters, respectively.

Keywords: Burrows–Wheeler transform · Parameterized string ·
Online algorithm

1 Introduction

The *parameterized matching problem (p-matching problem)* [2] is a generalization
of the classic pattern matching problem in the sense that we here consider two
disjoint alphabets, the set Σ of *static characters* and the set Π of *parameter
characters*. We call a string over $\Sigma \cup \Pi$ a *parameterized string (p-string)*. Two
equal-length p-strings X and Y are said to *parameterized match (p-match)* if
there is a bijection that renames the parameter characters in X so X becomes
equal to Y. The *p-matching problem* is, given a text p-string T and pattern
p-string P, to output the positions of all substrings of T that p-match P. The p-
matching problem is motivated by applications in the software maintenance [1,2],
the plagiarism detection [5], the analysis of gene structures [17], and so on. There

© The Author(s), under exclusive license to Springer Nature Switzerland AG 2022
D. Arroyuelo and B. Poblete (Eds.): SPIRE 2022, LNCS 13617, pp. 70–85, 2022.
https://doi.org/10.1007/978-3-031-20643-6_6

exist indexing structures that support p-matching, such as parameterized suffix trees [1,17], parameterized suffix arrays [7,10], and so on [4,6,13,14]; see also [12] for a survey. A drawback of these indexing structures is that they have high space requirements.

A more space-efficient indexing structure, the *parameterized Burrows–Wheeler transform (pBWT)*, was proposed by Ganguly et al. [8]. The pBWT is a variant of the *Burrows–Wheeler transform (BWT)* [3] that can be used as an indexing structure for p-matching using only $o(n \log n)$ bits of space. Later on, Kim and Cho [11] improved this indexing structure by changing the encoding of p-strings used for defining the pBWT. Recently, Ganguly et al. [9] augmented this index with capabilities of a suffix tree while keeping the space within $o(n \log n)$ bits. However, as far as we are aware of, none research related to the pBWT [8,9,11,18] has discussed how to construct their pBWT-based data structures in detail. Their construction algorithms mainly rely on the parameterized suffix tree. Given the parameterized suffix tree of a p-string T of length n, the pBWT of T can be constructed in $O(n \log(|\Sigma| + |\Pi|))$ time offline.

In this paper, we propose an algorithm for constructing pBWTs and related data structures used for indexing structures of p-matching. Our algorithm constructs the data structures directly in an online manner by reading the input text from right to left. The algorithm uses the dynamic array data structures of Navarro and Nekrich [15] to maintain our growing arrays. For each character read, our algorithm takes $O(|\Pi| \log n / \log \log n)$ amortized time, where n is the size of input string. Therefore, we can compute pBWT of a p-string T of length n in $O(n|\Pi| \log n / \log \log n)$ time in total. In comparison, computing the standard BWT on a string T (i.e., the pBWT on a string having no parameter characters) can be done in $O(n \log n / \log \log n)$ time with the dynamic array data structures [15] (see [16] for a description of this online algorithm). Looking at our time complexity, the factor $|\Pi|$ also appears in the time complexity of an offline construction algorithm of parameterized suffix arrays [7] as $O(n|\Pi|)$ and a right-to-left online construction algorithm of parameterized suffix trees [13] as $O(n|\Pi| \log(|\Pi| + |\Sigma|))$. This suggests it would be rather hard to improve the time complexity of the online construction of pBWT to be independent of $|\Pi|$.

2   Preliminaries

We denote the set of nonnegative integers by $\mathbb{N}$ and let $\mathbb{N}_+ = \mathbb{N} \setminus \{0\}$ and $\mathbb{N}_\infty = \mathbb{N}_+ \cup \{\infty\}$. The set of strings over an alphabet A is denoted by A^*. The empty string is denoted by ε. The length of a string $W \in A^*$ is denoted by $|W|$. For a subset $B \subseteq A$, the set of elements of B occurring in $W \in A^*$ is denoted by $B \upharpoonright W$. We count the number of occurrences of characters of B in a string W by $|W|_B$. So, $|W|_A = |W|$. When B is a singleton of b, i.e., $B = \{b\}$, we often write $|W|_b$ instead of $|W|_{\{b\}}$. When W is written as $W = XYZ$, X, Y, and Z are called *prefix*, *factor*, and *suffix* of W, respectively. The i-th character of W is denoted by $W[i]$ for $1 \le i \le |W|$. The factor of W that begins at position i and ends at position j is $W[i : j]$ for $1 \le i \le j \le |W|$. For convenience, we abbreviate $W[1 : i]$ to $W[: i]$ and $W[i : |W|]$ to $W[i :]$ for $1 \leq i \leq |W|$. Let $Rot(W, 0) = W$ and $Rot(W, i+1) = Rot(W, i)[|W|]Rot(W, i)[: |W| - 1]$ be the i-th right rotation of W. Note that $Rot(W, i) = Rot(W, i + |W|)$. For convenience we denote $W_i = Rot(W, i)$. Let $Left_W(a)$ and $Right_W(a)$ be the leftmost and rightmost positions of a character $a \in A$ in W, respectively. If a does not occur in W, define $Left_W(a) = Right_W(a) = 0$.

2.1   Parameterized Burrows–Wheeler Transform

Throughout this paper, we fix two disjoint ordered alphabets Σ and Π. We call elements of Σ *static characters* and those of Π *parameter characters*. Elements of Σ^* and $(\Sigma \cup \Pi)^*$ are called *static strings* and *parameterized strings* (or *p-strings* for short), respectively.

Two p-strings S and T of the same length are a *parameterized match (p-match)*, denoted by $S \approx T$, if there is a bijection f on $\Sigma \cup \Pi$ such that $f(a) = a$ for any $a \in \Sigma$ and $f(S[i]) = T[i]$ for all $1 \leq i \leq |T|$ [2]. We use Kim and Cho's version of p-string encoding [11], which replaces 0 in Baker's encoding [1] by ∞. The *prev-encoding* $\langle T \rangle$ of T is the string over $\Sigma \cup \mathbb{N}_\infty$ of length $|T|$ defined by

$$\langle T \rangle[i] = \begin{cases} T[i] & \text{if } T[i] \in \Sigma, \\ \infty & \text{if } T[i] \in \Pi \text{ and } Right_{T[:i-1]}(T[i]) = 0, \\ i - Right_{T[:i-1]}(T[i]) & \text{if } T[i] \in \Pi \text{ and } Right_{T[:i-1]}(T[i]) \neq 0 \end{cases}$$

for $1 \leq i \leq |T|$. When $T[i] \in \Pi$, $\langle T \rangle[i]$ represents the distance between i and the previous occurrence position of the same parameter character. If $T[i]$ does not occur before the position i, the distance is assumed to be ∞. We call a string $W \in (\Sigma \cup \mathbb{N}_\infty)^*$ a *pv-string* if $W = \langle T \rangle$ for some p-string T. For any p-strings S and T, $S \approx T$ if and only if $\langle S \rangle = \langle T \rangle$ [2]. For example, given $\Sigma = \{a, b\}$ and $\Pi = \{u, v, x, y\}$, $S = \text{uvvauvb}$ and $T = \text{xyyaxyb}$ are a p-match by f with $f(u) = x$ and $f(v) = y$, where $\langle S \rangle = \langle T \rangle = \infty\infty 1a43b$.

For defining pBWT, we use another encoding $[\![T]\!]$ given by

$$[\![T]\!][i] = \begin{cases} T[i] & \text{if } T[i] \in \Sigma, \\ |\Pi \upharpoonright T_{n-i}[1 : Left_{T_{n-i}}(T[i])]| & \text{if } T[i] \in \Pi \end{cases}$$

for $1 \leq i \leq |T|$. When $T[i] \in \Pi$, $[\![T]\!][i]$ counts the number of distinct parameter characters in T between i and the next occurrence of $T[i]$, if $T[i]$ occurs after i. If i is the rightmost occurrence position of $T[i]$, then we continue counting parameter characters from the left end to the right until we find $T[i]$. Since $T[i]$ occurs in T_{n-i} as the last character, $[\![T]\!][i]$ cannot be zero. Note that $[\![Rot(T, i)]\!] = Rot([\![T]\!], i)$ by definition. It is not hard to see that for any p-strings S and T, $S \approx T$ if and only if $[\![S]\!] = [\![T]\!]$ (see Proposition 1 in the appendix). For example, the two strings S and T given above are encoded as $[\![S]\!] = [\![T]\!] = 212a22b$.

Hereafter in this section, we fix a p-string T of length n which ends with a special static character \$ which occurs nowhere else in T. We extend the linear

Table 1. The pBWT $pBWT(T) = L_T = $ a33131\$22aa of the example string $T = $ xayzzazyza\$ with related arrays, where $\Sigma = \{a\}$ and $\Pi = \{x, y, z\}$.

i	T_i	$\langle T_i \rangle$	$RA_T[i]$	$LCP_T^\infty[i]$	$\langle T_{RA_T[i]} \rangle$	$F_T[i]$	$\llbracket T_{RA_T[i]} \rrbracket$	$L_T[i]$
1	\$xayzzazyza	\$∞a∞∞1a252a	1	0	\$∞a∞∞1a252a	\$	\$3a211a233a	a
2	a\$xayzzazyz	a\$∞a∞∞1a252	2	0	a\$∞a∞∞1a252	a	a\$3a211a233	3
3	za\$xayzzazy	∞a\$∞a∞61a25	10	2	a∞∞1a252a\$∞	a	a211a233a\$3	3
4	yza\$xayzzaz	∞∞a\$∞a661a2	6	0	a∞∞2a\$∞a661	a	a233a\$3a211	1
5	zyza\$xayzza	∞∞2a\$∞a661a	3	1	∞a\$∞a∞61a25	3	3a\$3a211a23	3
6	azyza\$xayzz	a∞∞2a\$∞a661	7	1	∞a2∞2a\$∞a66	1	1a233a\$3a21	1
7	zazyza\$xayz	∞a2∞2a\$∞a66	11	1	∞a∞∞1a252a\$	3	3a211a233a\$	\$
8	zzazyza\$xay	∞1a2∞2a\$∞a6	8	1	∞1a2∞2a\$∞a6	1	11a233a\$3a2	2
9	yzzazyza\$xa	∞∞1a252a\$∞a	4	2	∞∞a\$∞a661a2	3	33a\$3a211a2	2
10	ayzzazyza\$x	a∞∞1a252a\$∞	9	2	∞∞1a252a\$∞a	2	211a233a\$3a	a
11	xayzzazyza\$	∞a∞∞1a252a\$	5	0	∞∞2a\$∞a661a	2	233a\$3a211a	a

order over Σ to $\Sigma \cup \mathbb{N}_\infty$ by letting $\$ < a < i < \infty$ for any $a \in \Sigma \setminus \{\$\}$ and $i \in \mathbb{N}_+$. The order over $\mathbb{N}_+$ coincides with the usual numerical order.

The pBWT of T is defined through sorting $\llbracket T_p \rrbracket$ for $p = 1, \ldots, n$ using $\langle T_p \rangle$ as keys.

Definition 1 (Parameterized rotation array). The parameterized rotation array RA_T of T is an array of size n such that $RA_T[i] = p$ with $1 \le p \le n$ if and only if $\langle T_p \rangle$ is the i-th lexicographically smallest string in $\{ \langle T_p \rangle \mid 1 \le p \le n \}$. We denote its inverse by RA_T^{-1}, i.e., $RA_T^{-1}[p] = i$ iff $RA_T[i] = p$.

Note that RA_T and RA_T^{-1} are well-defined and bijective due to the presence of \$ in T. Here, we have $RA_T[i] = n - pSA_T[i] + 1$, where pSA_T refers to the suffix array pSA_∞ in [11]. The array gives an $n \times n$ square matrix $(\llbracket T_{RA_T[i]} \rrbracket)_{i=1}^{n}$, which we call the *rotation sort matrix* of T, whose (i, p) entry is $\llbracket T_{RA_T[i]} \rrbracket[p]$. The pBWT of T is formed by the characters in the last column of the matrix.

Definition 2 (pBWT [11]). The parameterized Burrows–Wheeler transform (pBWT) of a p-string T, denoted by $pBWT(T)$, is a string of length n such that $pBWT(T)[i] = \llbracket T_{RA[i]} \rrbracket[n]$.

An example pBWT can be found in Table 1. We will use L_T as a synonym of $pBWT(T)$, since it represents the *last* column of the matrix $(\llbracket T_{RA_T[i]} \rrbracket)_{i=1}^{n}$. When picking up the characters from the *first* column, we obtain another array F_T. That is, $F_T[i] = \llbracket T_{RA_T[i]} \rrbracket[1]$ for all $i \in \{1, \ldots, n\}$. Those arrays L_T and F_T are "linked" by the following mapping.

Definition 3 (LF mapping). The LF mapping $LF_T : \{1, \ldots, n\} \to \{1, \ldots, n\}$ for T is defined as $LF_T(i) = j$ if $T_{RA[i]+1} = T_{RA[j]}$.

By rotating T_p to the right by one, the last character moves to the first position in T_{p+1}. Roughly speaking, $L_T[i]$ and $F_T[LF_T(i)]$ "originate" in the same character occurrence of T, which implies $\mathsf{L}_T[i] = \mathsf{F}_T[LF_T(i)]$ in particular. One can recover $\llbracket T \rrbracket$ as $\llbracket T \rrbracket[p] = \mathsf{L}_T[LF^{-p}(k_T)]$ for $1 \leq p \leq n$ where $k_T = RA_T^{-1}[n]$. L_T, F_T, and LF_T are used for pattern matching based on pBWT. See [11] for the details.

Our pBWT construction algorithm maintains neither RA_T nor LF_T, but involves some helper data structures in addition to L_T and F_T. Among those, the array LCP_T^{∞} is worth explaining before going into the algorithmic details. For two pv-strings X and Y, let $lcp^{\infty}(X, Y) = |W|_{\infty}$ be the number of ∞'s in the longest common prefix W of X and Y. The following array counts the number of ∞'s in the longest common prefixes of two adjacent rows in $(\langle T \rangle_{RA_T[i]})_{i=1}^{n}$.

Definition 4 (∞-LCP array). The ∞-LCP array LCP_T^{∞} of T is an array of size n such that $\mathsf{LCP}_T^{\infty}[n] = 0$ and $\mathsf{LCP}_T^{\infty}[i] = lcp^{\infty}(\langle T_{RA_T[i]} \rangle, \langle T_{RA_T[i+1]} \rangle)$ for $1 \leq i < n$.

Table 1 shows an example of a pBWT and related (conceptual) data structures. We can compute $lcp^{\infty}(\langle T_{RA_T[i]} \rangle, \langle T_{RA_T[j]} \rangle)$ using LCP_T^{∞} as follows.

Lemma 1. For $1 \leq i < j \leq n$, $lcp^{\infty}(\langle T_{RA_T[i]} \rangle, \langle T_{RA_T[j]} \rangle) = \min_{i \leq k < j} \mathsf{LCP}_T^{\infty}[k]$.

Kim and Cho [11] showed some basic relations among L_T, LF_T, and lcp^{∞}. We rephrase Lemma 3 of [11] into a form convenient for our discussions.

Lemma 2. Consider i and j with $1 \leq i < j \leq n$ and $T_{RA_T[i]}[n], T_{RA_T[j]}[n] \in \Pi$. Then, $LF_T(i) < LF_T(j)$ iff $\min\{\mathsf{L}_T[i] - 1, \ lcp^{\infty}(\langle T_{RA_T[i]} \rangle, \langle T_{RA_T[j]} \rangle)\} < \mathsf{L}_T[j]$.

Corollary 1 ([11]). If $i < j$ and $\mathsf{L}_T[i] = \mathsf{L}_T[j]$, then $LF_T(i) < LF_T(j)$.

To maintain L_T, F_T, and LCP_T^{∞} dynamically, our algorithm uses the data structure for dynamic arrays by Navarro and Nekrich [15] that supports the following operations on an array Q of size m in $O(\frac{\log m}{\log \log m})$ amortized time.

1. $\mathsf{access}(Q, i)$: returns $Q[i]$ for $1 \leq i \leq m$;
2. $\mathsf{rank}_a(Q, i)$: returns $|Q[:i]|_a$ for $1 \leq i \leq m$;
3. $\mathsf{select}_a(Q, i)$: returns i-th occurrence position of a for $1 \leq i \leq \mathsf{rank}_a(Q, m)$;
4. $\mathsf{insert}_a(Q, i)$: inserts a between $Q[i-1]$ and $Q[i]$ for $1 \leq i \leq m + 1$;
5. $\mathsf{delete}(Q, i)$: deletes $Q[i]$ from Q for $1 \leq i \leq m$.

Corollary 1 implies that we can compute $LF_T(i)$ and its inverse $LF_T^{-1}(j)$ by

$$LF_T(i) = \mathsf{select}_x(\mathsf{F}_T, \mathsf{rank}_x(\mathsf{L}_T, i)) \quad \text{where} \quad x = \mathsf{L}_T[i],$$

$$LF_T^{-1}(j) = \mathsf{select}_y(\mathsf{L}_T, \mathsf{rank}_y(\mathsf{F}_T, j)) \quad \text{where} \quad y = \mathsf{F}_T[j].$$

3   Computing pBWT Online

This section introduces our algorithm computing $pBWT_T$ in an online manner by reading a p-string T from right to left. Let $T = cS$ for $c \in \Sigma \cup \Pi \setminus \{\$\}$ and $n = |S| \geq 1$. We consider updating L_S to L_T. Hereafter, we assume that Σ is known and $|\Sigma| \leq |T|$ as in [16]. Among the rows of the rotation matrices of S and T, the rows of $\llbracket S \rrbracket = \llbracket S_n \rrbracket$ and $\llbracket T \rrbracket = \llbracket T_{n+1} \rrbracket$ play important roles when updating. Let $k_S = RA_S^{-1}[n]$ and $k_T = RA_T^{-1}[n+1]$. We note that $\mathsf{L}_S[k_S] = \mathsf{L}_T[k_T] = \$$.

First, we observe RA_T is obtained from RA_S just by "inserting" $n + 1$ at k_T.

Algorithm 1: PBWT update algorithm

---

1 **Function** UpdateAll($c, n, \mathsf{L}, \mathsf{F}, \mathsf{Right}, \mathsf{Left}, \mathsf{RM}, \mathsf{C}, \mathsf{LCP}^{\infty}$)
2    $k = \mathsf{select}_{\$}(\mathsf{L}, 1)$; $// = k_S$
3    $\mathsf{L}, \mathsf{F}, \mathsf{Right}, \mathsf{Left}, \mathsf{RM} = \mathsf{UpdateLF}(c, n, \mathsf{L}, \mathsf{F}, \mathsf{Right}, \mathsf{Left}, \mathsf{RM}, k)$;
    $// = \mathsf{L}_T^{\circ}, \mathsf{F}_T^{\circ}, \mathsf{Right}_T, \mathsf{Left}_T, \mathsf{RM}_T^{\circ}$
4    $\mathsf{L}, \mathsf{F}, \mathsf{C}, k' = \mathsf{InsertRow}(n, \mathsf{L}, \mathsf{F}, \mathsf{C}, k)$; $// = \mathsf{L}_T, \mathsf{F}_T, \mathsf{C}_T, k_T$
5   **foreach** $a \in \Pi$ **do**
6   **if** $\mathsf{RM}[a] \geq k'$ **then** $\mathsf{RM}[a] = \mathsf{RM}[a] + 1$; $// = \mathsf{RM}_T[a]$
7    $x = \mathsf{UpdateLCP}(\mathsf{L}, \mathsf{F}, \mathsf{LCP}^{\infty}, k')$; $// = \mathsf{LCP}_T^{\infty}[k_T]$
8    $\mathsf{LCP}^{\infty}[k' - 1] = \mathsf{UpdateLCP}(\mathsf{L}, \mathsf{F}, \mathsf{LCP}^{\infty}, k' - 1)$; $// = \mathsf{LCP}_T^{\infty}[k_T - 1]$
9    $\mathsf{insert}_x(\mathsf{LCP}^{\infty}, k')$;
10   **return** $n + 1, \mathsf{L}, \mathsf{F}, \mathsf{Right}, \mathsf{Left}, \mathsf{RM}, \mathsf{C}, \mathsf{LCP}^{\infty}$;

Lemma 3. *For* $1 \leq i \leq n + 1$,

$$RA_T[i] = \begin{cases} RA_S[i] & \textit{if } i < k_T, \\ n + 1 & \textit{if } i = k_T, \\ RA_S[i - 1] & \textit{if } i > k_T. \end{cases}$$

In the BWT, where S and T have no parameter characters, this implies that $\mathsf{L}_T[i] = T_{RA_T[i]}[n + 1] = S_{RA_S[i]}[n] = \mathsf{L}_S[i]$ for $i < k_T$ and $\mathsf{L}_T[i + 1] = T_{RA_T[i+1]}[n + 1] = S_{RA_S[i]}[n] = \mathsf{L}_S[i]$ for $i > k_T$, except when $i = k_S$. Therefore, for computing L_T from L_S, we only need to update $\mathsf{L}_S[k_S] = \$$ to c and to find the position $k_T = RA_T^{-1}[n + 1]$ where $\$$ should be inserted. However in the pBWT, $RA_T[i] = RA_S[i]$ does not necessarily imply that the values $\mathsf{L}_T[i] = \llbracket T_{RA_T[i]} \rrbracket[n + 1]$ and $\mathsf{L}_S[i] = \llbracket S_{RA_S[i]} \rrbracket[n]$ coincide, since it is not always true that $\llbracket S \rrbracket = \llbracket T \rrbracket[2:]$. So we also need to update the values of the encoding.

Algorithm 1 shows our update procedure, which maintains the array F and other auxiliary data structures in addition to L. After getting the key position k_S as the unique occurrence position of $\$$ in L_S at Line 2, to update the values of L and F from L_S and F_S to L_T and F_T, respectively, we compute intermediate arrays L_T° and F_T° of length n, which satisfy

$$\mathsf{L}_T^{\circ}[i] = \llbracket T_{RA_S[i]} \rrbracket[n + 1] \quad \text{and} \quad \mathsf{F}_T^{\circ}[i] = \llbracket T_{RA_S[i]} \rrbracket[1]$$

for $1 \leq i \leq n$ using UpdateLF at Line 3. In other words, L_T° and F_T° are extracted from the last and the first columns of the $n \times (n + 1)$ matrix $(\llbracket T_{RA_S[i]} \rrbracket)_{i=1}^{n}$, respectively, which can conceptionally be obtained by deleting the k_T-th row of the rotation sort matrix of T. We then find the other key position k_T and inserts appropriate values into L_T° and F_T° at k_T to turn them into L_T and F_T, respectively, by InsertRow at Line 4. The rest of the algorithm is devoted to maintaining some of the helper arrays. Particularly, a dedicated function UpdateLCP is used to update the ∞-LCP array. In the remainder of this section, we will explain those functions and involved auxiliary data structures in respective subsections. Table 2 shows an example of our 2-step update.

Table 2. An example of our update step for $S = \textbf{xayzzazyza\$}$ and $T = yS$. The updated and inserted values are highlighted. In the arrays $\langle T_{RA_S[i]}\rangle$, updated/inserted values appear only after \$. Lemmas 3 and 8 are immediate consequences of this observation.

$F_S[i]$	$\langle S_{RA_S[i]}\rangle$	$L_S[i]$
\$	\$∞a∞∞1a252a	a
a	a\$∞a∞∞1a252	3
a	a∞∞1a252a\$∞	3
a	a∞∞2a\$∞a661	1
3	∞a\$∞a∞61a25	3
1	∞a2∞2a\$∞a66	1
3	∞a∞∞1a252a\$	\$
1	∞1a2∞2a\$∞a6	2
3	∞∞a\$∞a661a2	2
2	∞∞1a252a\$∞a	a
2	∞∞2a\$∞a661a	a

$F_T^\circ[i]$	$\langle T_{RA_S[i]}\rangle$	$L_T^\circ[i]$
\$	\$∞∞a3∞1a252a	a
a	a\$∞∞a3∞1a252	3
a	a∞∞1a252a\$4∞	3
a	a∞∞2a\$4∞a361	1
3	∞a\$∞∞a361a25	2
1	∞a2∞2a\$4∞a36	1
3	∞a∞∞1a252a\$4	2
1	∞1a2∞2a\$4∞a3	2
2	∞∞a\$4∞a361a2	2
2	∞∞1a252a\$4∞a	a
2	∞∞2a\$4∞a361a	a

$F_T[i]$	$\langle T_{RA_T[i]}\rangle$	$L_T[i]$
\$	\$∞∞a3∞1a252a	a
a	a\$∞∞a3∞1a252	3
a	a∞∞1a252a\$4∞	3
a	a∞∞2a\$4∞a361	1
3	∞a\$∞∞a361a25	2
1	∞a2∞2a\$4∞a36	1
3	∞a∞∞1a252a\$4	2
1	∞1a2∞2a\$4∞a3	2
2	∞∞a\$4∞a361a2	2
2	∞∞a3∞1a252a\$	\$
2	∞∞1a252a\$4∞a	a
2	∞∞2a\$4∞a361a	a

3.1 Step 1: UpdateLF Computes $L_T^\circ[i]$ and $F_T^\circ[i]$

When $c \in \Sigma$, computing L_T° and F_T° from L_S and F_S, respectively, is easy.

Lemma 4. *If $c \in \Sigma$, then for any $i \in \{1, \ldots, n\}$, $F_T^\circ[i] = F_S[i]$ and $L_T^\circ[i] = L_S[i]$ except for $L_T^\circ[k_S] = c$.*

Concerning the case $c \in \Pi$, first let us express the values of $[\![T]\!]$ using $[\![S]\!]$.

Lemma 5. *Suppose $c \in \Pi$.*

$$[\![T]\!][1] = \begin{cases} |\Pi \restriction S| + 1 & \text{if } Left_S(c) = 0, \\ |\Pi \restriction S[1 : Left_S(c)]| & \text{otherwise.} \end{cases}$$

For $1 \leq p \leq n$, if $S[p] \in \Sigma$ or $p \neq Right_S(S[p])$, then $[\![T]\!][p+1] = [\![S]\!][p]$. If $S[p] = a \in \Pi$ and $p = Right_S(a)$, then

$$[\![T]\!][p+1] = \begin{cases} |\Pi \restriction S[p+1:n]| + 1 & \text{if } a = c, \\ [\![S]\!][p] + 1 & \text{if } Left_S(c) = 0 \text{ or} \\ & \quad Left_S(a) < Left_S(c) \leq Right_S(c) < Right_S(a), \\ [\![S]\!][p] & \text{otherwise.} \end{cases}$$

Based on Lemmas 4 and 5, Algorithm 2 computes $F_T^\circ[i]$ and $L_T^\circ[i]$ from $F_S[i]$ and $L_S[i]$, as well as other auxiliary data structures. Note that, since the intermediate matrix $(T_{RA_S[i]})_{i=1}^n$ misses a row corresponding to $[\![T]\!]$, the value $[\![T]\!][1]$ does not matter for F_T°, whereas it appears as $L_T^\circ[k_S] = L_T^\circ[RA_S^{-1}[n]]$. When $c \in \Pi$, Lemma 5 implies that, other than $L_T^\circ[k_S] = [\![T]\!][1]$, we only need to update the values at the positions in L and F corresponding to the rightmost

---

Algorithm 2: Computing L_T° and F_T°

---

1 **Function** UpdateLF$(c, n, L, F, \text{Right}, \text{Left}, \text{RM}, k)$
2 $\quad$ **if** $c \in \Sigma$ **then** $L[k] = c$;
3 $\quad$ **else**
4 $\quad\quad$ **foreach** $a \in \Pi$ with Left$[a] \neq 0$ **do**
   $\quad\quad\quad$ // *Computing* $L_T^\circ[\text{RM}[a]] = F_T^\circ[LF_S(\text{RM}[a])]$
5 $\quad\quad\quad$ $i = \text{RM}[a]$;
6 $\quad\quad\quad$ $j = \text{select}_{L[i]}(F, \text{rank}_{L[i]}(L, i))$; $\qquad\qquad$ // $j = LF_S(i)$
7 $\quad\quad\quad$ **if** $a = c$ **then**
8 $\quad\quad\quad\quad$ $cnt = 0$;
9 $\quad\quad\quad\quad$ **foreach** $b \in \Pi$ with Left$[b] \neq 0$ **do**
10 $\quad\quad\quad\quad\quad$ **if** Left$[a] \geq$ Right$[b]$ **then** $cnt = cnt + 1$;
11 $\quad\quad\quad$ **else**
12 $\quad\quad\quad\quad$ $cnt = L[i]$;
13 $\quad\quad\quad\quad$ **if** Left$[c] = 0$ or Left$[a] >$ Left$[c] \geq$ Right$[c] >$ Right$[a]$ **then**
14 $\quad\quad\quad\quad\quad$ $cnt = cnt + 1$;
15 $\quad\quad\quad$ $L[i] = cnt$; $F[j] = cnt$;
   $\quad\quad$ // *Computing* $L_T^\circ[k_S] = [\![T]\!][1]$
16 $\quad\quad$ $cnt = 1$;
17 $\quad\quad$ **if** Left$[c] = 0$ **then**
18 $\quad\quad\quad$ **foreach** $a \in \Pi$ with Left$[a] \neq 0$ **do** $cnt = cnt + 1$;
19 $\quad\quad\quad$ Right$[c] = n + 1$; Left$[c] = n + 1$; RM$[c] = k$;
20 $\quad\quad$ **else**
21 $\quad\quad\quad$ **foreach** $a \in \Pi$ with Left$[a] >$ Left$[c]$ **do** $cnt = cnt + 1$;
22 $\quad\quad\quad$ Left$[c] = n + 1$;
23 $\quad\quad$ $L[k] = cnt$;
24 $\quad$ **return** L, F, Right, Left, RM;

---

occurrence position $p = Right_S(a)$ of each parameter character $a \in \Pi$ in S. By rotating S to the right by $n - p$, that occurrence comes to the right end and appears in the pBWT. That is, the array L needs to be updated only at i such that $RA_S[i] = n - Right_S(a)$. The algorithm maintains such position i as $\text{RM}_S[a]$ for each $a \in \Pi \restriction S$, i.e. $\text{RM}_S[a] = RA_S^{-1}[n - Right_S(a)]$. Similarly, we only need to update F at $LF_S(\text{RM}_S[a])$, where $F_T^\circ[LF_S(\text{RM}_S[a])] = L_T^\circ[\text{RM}_S[a]]$. In our algorithm, as alternatives of $Left_S$ and $Right_S$, we maintain two arrays Left and Right that store the leftmost and rightmost occurrence positions of parameter characters counting *from the right end*, respectively, i.e., $\text{Left}_S[a] = Right_{\overline{S}}(a)$ and $\text{Right}_S[a] = Left_{\overline{S}}(a)$ for each $a \in \Pi$, where $\overline{S}$ is the reverse of S.

Algorithm 2 also updates RM to RM_T°, which indicates the row of L_T° corresponding to the rightmost occurrence of each parameter character in T. That is, $\text{RM}_T^\circ[a] = i$ iff $RA_S[i] = \text{Right}_T[a]$, as long as a occurs in T. When $c \in \Pi$ and it appears in the text for the first time, we have $\text{RM}_T^\circ[c] = k_S$ (Line 19). Other than that, $\text{RM}_T^\circ[a] = \text{RM}_S[a]$ for every $a \in \Pi$.

Lemma 6. *Algorithm 2 computes* $\mathsf{L}_T^\circ[i]$, $\mathsf{F}_T^\circ[i]$, Right_T, Left_T, *and* RM_T° *in* $O(|\Pi|\frac{\log n}{\log\log n})$ *amortized time.*

3.2 Step 2: InsertRow Computes L_T and F_T

To transform F_T° and L_T° into F_T and L_T, we insert the values $[\![T]\!][1]$ and $[\![T]\!][n+1]$ at the position k_T, respectively. We know those values as $[\![T]\!][1] = \mathsf{L}_T^\circ[k_S]$ and $[\![T]\!][n+1] = \$$. Therefore, it is enough to discuss how to find the position k_T.

In the case $c \in \Sigma$, the position k_T can be calculated similarly to the case of BWT for static strings thanks to Corollary 1. Define $\Sigma_{<b} = |\{\, a \in \Sigma \mid a < b \,\}|$.

Lemma 7. *If* $c \in \Sigma$, $k_T = |T|_{\Sigma_{<c}} + |\{\, i \mid \mathsf{L}_T^\circ[i] = c,\ 1 \le i \le k_S \,\}|$.

In the case $c \in \Pi$, we will use Lemma 2 for finding k_T in Lemma 9 below. We first observe that one can use LCP_S^∞ to calculate $lcp^\infty(\langle T_p\rangle, \langle T_q\rangle)$ for most cases.

Lemma 8. *For* $1 \le p < q \le n$, $lcp^\infty(\langle T_p\rangle, \langle T_q\rangle) = lcp^\infty(\langle S_p\rangle, \langle S_q\rangle)$.

Lemma 9. *Suppose* $c \in \Pi$. *Let* $\ell_i = lcp^\infty(\langle S_{RA_S[i]}\rangle, \langle S_{RA_S[k_S]}\rangle)$ *for* $1 \le i \le n$. *Then,*

$$
\begin{aligned}
k_T = 1 + |T|_\Sigma & \\
+ |\{\, i \mid 1 \le \mathsf{L}_T^\circ[i] \le \mathsf{L}_T^\circ[k_S],\ 1 \le i < k_S \,\}| & \qquad (1) \\
+ |\{\, i \mid \ell_i < \mathsf{L}_T^\circ[k_S] < \mathsf{L}_T^\circ[i],\ 1 \le i < k_S \,\}| & \qquad (2) \\
+ |\{\, i \mid 1 \le \mathsf{L}_T^\circ[i] \le \min\{\mathsf{L}_T^\circ[k_S] - 1, \ell_i\},\ k_S < i \le n \,\}|. & \qquad (3)
\end{aligned}
$$

Proof. By definition,

$$
k_T = 1 + |T|_\Sigma + |\{\, j \mid \mathsf{F}_T[j] \in \mathbb{N}_+,\ 1 \le j < k_T \,\}|,
$$

of which we focus on the last term. Let $h = LF_T^{-1}(k_T)$ and $m_i = lcp^\infty(\langle T_{RA_T[i]}\rangle, \langle T_{RA_T[h]}\rangle)$ for $1 \le i \le n+1$. By Lemma 2, $\mathsf{F}_T[j] \in \mathbb{N}_+$ and $1 \le j < k_T$ iff for $i = LF_T^{-1}(j)$, either

1. $1 \le i < h$ and $1 \le \mathsf{L}_T[i] \le \mathsf{L}_T[h]$,
2. $1 \le i < h$ and $m_i < \mathsf{L}_T[h] < \mathsf{L}_T[i]$, or
3. $h < i \le n+1$ and $1 \le \mathsf{L}_T[i] \le \min\{\mathsf{L}_T[h] - 1, m_i\}$.

Those three cases are mutually exclusive. Let $m_i^\circ = lcp^\infty(\langle T_{RA_S[i]}\rangle, \langle T_{RA_S[k_S]}\rangle)$. Counting each of the above cases is equivalent to counting i such that

1. $1 \le i < k_S$ and $1 \le \mathsf{L}_T^\circ[i] \le \mathsf{L}_T^\circ[k_S]$,
2. $1 \le i < k_S$ and $m_i^\circ < \mathsf{L}_T^\circ[k_S] < \mathsf{L}_T^\circ[i]$, or
3. $k_S < i \le n$ and $1 \le \mathsf{L}_T^\circ[i] \le \min\{\mathsf{L}_T^\circ[k_S] - 1, m_i^\circ\}$.

Algorithm 3: Inserting $[\![T]\!][1]$ to F and $[\![T]\!][n+1]$ to L

1 **Function** InsertRow$(n, \mathsf{L}, \mathsf{F}, \mathsf{C}, k)$
2 $\quad x = \mathsf{L}[k];$ $\qquad\qquad\qquad\qquad\qquad\qquad\qquad$ $// = [\![T]\!][1]$
3 $\quad$ **if** $x \in \Sigma$ **then**
4 $\quad\quad k' = \mathsf{select}_x(\mathsf{C}, 1) - |\Sigma_{<x}| - 1 + \mathsf{rank}_x(\mathsf{L}, k);$
  $\quad\quad$ $// \; k_T = |S|_{\Sigma_{<c}} + |\{\, i \mid \mathsf{L}_T^{\circ}[i] = c,\ 1 \le i \le k_S \,\}|$
5 $\quad\quad \mathsf{insert}_x(\mathsf{C}, \mathsf{select}_x(\mathsf{C}, 1));$
6 $\quad$ **else**
7 $\quad\quad k' = 1 + |\mathsf{C}| - |\Sigma|;$ $\qquad\qquad\qquad\qquad\quad$ $// = 1 + |S|_{\Sigma}$
8 $\quad\quad$ **for** $y = 1$ **to** x **do** $k' = k' + \mathsf{rank}_y(\mathsf{L}, k-1);$ $\qquad$ $// \; Term\ (1)$
9 $\quad\quad j = 0;$
10 $\quad\quad$ **for** $y = 0$ **to** $x - 1$ **do**
11 $\quad\quad\quad$ **if** $\mathsf{rank}_y(\mathsf{LCP}^{\infty}, k-1) \ne 0$ **then**
12 $\quad\quad\quad\quad j = \max\{j, \mathsf{select}_y(\mathsf{LCP}^{\infty}, \mathsf{rank}_y(\mathsf{LCP}^{\infty}, k-1))\};$
  $\quad\quad\quad\quad$ $// \; j = \max(\{j\} \cup \{\, i \mid \mathsf{LCP}^{\infty}[i] = y \; and \; 1 \le i < k_S \,\})$
13 $\quad\quad$ **for** $y = x+1$ **to** $|\Pi|$ **do** $k' = k' + \mathsf{rank}_y(\mathsf{L}, j);$ $\qquad$ $// \; Term\ (2)$
14 $\quad\quad j = n;$ $\qquad\qquad\qquad\qquad\qquad\qquad\qquad\qquad\quad$ $// \; j_0 = n$
15 $\quad\quad$ **for** $y = 1$ **to** $x - 1$ **do** $\qquad\qquad\qquad\qquad\qquad$ $// \; Term\ (3)$
16 $\quad\quad\quad$ **if** $\mathsf{rank}_{y-1}(\mathsf{LCP}^{\infty}, k-1) < \mathsf{rank}_{y-1}(\mathsf{LCP}^{\infty}, n)$ **then**
17 $\quad\quad\quad\quad j = \min\{j, \mathsf{select}_{y-1}(\mathsf{LCP}^{\infty}, \mathsf{rank}_{y-1}(\mathsf{LCP}^{\infty}, k-1) + 1)\};$
  $\quad\quad\quad\quad$ $// \; j_y = \min(\{j_{y-1}\} \cup \{\, i \mid \mathsf{LCP}^{\infty}[i] = y - 1 \; and \; k_S \le i \le n \,\})$
18 $\quad\quad\quad k' = k' + \mathsf{rank}_y(\mathsf{L}, j-1) - \mathsf{rank}_y(\mathsf{L}, k);$
19 $\quad \mathsf{insert}_\$(\mathsf{L}, k'); \; \mathsf{insert}_x(\mathsf{F}, k');$
20 $\quad$ **return** $\mathsf{L}, \mathsf{F}, \mathsf{C}, k';$

This is because the matrix $([\![T_{RA_S[i]}]\!])_{i=1}^{n}$ can conceptionally be obtained by removing the k_T-th row of the matrix of $([\![T_{RA_T[i]}]\!])_{i=1}^{n+1}$, where the row k_S of $([\![T_{RA_S[i]}]\!])_{i=1}^{n}$ corresponds to the row h of $([\![T_{RA_T[i]}]\!])_{i=1}^{n+1}$ in particular $(RA_S[k_S] = RA_T[h] = n)$, and $i = k_T = RA_T^{-1}[n+1]$ is not counted due to $T[n+1] = \$ \in \Sigma$.

Lemma 8 implies $m_i^{\circ} = \ell_i$, which completes the proof. $\qquad\qquad\qquad$ $\square$

Based on Lemmas 7 and 9, Algorithm 3 finds the key position k_T.

For handling the case $c \in \Sigma$, we maintain a dynamic array C by which one can obtain the value $|T|_{\Sigma_{<c}} = |S|_{\Sigma_{<c}}$ quickly. The array C_S can be seen as a string of the form $\mathsf{C}_S = a_1^{|S|_{a_1}+1} \dots a_\sigma^{|S|_{a_\sigma}+1}$, where $a_1, \dots, a_\sigma$ enumerate the static characters of Σ in the lexicographic order ($\sigma = |\Sigma|$) and a^s denotes the sequence of a of length s. Then, $|T|_{\Sigma_{<c}} = \mathsf{select}_c(\mathsf{C}_S, 1) - |\Sigma_{<c}| - 1$. The other term $|\{\, i \mid \mathsf{L}_T^{\circ}[i] = c,\ 1 \le i \le k_S \,\}|$ in Lemma 7 is calculated as $\mathsf{rank}_c(\mathsf{L}_T^{\circ}, k_S)$. We remark C_S has $a^{|S|_a+1}$ rather than $a^{|S|_a}$ so that $\mathsf{select}_c(\mathsf{C}, 1)$ is always defined.

Suppose $c \in \Pi$. The term $|T|_{\Sigma}$ of the equation of Lemma 9 is calculated as $|T|_{\Sigma} = |S|_{\Sigma} = |\mathsf{C}| - |\Sigma|$. Let $x = \mathsf{L}_T^{\circ}[k_S]$. Term (1) is obtained at Line 8 by

$$(1) = \sum_{y=1}^{x} \mathsf{rank}_y(\mathsf{L}_T^{\circ}, k_S - 1).$$

Concerning Term (2), we first find the range of $i < k_S$ satisfying $\ell_i < x$. By Lemma 1, $\ell_i = \min_{i \leq j < k_S} \mathsf{LCP}_S^\infty[j]$. Thus, for any $i < k_S$, $\ell_i < x$ iff $i \leq j_* = \max\{\, j \mid \mathsf{LCP}_S^\infty[j] < x,\ j < k_S \,\}$. The **for** loop of Line 10 computes such j_*. Then, (2) is computed at Line 13 as

$$(2) = |\{\, i \mid x < \mathsf{L}_T^\circ[i],\ 1 \leq i \leq j_* \,\}| = \sum_{y=x+1}^{|\Pi|} \mathsf{rank}_y(\mathsf{L}_T^\circ, j_*)\,.$$

We compute Term (3) by summing up the numbers of positions $i > k_S$ such that $\mathsf{L}_T^\circ[i] = y \leq \ell_i$ for all $y = 1, \ldots, x-1$ in the **for** loop of Line 15. To this end, we find the range of $i > k_S$ such that $\ell_i \geq y$. By Lemma 1, $\ell_i = \min_{k_S \leq j < i} \mathsf{LCP}_S^\infty[j]$. Thus, for every $i > k_S$, $\ell_i \geq y$ iff $i < j_y = \min\{\, j \mid \mathsf{LCP}_S^\infty[j] < y,\ k_S \leq j \leq n \,\}$. Note that $j_y = \min(\{j_{y-1}\} \cup \{\, j \mid \mathsf{LCP}_S^\infty[j] = y-1,\ k_S \leq j \leq n \,\})$ for any $y \geq 1$ assuming $j_0 = n$. Line 17 computes j_y as the first occurrence of $y-1$ after those in $\mathsf{LCP}^\infty[1 : k_S - 1]$. Then, (3) is calculated by

$$(3) = |\{\, i \mid 1 \leq \mathsf{L}_T^\circ[i] \leq x-1,\ k_S < i < j_y \,\}|$$
$$= \sum_{y=1}^{x-1} \left(\mathsf{rank}_y(\mathsf{L}_T^\circ, j_y - 1) - \mathsf{rank}_y(\mathsf{L}_T^\circ, k_S) \right).$$

Lemma 10. *Algorithm 3 computes* L_T, F_T, C_T, *and* k_T *in* $O(|\Pi| \frac{\log n}{\log \log n})$ *amortized time.*

3.3   Step 3: Updating LCP$^\infty$ by UpdateLCP

What remains to do is updating the arrays RM and LCP$^\infty$. On the one hand, updating RM from RM_T° to RM_T is easy. $\mathsf{RM}_T^\circ[a]$ should be incremented by one just if $\mathsf{RM}_T^\circ[a] \geq k_T$. Otherwise, $\mathsf{RM}_T[a] = \mathsf{RM}_T^\circ[a]$. On the other hand, Lemma 8 implies LCP_T^∞ is almost identical to LCP_S^∞.

Corollary 2. $\mathsf{LCP}_T^\infty[i] = \mathsf{LCP}_S^\infty[i]$ *if* $i < k_T - 1$, *and* $\mathsf{LCP}_T^\infty[i] = \mathsf{LCP}_S^\infty[i-1]$ *if* $i > k_T$.

By Corollary 2, we only need to compute $\mathsf{LCP}_T^\infty[k_T - 1]$ and $\mathsf{LCP}_T^\infty[k_T]$, to which Lemma 8 cannot directly be applied. The following lemma allows us to reduce the calculation of $\mathsf{LCP}_T^\infty[k] = lcp^\infty(\langle T_{RA_T[k]} \rangle, \langle T_{RA_T[k+1]} \rangle)$ to that of $lcp^\infty(\langle T_{RA_T[LF_T^{-1}(k)]} \rangle, \langle T_{RA_T[LF_T^{-1}(k+1)]} \rangle)$, to which Lemma 8 may be applied.

Lemma 11. *Let* $1 \leq i, j \leq n+1$, $p = RA_T[i]$, $q = RA_T[j]$, $\ell = lcp^\infty(\langle T_p \rangle, \langle T_q \rangle)$, $i' = LF_T^{-1}(i)$, $j' = LF_T^{-1}(j)$, $p' = RA_T[i']$, $q' = RA_T[j']$, *and* $\ell' = lcp^\infty(\langle T_{p'} \rangle, \langle T_{q'} \rangle)$.

1. *If* $\mathsf{F}_T[i] = \mathsf{F}_T[j] \in \Sigma$, *then* $\ell = \ell'$.
2. *If* $\mathsf{F}_T[i] \neq \mathsf{F}_T[j]$ *and either* $\mathsf{F}_T[i] \in \Sigma$ *or* $\mathsf{F}_T[j] \in \Sigma$, *then* $\ell = 0$.
3. *If* $\mathsf{F}_T[i], \mathsf{F}_T[j] \in \mathbb{N}_+$, *then*

$$\ell = \begin{cases} \ell' + 1 & \text{if } \ell' < \min\{\mathsf{F}_T[i], \mathsf{F}_T[j]\}, \\ \ell' & \text{if } \ell' \geq \mathsf{F}_T[i] = \mathsf{F}_T[j], \\ \min\{\mathsf{F}_T[i], \mathsf{F}_T[j]\} & \text{otherwise.} \end{cases}$$

Algorithm 4: Updating $\mathsf{LCP}^\infty[i]$

---

1 **Function** UpdateLCP($\mathsf{L}, \mathsf{F}, \mathsf{LCP}^\infty, i$)
2   $j = i + 1$; $x = 0$;
3   **if** $\mathsf{F}[i] = \mathsf{F}[j]$ or $\mathsf{F}[i], \mathsf{F}[j] \in \mathbb{N}_+$ **then**
4    $i' = \mathsf{select}_{\mathsf{F}[i]}(\mathsf{L}, \mathsf{rank}_{\mathsf{F}[i]}(\mathsf{F}, i))$; $// \ i' = LF_T^{-1}(i)$
5    $j' = \mathsf{select}_{\mathsf{F}[j]}(\mathsf{L}, \mathsf{rank}_{\mathsf{F}[j]}(\mathsf{F}, j))$; $// \ j' = LF_T^{-1}(i+1)$
6   **for** $y = |\Pi|$ **downto** 0 **do**
7     $\lfloor$ **if** $\mathsf{rank}_y(\mathsf{LCP}^\infty, i' - 1) \neq \mathsf{rank}_y(\mathsf{LCP}^\infty, j' - 1)$ **then** $x = y$;
    $// \ x = lcp^\infty(\langle S_{RAS[i']}\rangle, \langle S_{RAS[j']}\rangle)$
8   **if** $\mathsf{F}[i], \mathsf{F}[j] \in \mathbb{N}_+$ **then**
9   **if** $x < \min\{\mathsf{F}[i], \mathsf{F}[j]\}$ **then** $x = x + 1$;
10   **else if** $\mathsf{F}[i] \neq \mathsf{F}[j]$ **then** $x = \min\{\mathsf{F}[i], \mathsf{F}[j]\}$;
11   **return** x;

Proof. Let $T_p = aU$, $T_q = bV$, $T_{p'} = Ua$ and $T_{q'} = Vb$.

1. If $a = b \in \Sigma$, then $\ell = \ell' = lcp^\infty(\langle U \rangle, \langle V \rangle)$.
2. In the case $a \neq b$ and $\{a, b\} \cap \Sigma \neq \emptyset$, clearly $\langle T_p \rangle[1] \neq \langle T_q \rangle[1]$. Thus $\ell = 0$.
3. In the case $a, b \in \Pi$, let W be the longest common prefix of $\langle T_{p'} \rangle$ and $\langle T_{q'} \rangle$, u and v be the first occurrence positions of a in $T_{p'}$ and b in $T_{q'}$, respectively, and w be the ℓ'-th occurrence position of ∞ in W.

Suppose $\ell' < \min\{\mathsf{F}_T[i], \mathsf{F}_T[j]\} = \min\{\mathsf{L}_T[i'], \mathsf{L}_T[j']\}$. That is, $|W|_\infty < \min\{|\langle T_{p'} \rangle[: u]|_\infty, |\langle T_{q'} \rangle[: v]|_\infty\}$. This means $|W| < \min\{u, v\}$ and thus ∞W is the longest common prefix of $\langle T_p \rangle$ and $\langle T_q \rangle$. Thus, we have $\ell = |\infty W|_\infty = \ell' + 1$.

Suppose $\ell' \geq \mathsf{F}_T[i] = \mathsf{F}_T[j]$, i.e., $|W|_\infty \geq |\langle T_{p'} \rangle[: u]|_\infty = |\langle T_{q'} \rangle[: v]|_\infty$. Then $|W[: u]|_\infty = |W[: v]|_\infty$ and $W[u] = W[v] = \infty$ implies $u = v$. Let Z be the longest common prefix of $\langle T_p \rangle$ and $\langle T_q \rangle$. Then, W and Z can be written as $W = X\infty Y$ and $Z = \infty X u Y$, where $|X| = u - 1$. Therefore, $\ell = \ell'$.

Otherwise, $\mathsf{F}_T[i] \neq \mathsf{F}_T[j]$ and $\ell' \geq \min\{\mathsf{F}_T[i], \mathsf{F}_T[j]\}$. Assume $\mathsf{F}_T[i] < \mathsf{F}_T[j]$ (the case $\mathsf{F}_T[j] < \mathsf{F}_T[i]$ is symmetric). Then $u \leq |W|$. Moreover, we have $u < v$, since otherwise, $\langle T_{q'} \rangle[: v]$ had to be a prefix of $\langle T_{p'} \rangle[: u]$, which is impossible by $\mathsf{F}_T[i] < \mathsf{F}_T[j]$. Let $\langle T_p[: |W| + 1] \rangle = \infty X u Y$, where $|X| = u - 1$. Then we have $\langle T_q[: |W| + 1] \rangle = \infty X \infty Y'$ for some $Y' \in (\Sigma \cup \mathbb{N}_\infty)^*$. Thus $\ell = |\infty X|_\infty = \mathsf{F}_T[i]$. □

One can compute $\ell' = lcp^\infty(\langle T_{p'} \rangle, \langle T_{q'} \rangle)$ in Lemma 11 for $1 \leq p' < q' \leq n$ using Lemmas 8 and 1 as

$$lcp^\infty(\langle T_{p'} \rangle, \langle T_{q'} \rangle) = lcp^\infty(\langle S_{p'} \rangle, \langle S_{q'} \rangle) = \min\{\ \mathsf{LCP}_S^\infty[h] \mid i' \leq h < j'\ \}$$
$$= \min(\{0\} \cup \{\ y \mid \mathsf{rank}_y(\mathsf{LCP}_S^\infty, i' - 1) \neq \mathsf{rank}_y(\mathsf{LCP}_S^\infty, j' - 1)\ \}).$$

Finally, when $q' = n + 1$, we have $\mathsf{F}_T[j] = \$ \neq \mathsf{F}_T[i]$, and thus $lcp^\infty(\langle T_p \rangle, \langle T_q \rangle) = 0$. Algorithm 4 computes $\mathsf{LCP}_T^\infty[i]$ using F_T, L_T, and LCP_S^∞.

Lemma 12. *Algorithm 4 computes* $\mathsf{LCP}_T^\infty[i]$ *in* $O(|\Pi| \frac{\log n}{\log \log n})$ *amortized time.*

By Lemmas 6, 10, and 12, we have the following theorem.

Theorem 1. *Given $c \in \Sigma \cup \Pi$, $n = |S|$, $L = L_S$, $F = F_S$, Right = Right$_S$, Left = Left$_S$, RM = RM$_S$, C = C$_S$, and LCP$^\infty$ = LCP$_S^\infty$ for some $S \in (\Sigma \cup \Pi)^*$, Algorithm 1 computes $|T|$, L_T, F_T, Right$_T$, Left$_T$, RM$_T$, C$_T$, and LCP$_T^\infty$ for $T = cS$ in $O(|\Pi| \frac{\log n}{\log \log n})$ amortized time per input character.*

Corollary 3. *For a p-string T of length n, pBWT$_T$ can be computed in an online manner by reading T from right to left in $O(n|\Pi| \frac{\log n}{\log \log n})$ time.*

Acknowledgments. This work was supported by JSPS KAKENHI Grant Numbers JP19K20208 (DH), JP21K17701 (DK), JP21H05847 (DK), JP22H03551 (DK), JP18H04091 (RY), JP18K11150 (RY), JP20H05703 (RY), and JP21K11745 (AS).

A   Proofs

Proposition 1. *For any p-strings S and T, $S \approx T$ if and only if $[\![S]\!] = [\![T]\!]$.*

Proof. For simplicity, assume that S and T contain no static character. Suppose $S \approx T$. Since $S[i] = S[j]$ iff $T[i] = T[j]$ for any indices i, j, we have

$$[\![S]\!][i] = |\Pi \upharpoonright S_{n-i}[1 : Left_{S_{n-i}}(S[i])]| = |\Pi \upharpoonright T_{n-i}[1 : Left_{T_{n-i}}(T[i])]| = [\![T]\!][i].$$

for all i.

Suppose $S \not\approx T$. Let i be the leftmost position such that $\langle S \rangle[i] \neq \langle T \rangle[i]$. We may assume without loss of generality that $\langle S \rangle[i] < \langle T \rangle[i]$. Let $j = i - \langle S \rangle[i]$. Then,

$$[\![S]\!][j] = |\Pi \upharpoonright S[j+1 : i]| = |\Pi \upharpoonright S[j : i-1]| = |\Pi \upharpoonright T[j : i-1]|,$$

since $S[j : i-1] \approx T[j : i-1]$. The fact $S[j] \notin \Pi \upharpoonright S[j+1 : i-1]$ implies $T[j] \notin \Pi \upharpoonright T[j+1 : i-1]$. Moreover, $\langle S \rangle[i] < \langle T \rangle[i]$ implies $T[i] \notin \Pi \upharpoonright T[j : i-1]$. Hence,

$$[\![T]\!][j] \geq |(\Pi \upharpoonright T[j : i-1]) \cup \{T[i]\}| > |\Pi \upharpoonright T[j : i-1]| = [\![S]\!][j].$$

$\square$

Lemma 3 is a corollary to the following lemma.

Lemma 13. *For any i and j such that $1 \leq i < j \leq n$, $RA_S^{-1}[i] < RA_S^{-1}[j]$ iff $RA_T^{-1}[i] < RA_T^{-1}[j]$.*

Proof. Let $S_i = U\$V$ and $S_j = XY\$Z$, where $|U\$| = |X| = i < j \leq n$. We have $T_i = U\$cV$ and $T_j = XY\$cZ$. Since $\$$ does not occur in X, $\langle U\$\rangle \neq \langle X \rangle$. Thus,

$$RA_S^{-1}[i] < RA_S^{-1}[j] \iff \langle U\$\rangle < \langle X \rangle \iff RA_T^{-1}[i] < RA_T^{-1}[j].$$

$\square$

Lemma 4. *If $c \in \Sigma$, then for any $i \in \{1, \dots, n\}$, $F_T^\circ[i] = F_S[i]$ and $L_T^\circ[i] = L_S[i]$ except for $L_T^\circ[k_S] = c$.*

Proof. If $c \in \Sigma$, then $[\![S]\!] = [\![T]\!][2:]$ by definition. So, for any $i \in \{1, \ldots, n\}$,

$$\mathsf{F}_T^\circ[i] = [\![T_{RA_S[i]}]\!][1] = [\![S_{RA_S[i]}]\!][1] = \mathsf{F}_S[i]$$

and

$$\mathsf{L}_T^\circ[i] = [\![T_{RA_S[i]}]\!][n+1] = \begin{cases} [\![S_{RA_S[i]}]\!][n] = \mathsf{L}_S[i] & \text{if } RA_S[i] \neq n, \\ c & \text{if } RA_S[i] = n. \end{cases}$$

$\square$

Lemma 5. *Suppose $c \in \Pi$.*

$$[\![T]\!][1] = \begin{cases} |\Pi \upharpoonright S| + 1 & \text{if } Left_S(c) = 0, \\ |\Pi \upharpoonright S[1 : Left_S(c)]| & \text{otherwise.} \end{cases}$$

For $1 \leq p \leq n$, if $S[p] \in \Sigma$ or $p \neq Right_S(S[p])$, then $[\![T]\!][p+1] = [\![S]\!][p]$. If $S[p] = a \in \Pi$ and $p = Right_S(a)$, then

$$[\![T]\!][p+1] = \begin{cases} |\Pi \upharpoonright S[p+1 : n]| + 1 & \text{if } a = c, \\ [\![S]\!][p] + 1 & \text{if } Left_S(c) = 0 \text{ or} \\ & \quad Left_S(a) < Left_S(c) \leq Right_S(c) < Right_S(a), \\ [\![S]\!][p] & \text{otherwise.} \end{cases}$$

Proof. The claim on value of $[\![T]\!][1]$ is clear by definition. For $p \geq 1$, if $T[p+1] = S[p] \in \Sigma$, then $[\![T]\!][p+1] = [\![S]\!][p]$.

Let us consider the case $S[p] = a \in \Pi$. If $p \neq Right_S(a)$, then a occurs somewhere after p in S. Let $q > p$ be the first occurrence position of a after p in S. By definition, $[\![S]\!][p] = |\Pi \upharpoonright S[p+1 : q]| = |\Pi \upharpoonright T[p+2 : q+1]| = [\![T]\!][p+1]$.

Suppose $p = Right_S(a)$ for some $a \in \Pi$. If $a = c$, since $T[1] = c$ and $c \notin \Pi \upharpoonright S[p+1 : n]$, we have $[\![T]\!][p+1] = |\Pi \upharpoonright S[p+1 : n] \cup \{c\}| = |\Pi \upharpoonright S[p+1 : n]| + 1$. If $Right_S(c) = 0$ or $Left_S(a) < Left_S(c) \leq Right_S(c) < Right_S(a) = p$, $[\![S]\!][p]$ counts the number of distinct p-characters in $S[p + 1 :]S[: Left_S(a)]$, where c does not occur. On the other hand, $[\![T]\!][p+1]$ counts the ones in $S[p+1 :]cS[: Left_S(a)]$. That is, $[\![T]\!][p+1] = [\![S]\!][p] + 1$. Otherwise, if $Left_S(c) < Left_S(a)$ or $Right_S(a) < Right_S(c)$, we already have c in $S[p + 1 :]S[: Left_S(a)]$. Thus $[\![T]\!][p+1] = [\![S]\!][p]$.

$\square$

Lemma 7. *If $c \in \Sigma$, $k_T = |T|_{\Sigma_{<c}} + |\{ i \mid \mathsf{L}_T^\circ[i] = c, 1 \leq i \leq k_S \}|$.*

Proof. By definition, $k_T = |T|_{\Sigma_{<c}} + |\{ j \mid \mathsf{F}_T[j] = c, 1 \leq j \leq k_T \}|$. By Corollary 1 and the bijectivity of LF_T, the second term equals

$$|\{ i \mid \mathsf{L}_T[i] = c, 1 \leq i \leq LF_T^{-1}(k_T) \}|$$

and further more equals

$$|\{ i \mid \mathsf{L}_T^\circ[i] = c, 1 \leq i \leq k_S \}|$$

because L_T and L_T° are different only in that L_T has an extra element $\$ < c$, and the position $LF_T^{-1}(k_T)$ in L_T corresponds to the position k_S in L_T°. $\square$

Lemma 8. *For* $1 \leq p < q \leq n$, $lcp^{\infty}(\langle T_p \rangle, \langle T_q \rangle) = lcp^{\infty}(\langle S_p \rangle, \langle S_q \rangle)$.

Proof. Let $S_p = U\$V$ and $S_q = XY\$Z$, where $|U\$| = |X| = p < q = |XY\$|$. Then, $T_p = U\$cV$ and $T_q = XY\$cZ$. Since $\$$ does not appear in X,

$$lcp^{\infty}(\langle S_p \rangle, \langle S_q \rangle) = lcp^{\infty}(U\$, X) = lcp^{\infty}(\langle T_p \rangle, \langle T_q \rangle).$$

$\square$

References

1. Baker, B.S.: A theory of parameterized pattern matching: algorithms and applications. In: Proceedings of the Twenty-fifth Annual ACM Symposium on Theory of Computing (STOC 1993), pp. 71–80 (1993)
2. Baker, B.S.: Parameterized pattern matching: algorithms and applications. J. Comput. Syst. Sci. **52**(1), 28–42 (1996)
3. Burrows, M., Wheeler, D.: A block-sorting lossless data compression algorithm. Technical report 124, Digital Equipment Corporation (1994)
4. Diptarama, Katsura, T., Otomo, Y., Narisawa, K., Shinohara, A.: Position heaps for parameterized strings. In: Proceedings of the 28th Annual Symposium on Combinatorial Pattern Matching (CPM 2017), pp. 8:1–8:13 (2017)
5. Fredriksson, K., Mozgovoy, M.: Efficient parameterized string matching. Inf. Process. Lett. **100**(3), 91–96 (2006)
6. Fujisato, N., Nakashima, Y., Inenaga, S., Bannai, H., Takeda, M.: Right-to-left online construction of parameterized position heaps. In: Proceedings of the Prague Stringology Conference 2018 (PSC 2018), pp. 91–102 (2018)
7. Fujisato, N., Nakashima, Y., Inenaga, S., Bannai, H., Takeda, M.: Direct linear time construction of parameterized suffix and LCP arrays for constant alphabets. In: Proceedings of the 26th International Symposium on String Processing and Information Retrieval (SPIRE 2019), pp. 382–391 (2019)
8. Ganguly, A., Shah, R., Thankachan, S.V.: pBWT: achieving succinct data structures for parameterized pattern matching and related problems. In: Proceedings of the 28th Annual ACM-SIAM Symposium on Discrete Algorithms (SODA 2017), pp. 397–407 (2017)
9. Ganguly, A., Shah, R., Thankachan, S.V.: Fully functional parameterized suffix trees in compact space. In: Proceedings of the 49th International Colloquium on Automata, Languages, and Programming (ICALP 2022), pp. 65:1–65:18 (2022)
10. I, T., Deguchi, S., Bannai, H., Inenaga, S., Takeda, M.: Lightweight parameterized suffix array construction. In: Fiala, J., Kratochvíl, J., Miller, M. (eds.) IWOCA 2009. LNCS, vol. 5874, pp. 312–323. Springer, Heidelberg (2009). https://doi.org/10.1007/978-3-642-10217-2_31
11. Kim, S., Cho, H.: Simpler FM-index for parameterized string matching. Inf. Process. Lett. **165**, 106026 (2021)
12. Mendivelso, J., Thankachan, S.V., Pinzón, Y.J.: A brief history of parameterized matching problems. Discret. Appl. Math. **274**, 103–115 (2020)
13. Nakashima, K., et al.: DAWGs for parameterized matching: online construction and related indexing structures. In: Proceedings of the 31st Annual Symposium on Combinatorial Pattern Matching (CPM 2020), pp. 26:1–26:14 (2020)

14. Nakashima, K., Hendrian, D., Yoshinaka, R., Shinohara, A.: An extension of linear-size suffix tries for parameterized strings. In: SOFSEM 2020 Student Research Forum, pp. 97–108 (2020)
15. Navarro, G., Nekrich, Y.: Optimal dynamic sequence representations. SIAM J. Comput. **43**(5), 1781–1806 (2014)
16. Policriti, A., Prezza, N.: Fast online Lempel-Ziv factorization in compressed space. In: Proceedings of the the 22nd International Symposium on String Processing and Information Retrieval (SPIRE 2015), pp. 13–20 (2015)
17. Shibuya, T.: Generalization of a suffix tree for RNA structural pattern matching. Algorithmica **39**(1), 1–19 (2004)
18. Thankachan, S.V.: Compact text indexing for advanced pattern matching problems: parameterized, order-isomorphic, 2D, etc. (invited talk). In: Proceedings of the 33rd Annual Symposium on Combinatorial Pattern Matching (CPM 2022), pp. 3:1–3:3 (2022)

Accessing the Suffix Array via ϕ^{-1}-Forest

Christina Boucher[1], Dominik Köppl[2(✉)], Herman Perera[1],
and Massimiliano Rossi[1]

[1] Department of Computer and Information Science and Engineering,
Herbert Wertheim College of Engineering, University of Florida, Gainesville, FL, USA
{christinaboucher,hperera1,rossi.m}@ufl.edu
[2] Tokyo Medical and Dental University, M&D Data Science Center, Tokyo, Japan
koeppl.dsc@tmd.ac.jp

Abstract. Kärkkainen et al. (CPM, 2009) defined the concept of ϕ that
later became key to the construction of the r-index. Given a string
$S[1..n]$, its suffix array SA and its inverse suffix array ISA, we define
ϕ as the permutation of $\{1, \ldots, n\}$ such that $\phi(i) = \mathrm{SA}[\mathrm{ISA}[i] - 1]$ if
$\mathrm{ISA}[i] > 1$, and $\phi(i) = \mathrm{SA}[n]$ otherwise. Gagie et al. (JACM, 2020)
showed that it is possible to store $\mathcal{O}(r)$ words such that the permutations ϕ and ϕ^{-1} are evaluated in $\mathcal{O}(\log\log_w(n/r))$-time, which was
improved to $\mathcal{O}(1)$-time by Nishimoto and Tabei (ICALP, 2021). In this
paper, we introduce the concept of ϕ^{-1}-forest, which is a data structure
using sampled SA values to speed up random access to SA. We implemented our approach and compared its performance with respect to the
r-index.

Keywords: Compressed suffix array · r-index · ϕ function ·
Burrows–Wheeler transform

1 Introduction

Biological public datasets have become increasingly large, extensive and numer-
ous. Currently, for almost every biomedically or agriculturally interesting species,
there exists a sequencing consortium aimed at sequencing a large number of
individuals or cultivars of that species. For example, with relative ease you can
download over 5,000 human genomes, over 3 million SARS-COVID sequence
datasets, and over 5 petabytes sequence data from The Cancer Genome Atlas
(TCGA). Fortunately or unfortunately the majority of methods that aim to analyze these and other large datasets rely on the use of succinct data structures
that are capable of being constructed and stored in relatively small amount of
space and time, while performing efficient queries. In this paper, we focus on
developing a data structure that provides efficient access to the *suffix array*,
which is defined as follows: $\mathrm{SA}[1..n]$ is an array that is a permutation of $1, \ldots, n$
such that the suffixes of S starting at the consecutive positions indicated in SA
are in lexicographical order.

© The Author(s), under exclusive license to Springer Nature Switzerland AG 2022
D. Arroyuelo and B. Poblete (Eds.): SPIRE 2022, LNCS 13617, pp. 86–98, 2022.
https://doi.org/10.1007/978-3-031-20643-6_7

The suffix array, however, predates read alignment or even high-throughput sequencing as it was first introduced by Manber and Myers [12]. The lexicographical order implies that the starting positions of the longest match of a pattern P is within a contiguous range in SA. When combined with the Burrows–Wheeler Transform (BWT) of S [3], this range can be found in $2|P|$ rank queries via backward search. This is one of the main features that read alignment methods, such as Bowtie [9] and BWA [10], exploit in order to efficiently align reads to a database of genomes. Hence, given this application and the ever-increasing size of public, biological datasets, there has been significant interest in reducing the construction size and time of the SA while still allowing for efficient queries. There has been a significant amount of work in developing more efficient compressed suffix array representations and implementations. Most recently Puglisi and Zhukova [15] showed that the suffix array can be compressed via relative Lempel–Ziv (RLZ) dictionary compression in a manner that does not greatly affect the time to access the elements of the SA. In a different direction, Gagie et al. [7] showed that a single element of the SA for each run of the BWT is needed, and that all elements of the SA can be recovered via a small auxiliary data structure (e.g., ϕ). The resulting data structures of Gagie et al. is referred to as the r-index, where r stands for the number of single character runs in the BWT. Then Cobas et al. [5] improved upon the space usage of the r-index by a more careful sampling of the elements of the SA. Their resulting method, referred to as the subsampled r-index or the sr-index, was shown to be between 1.5 to 3.0 times smaller than the r-index in practice.

Although the methods of Gagie et al. and Cobas et al. are provably and practically space efficient, the time required to access the elements of the SA (which relies on ϕ or ϕ^{-1}) is slow in practice. In this paper, we revisit the problem of providing efficient access to the suffix array in the sampled suffix array of Gagie et al. via a small auxiliary data structure that exploits the iterative properties of ϕ. We implemented our method and compared it to the r-index on Chromosome 19 sequences from the 1000 Genomes Project [17] and the Pizza&Chili repetitive corpus [1]. We showed that the runtime for random access of our new method was favorable to that of the r-index. We report our new method was consistently faster on the Pizza&Chili, offering between 3x and 6x speed-up on all Pizza&Chili datasets except for one (`cere`) where both methods performed comparably. Our method is publicly available at https://github.com/koeppl/rasarindex.

2   Preliminaries

We define a string S as a finite sequence of characters $S = S[1..n] = S[1] \cdots S[n]$ over an alphabet $\Sigma = \{c_1, \ldots, c_\sigma\}$. We denote by ε the empty string, and the length of S as $|S|$. We denote by $S[i..j]$ the substring $S[i] \cdots S[j]$ of S starting in position i and ending in position j, with $S[i..j] = \varepsilon$ if $i > j$. For a string S and $1 \le i \le n$, $S[1..i]$ is called the i-th prefix of S, and $S[i..n]$ is called the i-th suffix of S.

Given an array $A[1..n]$ of n integers over a universe $\mathcal{U} = \{1, ..., |\mathcal{U}|\}$, for all $i \in \mathcal{U}$ we define $\text{pred}_A(i)$ as the *predecessor* of i in A, i.e., $\text{pred}_A(i) = \max\{A[j] \le$

Table 1. SA, BWT, and *rotations matrix* of the string $S = $ GATTACAT$GATACAT$GATTA
GATA#. The SA samples at the beginning and end of each run of the BWT are highlighted in bold. We have $r = 13$.

$$S = \text{GATTACAT\$GATACAT\$GATTAGATA\#}$$

i	SA	BWT	rotations matrix
0	**26**	A	#GATTACAT$GATACAT$GATTAGATA
1	**8**	T	$GATACAT$GATTAGATA#GATTACAT
2	16	T	$GATTAGATA#GATTACAT$GATACAT
3	25	T	A#GATTACAT$GATACAT$GATTAGAT
4	4	T	ACAT$GATACAT$GATTAGATA#GATT
5	12	T	ACAT$GATTAGATA#GATTACAT$GAT
6	**21**	T	AGATA#GATTACAT$GATACAT$GATT
7	**6**	C	AT$GATACAT$GATTAGATA#GATTAC
8	14	C	AT$GATTAGATA#GATTACAT$GATAC
9	**23**	G	ATA#GATTACAT$GATACAT$GATTAG
10	10	G	ATACAT$GATTAGATA#GATTACAT$G
11	1	G	ATTACAT$GATACAT$GATTAGATA#G
12	**18**	G	ATTAGATA#GATTACAT$GATACAT$G
13	**5**	A	CAT$GATACAT$GATTAGATA#GATTA
14	13	A	CAT$GATTAGATA#GATTACAT$GATA
15	**22**	A	GATA#GATTACAT$GATACAT$GATTA
16	**9**	$	GATACAT$GATTAGATA#GATTACAT$
17	**0**	#	GATTACAT$GATACAT$GATTAGATA#
18	**17**	$	GATTAGATA#GATTACAT$GATACAT$
19	**7**	A	T$GATACAT$GATTAGATA#GATTACA
20	15	A	T$GATTAGATA#GATTACAT$GATACA
21	**24**	A	TA#GATTACAT$GATACAT$GATTAGA
22	**3**	T	TACAT$GATACAT$GATTAGATA#GAT
23	11	A	TACAT$GATTAGATA#GATTACAT$GA
24	**20**	T	TAGATA#GATTACAT$GATACAT$GAT
25	**2**	A	TTACAT$GATACAT$GATTAGATA#GA
26	19	A	TTAGATA#GATTACAT$GATACAT$GA

$i \mid 1 \leq j \leq n\}$. Analogously, we define $\text{succ}_A(i)$ as the *successor* of i in A, i.e., $\text{succ}_A(i) = \min\{A[j] \geq i \mid 1 \leq j \leq n\}$.

Next, to define the suffix array for a string S of length n, we consider all rotations of S in lexicographical order. The index of the starting positions in S of each rotation defines the SA of S. Related to the suffix array, given a string

S and the suffix array of S, we define the *inverse suffix array* (ISA) of S as the indexes in the SA of every suffix of S, i.e., $\text{ISA}[i] = j$ if and only if $\text{SA}[j] = i$.

In what follows, we assume that S ends with a unique character $\#$ smaller than all other characters appearing in S. Then the BWT of S is a permutation of S such that $\text{BWT}[i] = S[\text{SA}[i]-1]$ and $\text{BWT}[i] = \#$ if $\text{SA}[i] = 1$. The BWT is often run-length compressed, i.e., representing maximal consecutive appearances of the same character, also called *runs*, by a single occurrence of this character and an exponent reflecting the length of this run. We denote the number of the runs in BWT by r. See Table 1 for an example.

Throughout this paper, we assume working in the RAM model with machine word size w. Let us fix the length n of a given input string S. Kärkkainen et al. [8] defined the concept of the ϕ function that later became key to the construction of the r-index. Given a string $S[1..n]$, its suffix array SA and its inverse suffix array ISA, we define ϕ as the permutation of $\{1, \ldots, n\}$ such that $\phi(i) = \text{SA}[\text{ISA}[i]-1]$ if $\text{ISA}[i] > 1$, and $\phi(i) = \text{SA}[n]$ otherwise. Later, Gagie et al. [7] showed that it is possible to store $\mathcal{O}(r)$ words such that the permutations ϕ and ϕ^{-1} are evaluated in $\mathcal{O}(\log \log_w(n/r))$-time. We note that this was later improved by Nishimoto and Tabei [13] to $\mathcal{O}(1)$ time. In particular, let $\mathcal{E}$ be the list of the SA samples at the end of each run of the BWT such that $\mathcal{E}[i]$ is the sample corresponding to the i-th run. Analogously, let $\mathcal{S}$ be the list of the SA samples at the beginning of each run such that $\mathcal{S}[i]$ is the sample corresponding to the i-th run. See Table 2 for an example of $\mathcal{E}$ and $\mathcal{S}$.

Given Gagie et al.'s representation of ϕ, we can implement random access queries to the suffix array by iterating the ϕ function as follows. Let i be a position in the suffix array, and let $j \geq i$ be a sampled position of the suffix array, then $\text{SA}[i] = \phi^{j-i}(\text{SA}[j])$. This solution works in $\mathcal{O}(r)$ space, since $\mathcal{E}$ and $\mathcal{S}$ each contain r SA samples. The running time can be bounded by the length of the longest run.

Table 2. Lists $\mathcal{S}$ and $\mathcal{E}$ of our running example with costs c_i and limits ℓ_i. $\text{BWT}[\mathcal{S}[x]..\mathcal{E}[x]]$ is a unary string, for $x \in [1..r]$.

i	1	2	3	4	5	6	7	8	9	10	11	12	13
$\mathcal{E}[i]$	26	21	14	18	22	9	0	17	24	3	11	20	19
$\mathcal{S}[i]$	26	8	6	23	5	9	0	17	7	3	11	20	2
c_i	5	3	1	2	0	0	0	4	0	0	0	0	2
ℓ_i	1	1	3	1	2	2	3	1	2	6	3	2	

3   Access Data Structures to SA

In what follows, we assume that we have the SA samples of $\mathcal{S}$ and $\mathcal{E}$ as defined by Gagie et al. [7], and describe a method for accessing the missing SA samples via the construction of an auxiliary data structure that exploits the iterative computation of ϕ^{-1}. We note that we selected to describe our methods in terms of ϕ^{-1}, and due to the symmetry of ϕ^{-1} and ϕ everything can be described analogously for ϕ.

The key observation of our technique is that the difference between SA[i] and SA[$i + 1$] keeps the same when iteratively exchanging i with the index j such that SA[j] = SA[i] $- 1$ as long as BWT[i] = BWT[$i + 1$]. This is also known as a part of the so-called toehold lemma [6, Lemma 3.1]. In Table 1, when starting with $i = 1$, we have SA[i] = 8 and SA[$i + 1$] $-$ SA[i] = 8. Then we exchange i with 19 such that SA[i] = 7 but SA[$i + 1$] $-$ SA[i] = 8. The difference is the same up until SA[i] = 3 with $i = 22$ (for SA[j] = 2 with $j = 25$ we have SA[$j + 1$] $-$ SA[j] = 17). We can also observe that the number of iterations from SA value 8 to SA value 3 is five, which is the longest common suffix of Rows 1 and 2 of the rotations matrix. In particular, among all SA values stored in $\mathcal{E}$, 3 is the preceding text position of 8. In other terms, this iteration ends whenever we visit the predecessor stored in $\mathcal{E}$ since at that time we know we are at a run boundary and therefore the next value in BWT has to differ. We can make use of this phenomenon for computing ϕ^{-1} as follows. First, having $\mathcal{E}$ and $\mathcal{S}$, we can compute ϕ^{-1} for all SA samples stored in $\mathcal{E}$ because $\phi^{-1}(\mathcal{E}[x]) = \mathcal{S}[x + 1]$ for all $x = [1..r - 1]$, and $\phi^{-1}(\mathcal{E}[r]) = \mathcal{S}[1]$. Now, given a text position $i \in [1..n]$, let $s = \text{pred}_\mathcal{E}(i) \in \mathcal{E}$ be an SA sample for which we can evaluate ϕ^{-1}. Then, by the above observation, the difference between s and s's succeeding value in SA is the same as the difference of i and i's succeeding value in SA. Hence, $\phi^{-1}(i) = \text{SA}[\text{ISA}[i] + 1] = \text{SA}[\text{ISA}[s] + 1] + (i - s) = \phi^{-1}(s) + (i - s)$.

3.1 Access via ϕ^{-1}-Graph

We first give two definitions before defining our auxiliary data structure.

Definition 1 (Costs and Limits). *For* $i \in [1..r - 1]$, *we define the* cost $c_i = \mathcal{S}[i + 1] - \text{pred}_\mathcal{E}(i + 1)$, *and the* limit *to be* $\ell_i = \text{succ}_\mathcal{E}(\mathcal{E}[i] + 1) - \mathcal{E}[i]$.

Informally, costs reflect the distances in our above key observation. Starting with an initial cost of zero, by adding up the costs and moving to elements of $\mathcal{E}$, we can simulate Φ^{-1} as long as the costs are bounded by the limits. The limits reflect the fact that our key observation only works up to the point that the predecessor in $\mathcal{E}$ keeps the same, therefore when the distances become too large, we would move into the next run with predecessor $\text{succ}_\mathcal{E}(\mathcal{E}[i] + 1)$. Using costs and limits, we define the following directed graph with edge labels.

Definition 2 (ϕ^{-1}-Graphs). *Given* $\mathcal{E}$ *and* $\mathcal{S}$, *we build a directed graph* $G = (V, E)$ *such that there exists a node in* G *for each sample at the end of each BWT run, i.e.,* $V = \mathcal{E}$, *and an edge from* v_i *(corresponding to* $\mathcal{E}[i]$) *to* v_j *(corresponding to* $\mathcal{E}[j]$) *if and only if* $\mathcal{E}[j]$ *is the predecessor of* $\mathcal{S}[i + 1]$ *in* $\mathcal{E}$. *Lastly, we label edge* (v_i, v_j) *with the cost and limit* $< c_i, \ell_i >$. *We refer to this as the* ϕ^{-1}-graph.

An example of the ϕ^{-1}-graph is depicted in Fig. 1. We observe that all nodes of the ϕ^{-1}-graph have at most one outgoing edge due to the uniqueness of the predecessor function, and the only node with no outgoing edges is the node corresponding to $\mathcal{E}[r]$ since $\mathcal{S}[r+1]$ is not defined. In what follows, we refer to the node of the ϕ^{-1}-graph corresponding to $\mathcal{E}[i]$ as v_i, and the edge outgoing from v_i to v_j as $e_i = (v_i, v_j)$. This induces an injective mapping from edges e_i to nodes v_i (we identify an edge by its outgoing node).

Given the ϕ^{-1}-graph G, we can compute recursive applications of ϕ^{-1} as follows. Given a position $1 \leq s \leq n$ in the suffix array, let $v_i \in V$ be the node corresponding to $\text{pred}_{\mathcal{E}}(\text{SA}[s])$ and let $e_i = (v_i, v_j)$ be the edge outgoing from v_i, labeled with $< c_i, \ell_i >$. First, we consider the simplest case where $\text{SA}[s]$ is the end of the i-th run. In this case, we can compute ϕ^{-1} directly with the equation

$$\phi^{-1}(\text{SA}[s]) = \text{SA}[s+1] = \mathcal{S}[i+1] = \mathcal{E}[j] + \mathcal{S}[i+1] - \mathcal{E}[j] = \mathcal{E}[j] + c_i.$$

In our example, if we start from the node corresponding to the SA sample 26, by following the edge connecting 26 with 3, we have that $\phi^{-1}(26) = 3 + 5 = 8$.

Next, suppose we want to compute $\phi^{-1}(\text{SA}[s+1])$. Let $e_j = (v_j, v_k)$ be the edge outgoing from v_j, labeled with $< c_j, \ell_j >$. Hence,

$$\phi^{-1}(\text{SA}[s+1]) = \phi^{-1}(\phi^{-1}(\text{SA}[s])) = \phi^{-1}(\mathcal{S}[i+1]) = \phi^{-1}(\mathcal{E}[j] + c_i).$$

If $\text{pred}_{\mathcal{E}}(\mathcal{S}[i+1]) = \mathcal{E}[j]$ then $\phi^{-1}(\mathcal{E}[j] + c_i) = \phi^{-1}(\mathcal{E}[j]) + c_i = \mathcal{E}[k] + c_j + c_i$. This shows that as long as the condition $\text{pred}_{\mathcal{E}}(\mathcal{S}[i+1]) = \mathcal{E}[j]$ holds, we can compute iterated applications of ϕ^{-1} by traversing the graph and summing the costs stored at the edge labels.

Back to our example, we continue at the node corresponding to the SA sample 3, having already gathered a cumulative cost of 5. We now follow the edge connecting 3 with 11, and obtain $\phi^{-1}(8) = \phi^{-1}(3)+5 = 16$. Note that we cannot compute $\phi^{-1}(16)$ by following the edge connecting 11 and 20, since $\text{pred}_{\mathcal{E}}(16) = 14 \neq 11$.

In the case where $\text{SA}[s]$ is not at the end of a run, let $\mathcal{E}[i] = \text{pred}_{\mathcal{E}}(\text{SA}[s])$. We can write $\phi^{-1}(\text{SA}[s]) = \mathcal{E}[j] + c_i + c_0 = \mathcal{S}[i+1] + c_0$, where $c_0 = \text{SA}[s] - \mathcal{E}[i]$ by definition of ϕ^{-1}. We refer to c_0 as the *initial cost*.

Therefore, given $\text{SA}[s]$, where $\text{pred}_{\mathcal{E}}(\text{SA}[s]) = \mathcal{E}[i_1]$, we let $c_0 = \text{SA}[s] - \mathcal{E}[i_1]$ and $v_{i_1}, v_{i_2}, \ldots, v_{i_m}$ be the m-length path outgoing from v_i. If $\text{pred}_{\mathcal{E}}(\mathcal{E}[i_j] + \sum_{k=1}^{j-1} c_{i_k} + c_0) = \mathcal{E}[i_{j+1}]$ for all $j = 2, ..., m-1$, then after m iterations of ϕ^{-1}, we have

$$\phi^{-m}(\text{SA}[s]) = \mathcal{S}[i_m + 1] = \mathcal{E}[i_m] + \sum_{k=1}^{m-1} c_{i_k} + c_0.$$

We note that checking that $\text{pred}_{\mathcal{E}}(\mathcal{E}[i_j] + \sum_{k=1}^{j-1} c_{i_k} + c_0) = \mathcal{E}[i_{j+1}]$ for all $j = 2, ..., m-1$ is equivalent to checking that

$$\mathcal{E}[i_j] + \sum_{k=1}^{j-1} c_{i_k} + c_0 = \mathcal{S}[i_{j-1} + 1] + c_0 < \text{succ}_{\mathcal{E}}(\mathcal{E}[i_j] + 1) = \mathcal{E}[i_j] + \ell_{i_j},$$

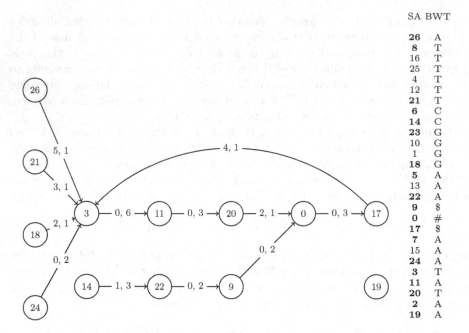

SA	BWT
26	A
8	T
16	T
25	T
4	T
12	T
21	T
6	C
14	C
23	G
10	G
1	G
18	G
5	A
13	A
22	A
9	$
0	#
17	$
7	A
15	A
24	A
3	T
11	A
20	T
2	A
19	A

Fig. 1. The ϕ^{-1}-graph corresponding to the string $S = $ GATTACAT\$GATACAT\$GATTA GATA#. Each node is labeled with the SA value at the end of each run. Each edge is labeled with its cost and its limit, which are shown in red. On the right we have the SA and BWT of the string S as derived in Table 1. (Color figure online)

that is equivalent to determining that for all $j = 2, ..., m - 1$

$$\sum_{k=1}^{j-1} c_{i_k} + c_0 < \ell_j.$$

Hence, it follows that we can access the SA samples via computing the costs in G with the condition that it is less than the limits at all visited edges.

On our running example, we can follow the edge connecting 3 and 11 since $\mathrm{pred}_\varepsilon(8) = 3$, which can be obtained checking if the cost 5 is smaller than the limit on the edge that is 6. However, we cannot traverse the edge between 11 and 20 starting with a total cost of 5, since $\mathrm{pred}_\varepsilon(16) = 14 \neq 11$ which is witnessed by the limit on the edge being 3, which is smaller than the cumulative cost of 5, given by the sum of costs on the edges between 26 and 3, and between 3 and 11.

3.2 Access via ϕ^{-1}-Forest

After we have the ϕ^{-1}-graph G, we can extract (long) paths from G and use them to speed up the computation for random access by allowing us to skip iterations of ϕ^{-1} to find a missing SA sample. We refer to this as *fast-forwarding* the random access. Given a path of length m in G, say $v_{i_1}, v_{i_2}, \ldots, v_{i_m}$, we want

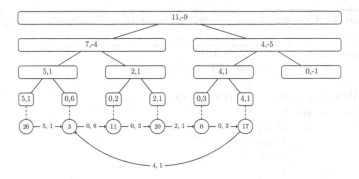

Fig. 2. Example of a balanced binary tree described in Sect. 3.2 on the path $[26, 3, 11, 20, 0, 17, 3]$ of the ϕ^{-1}-graph of Fig. 1. By adding dummy nodes like the right-most one with label $(0, -1)$, we make the tree full binary. This is not a requirement, but used in our implementation to keep all leaves on the same level such that we can store them consecutively in an array for leveraging data locality. That is because we alternatively can scan this array of leaves instead of walking through the tree if we know that we will only perform a short traversal. The downside is that we need to take care of the dummy nodes. The negative limits $(0, -1)$ of the dummy nodes prevents the algorithm from exploring the non-existent children of this node.

to build a data structure that solves the following problem: for a given $1 \leq j \leq m$ and a value c_0, find the largest index k such that $\sum_{x=j}^{h-1} c_{i_x} + c_0 < \ell_h$, for all $h = j, ..., k - 1$.

To solve this problem, we build a balanced binary tree, where the leaves represent the edges $e_{i_1}, e_{i_2}, \ldots, e_{i_{m-1}}$ of the path. Each node p of the tree stores a pair of integers $< c_p, \ell_p >$ such that p is the j-th leaf $< c_p, \ell_p >$ corresponding to the edge label of e_{i_j} if and only if $c_p = c_{i_j}$ and $\ell_p = \ell_{i_j}$. Otherwise, if p is an internal node, we let u and v be the left and right children of p respectively, and $c_p = c_u + c_v$ and $\ell_p = \min(\ell_u, \ell_v - c_u)$. We note that the definition of the costs and limits for the internal nodes guarantee that we can traverse all the edges in the subtree of the node p if we start from the leftmost leaf in the subtree of p with a cost $c < \ell_p$. See Fig. 2 for an example.

Given a suffix array sample $SA[s]$, we can query the tree as follows. Let i be such that $\mathcal{E}[i] = \text{pred}_{\mathcal{E}}(SA[s])$, $c_0 = SA[s] - \mathcal{E}[i]$, and let v be the leaf corresponding to the edge $e_i = e_{i_j}$. In order to find the largest index k such that $\sum_{x=j}^{h-1} c_{i_x} + c_0 < \ell_h$ for all $h = j, ..., k - 1$, we divide the query into two phases.

- In the first phase (Line 4 of Algorithm 1), we traverse the tree from the leaf v towards the root as follows. We consider a cumulative cost c, which we initialize to c_0. Now let p be the parent of v. If v is a right child of p we move to the parent of p. If v is a left child of p let u be the right child of p. If $c < \ell_u$ then we add c_u to c and we move to the parent of p. Otherwise, if $c \geq \ell_u$ or v is the root, we stop and continue with the second phase.
- In the second phase (Line 13 of Algorithm 1), we begin descending the tree starting from the right child u of p. Let q be the left child of u. If $c < \ell_q$ we

---

Algorithm 1. Computes the largest number of ϕ^{-1} steps in ϕ^{-1}-graph via ϕ^{-1}-forest

---

1: **function.** FAST-FORWARD($\mathcal{T}$, v, c_0)
2:    Let $v \in \mathcal{T}$ be the j-th leaf of $\mathcal{T}$, i.e., $v = v_{i_j}$.
3:    $c \leftarrow c_0$.
4:    **while** $c < \ell_v$ and v is not the root **do**
5:        $p \leftarrow$ parent(v).
6:        **if** $v =$ left-child(p) **then**
7:            $u \leftarrow$ right-child(p).
8:            **if** $c < \ell_u$ **then** $c \leftarrow c + c_u$.
9:            **else** $p \leftarrow u$.
10:       $v \leftarrow p$.
11:    **if** v is not leaf **then**
12:        $v \leftarrow$ right-child(v).
13:    **while** v is not leaf **do**
14:        $q \leftarrow$ left-child(v).
15:        **if** $c < \ell_q$ **then**
16:            $c \leftarrow c + c_q$.
17:            $v \leftarrow$ right-child(v).
18:        **else** $v \leftarrow q$.
19:    Let i_{k-1} be the index of v in $\mathcal{T}$.
20:    **return** (i_k, c, $k - j$)

---

move to the right child of u and update the value of c to $c + c_q$; otherwise we move to q. We repeat this procedure as long as u is not a leaf. At the end of the procedure we arrive at the leaf corresponding to the edge $e_{i_{k-1}}$. In addition to the node v_{i_k} that corresponds to the sample $\mathcal{E}[i_k]$, we report also the total cost c and the number of traversed edges $d = k - j$ so that we can recover the suffix array sample $\mathrm{SA}[s + d] = \mathcal{E}[i_k] + c$.

The steps of both phases are summarized in Algorithm 1 and we refer to them as *fast-forward* query. A trivial case is when all costs are less than the limits; in such a case we would ascend to the root and then descend to the rightmost leaf stored in the tree.

We can easily extend the fast-forward query by including a constraint on the total number of leaves that can be traversed, i.e., solving the following problem: given $1 \leq j \leq m$, a value c_0, and an integer d, find the largest index $k \leq d$ such that $\sum_{x=j}^{h-1} c_{i_x} + c_0 < \ell_h$, for all $h = j, ..., k - 1$. We refer to this function as *bounded fast-forward*. Thus, applying the above procedures to perform SA access is straightforward when we have a limit on the total number of leaves that can be traversed. We decompose the ϕ^{-1}-graph into non-overlapping paths, i.e., that no pair of paths of the decomposition shares an edge of the graph, in order to obtain a set of trees. We can build one ϕ^{-1}-tree for each path in the decomposition and use the trees collectively to compute ϕ^{-1}. Lastly, we note that we can choose any set of non-overlapping paths of the ϕ^{-1}-graph. In our implementation, we omit

Algorithm 2. SA access via ϕ^{-1}-forest

1: **function** SA ACCESS(SA, i)
2:     Let j be the end of a BWT run preceding i.
3:     Let $d \leftarrow i - j$ and let v be the node corresponding to $SA[j]$.
4:     $s \leftarrow$ SA$[j]$.
5:     **while** $d > 0$ **do**
6:         Let v be the node in G corresponding to $\text{pred}_{\mathcal{E}}(s)$.
7:         Let $c_0 = s - \text{pred}_{\mathcal{E}}(s)$.
8:         **if** v in one of ϕ^{-1}-forest **then**
9:            Let $\mathcal{T}$ be the tree containing v.
10:           $(i_k, c, t) \leftarrow$ BOUNDED FAST-FORWARD($\mathcal{T}$, v, c_0, d).
11:           $s \leftarrow \mathcal{E}[i_k] + c$.
12:         **else**
13:           $s \leftarrow \text{pred}_{\mathcal{E}}(s) + c_0$.
14:           $t \leftarrow 1$.
15:         $d \leftarrow d - t$.
16:     **return** s

---

paths that are too short to observe speedups gained by a fast-forward query. Therefore, we deal with nodes not present in the ϕ^{-1}-forest separately.

Given the set of trees (i.e., the forest) built on the non-overlapping paths, we can use them to speed up the random access computation as follows. Let i be a position in the suffix array. We first compute the position j of the end of a BWT run immediately preceding i. The number of iterations of ϕ^{-1} to be applied to SA$[j]$ are $d = i - j$. We start from $s =$ SA$[j]$, and find the node v in the ϕ^{-1}-graph corresponding to $\text{pred}_{\mathcal{E}}(s)$. Subsequently, we set $c_0 = s - \text{pred}_{\mathcal{E}}(s)$. Next, we check if v is stored in ϕ^{-1}-forest.

- If v is stored in a tree of ϕ^{-1}-forest (Line 8 in Algorithm 2), we perform a bounded fast-forward query on this tree, starting from the node v, with initial cost c_0, and a limit to the total number of leaves to be traversed to be d. The bounded fast-forward query will return the index i_k in the graph of the reached leaf, the final cost c, and the number of traversed edges t. Hence, we compute SA$[i - d + t] = \mathcal{E}[i_k] + c$, and update the remaining steps d to $d - t$.
- Otherwise (Line 12 in Algorithm 2), we apply the standard computation of ϕ^{-1} to obtain SA$[i - d + 1] = \text{pred}_{\mathcal{E}}(s) + c_0$. Subsequently, we decrement the remaining steps d by one.

We iterate this procedure until no further steps are required to be computed, i.e., until we obtain $d = 0$. We summarize this procedure in Algorithm 2. For simplicity, we assume there that we are not in the first run. Otherwise, we set $j = n = \mathcal{E}[r]$ and write $d \leftarrow i - j + n \mod n$. In ϕ^{-1}-graph, $\mathcal{E}[r]$ is always represented by a node having no outgoing edge. Therefore, we initially run into the **else** branch in Line 12.

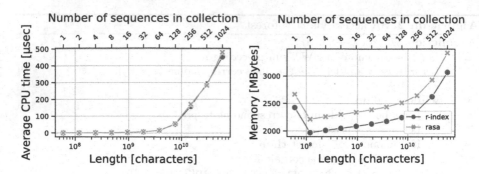

Fig. 3. Illustration of the SA access time (left) and space (right) of the `r-index` and `rasa` on increasingly larger numbers of Chromosome 19 sequences.

Theorem 1. *Given a string $S[1..n]$, the r-index of S augmented by ϕ^{-1}-forest, and an index $1 \leq i \leq n$, we can compute $SA[i]$ in $\mathcal{O}((i-j)(\log\log_w(n/r)+\log r))$ time, where j is the position of the end of a run preceding i and w is the machine word size, with $\mathcal{O}(r)$ additional space to the r-index.*

Proof. First, we note that our SA access only requires, in addition to the r-index, ϕ^{-1}-forest since all other components (including predecessor access) are part of the r-index. There exists at most one node for each SA entry in $\mathcal{E}$ in ϕ^{-1}-forest, and each node has at most one outgoing edge. Hence, it follows that ϕ^{-1}-forest requires $\mathcal{O}(r)$-space since the size of $\mathcal{E}$ is at most r. For the query time, in the case that no node is stored in any ϕ^{-1}-tree, we preform one predecessor query for each SA value between j and i, which takes $\mathcal{O}((i-j)\log\log_w(n/r))$ time. In the case that a tree is used, one bounded fast-forward query can be performed in $\mathcal{O}(\log r)$ time. Hence, we have $\mathcal{O}((i-j)(\log\log_w(n/r) + \log r))$ time.   □

4   Experiments

Experimental Details. We evaluated the performance on two different datasets. First, we compared the methods using Chromosome 19 sequences from the 1000 Genomes Project [17]. From this project, we created datasets consisting of 1, 2, 4, 8, 16, 32, 64, 128, 256, 512, and 1024 sequences, which we call `chr.x` in the following, where x is the number of assigned individuals. Next, we compared the methods using the Pizza&Chili repetitive corpus [1]. We performed all experiments on an AMD EPYC 75F3 32-core processor running at 2.95 GHz with 512 GB of RAM with 64-bit Linux. Time was measured using `std::chrono::system_clock` from the C++ standard library.

Competing Methods. We augmented the r-index implementation of Rossi et al. [16] with our data structure, which we refer to as `rasa`. We compared `rasa` to the standard ϕ implementation in the r-index of Rossi et al. We refer to this as `r-index`. We note that the sr-index of Cobas et al. [5] is up to 6x times smaller

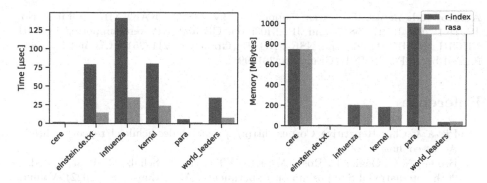

Fig. 4. Illustration of the SA access time (left) and space (right) of the `r-index` and `rasa` on Pizza&Chili repetitive corpus. We sampled 100,000 suffix array index positions between 1 and n at random and calculated the mean CPU time to perform an SA access.

but showed no significant difference in the SA access time than the `r-index` since it uses the same ϕ implementation. The block-tree CSA implementation of Cáceres and Navarro [4] and RLCSA of Mäkinen et al. [11] (`rlcsa`) used standard backward search. We wanted to compare against RLZCSA [14] but no implementation is available, even upon request.

Random Access. In order to evaluate our implementation of ϕ^{-1}, we randomly selected 100,000 SA entries and performed the SA random access 5 time for each entry. Figures 3 and 4 illustrate the average query times. On the Chromosome 19 datasets, the query times were comparable as they differed by at most half a microsecond. On Pizza&Chili the `r-index` the results were more pronounced as the `rasa` was between 3x and 6x times faster on all datasets except for `cere` where there was negligible difference in the running time of the methods. On all datasets, `rasa` required more memory, which is expected since it required the addition of the ϕ^{-1}-forest. However, even on the largest datasets, the additional memory was less than 3 GB.

5   Conclusion

In this paper, we introduced the concept of ϕ^{-1}-forest and demonstrated how it can be used to access the SA. Again, we note that it can be implemented analogously for ϕ. To the best of our knowledge, this is the third data structure for ϕ/ϕ^{-1} since the development of the r-index—with Nishimoto and Tabei [13] and sr-index [5] being the other two data structures. And although Brown et al. [2] implemented the LF data structure of Nishimoto and Tabei, the ϕ data structure of Nishimoto and Tabei has yet to be implemented. Our ϕ^{-1}-forest is competitive to the standard SA access of the r-index in practice, and also provides a graphical representation that we believe can be exploited to gain further insight into decreasing the theoretical time to access an entry of the SA if we store only sampled values.

Acknowledgments. DK was supported by JSPS KAKENHI (Grant No. JP21K17701, JP21H05847, and JP22H03551). CB and MR were supported by NIH NHGRI (R01HG011392) and NSF EAGER (Grant No. 2118251). CB and HP were funded by NSF SCH:INT (Grant No. 2013998).

References

1. Pizza & Chili Repetitive Corpus. http://pizzachili.dcc.uchile.cl/repcorpus.html.. Accessed June 2022
2. Brown, N.K., Gagie, T., Rossi, M.: RLBWT tricks. In Schulz, C., Uçar, B. (eds.) 20th International Symposium on Experimental Algorithms (SEA 2022), Volume 233 of Leibniz International Proceedings in Informatics (LIPIcs), pp. 16:1–16:16, Dagstuhl, Germany. Schloss Dagstuhl - Leibniz-Zentrum für Informatik (2022)
3. Burrows, M., Wheeler, D.: A block sorting lossless data compression algorithm. Technical report 124, Digital Equipment Corporation (1994)
4. Cáceres, M., Navarro, G.: Faster repetition-aware compressed suffix trees based on block trees. Inf. Comput. **285**, 104749 (2021)
5. Cobas, D., Gagie, T., Navarro, G.: A fast and small subsampled R-Index. In: 32nd Annual Symposium on Combinatorial Pattern Matching (CPM 2021). Schloss Dagstuhl-Leibniz-Zentrum für Informatik (2021)
6. Gagie, T., Navarro, G., Prezza, N.: Optimal-time text indexing in BWT-runs bounded space. In: Proceedings SODA, pp. 1459–1477 (2018)
7. Gagie, T., Navarro, G., Prezza, N.: Fully functional suffix trees and optimal text searching in BWT-runs bounded space. J. ACM **67**(1), 21–254 (2020)
8. Kärkkäinen, J., Manzini, G., Puglisi, S.J.: Permuted longest-common-prefix array. In: Kucherov, G., Ukkonen, E. (eds.) CPM 2009. LNCS, vol. 5577, pp. 181–192. Springer, Heidelberg (2009). https://doi.org/10.1007/978-3-642-02441-2_17
9. Langmead, B., Trapnell, C., Pop, M., Salzberg, S.L.: Ultrafast and memory-efficient alignment of short DNA sequences to the human genome. Genome Biol. **10**(3), 1–10 (2009)
10. Li, H., Durbin, R.: Fast and accurate long-read alignment with Burrows-Wheeler transform. Bioinformatics **26**(5), 589–595 (2010)
11. Mäkinen, V., Navarro, G., Sirén, J., Välimäki, N.: Storage and retrieval of highly repetitive sequence collections. J. Comput. Biol. **17**(3), 281–308 (2010)
12. Manber, U., Myers, E.W.: Suffix arrays: a new method for on-line string searches. SIAM J. Comput. **22**(5), 935–948 (1993)
13. Nishimoto, T., Tabei, Y.: Optimal-time queries on BWT-runs compressed indexes. In: Proceedings of the 48th International Colloquium on Automata, Languages, and Programming, (ICALP 2021), volume 198 of LIPIcs, pp. 101:1–101:15 (2021)
14. Puglisi, S.J., Zhukova, B.: Relative Lempel-Ziv compression of suffix arrays. In: Boucher, C., Thankachan, S.V. (eds.) SPIRE 2020. LNCS, vol. 12303, pp. 89–96. Springer, Cham (2020). https://doi.org/10.1007/978-3-030-59212-7_7
15. Puglisi, S.J., Zhukova, B.: Smaller RLZ-compressed suffix arrays. In: Proceedings of the 31st Data Compression Conference, (DCC 2021), pp. 213–222. IEEE (2021)
16. Rossi, M., Oliva, M., Langmead, B., Gagie, T., Boucher, C.: MONI: a pangenomic index for finding maximal exact matches. J. Comput. Biol. **29**(2), 169–187 (2022)
17. The 1000 Genomes Project Consortium. A global reference for human genetic variation. Nature **526** 68–74 (2015)

On the Optimisation of the GSACA Suffix Array Construction Algorithm

Jannik Olbrich[✉][iD], Enno Ohlebusch, and Thomas Büchler

University of Ulm, 89081 Ulm, Germany
{jannik.olbrich,enno.ohlebusch,thomas.buechler}@uni-ulm.de
https://www.uni-ulm.de/in/theo

Abstract. The suffix array is arguably one of the most important data structures in sequence analysis and consequently there is a multitude of suffix sorting algorithms. However, to this date the GSACA algorithm introduced in 2015 is the only known non-recursive linear-time suffix array construction algorithm (SACA). Despite its interesting theoretical properties, there has been little effort in improving the algorithm's subpar real-world performance. There is a super-linear algorithm DSH which relies on the same sorting principle and is faster than DivSufSort, the fastest SACA for over a decade. This paper is concerned with analysing the sorting principle used in GSACA and DSH and exploiting its properties in order to give an optimised linear-time algorithm. Our algorithm is not only significantly faster than GSACA but also outperforms DivSufSort and DSH.

Keywords: Suffix array · Suffix sorting · String algorithms

1 Introduction

The *suffix array* contains the indices of all suffixes of a string arranged in lexicographical order. It is arguably one of the most important data structures in *stringology*, the topic of algorithms on strings and sequences. It was introduced in 1990 by Manber and Myers for on-line string searches [9] and has since been adopted in a wide area of applications including text indexing and compression [12]. Although the suffix array is conceptually very simple, constructing it efficiently is not a trivial task.

When n is the length of the input text, the suffix array can be constructed in $\mathcal{O}(n)$ time and $\mathcal{O}(1)$ additional words of working space when the alphabet is linearly-sortable (i.e. the symbols in the string can be sorted in $\mathcal{O}(n)$ time) [7,8,10]. However, algorithms with these bounds are not always the fastest in practice. For instance, DivSufSort has been the fastest SACA for over a decade although having super-linear worst-case time complexity [3,5]. To the best of our knowledge, the currently fastest suffix sorter is libsais, which appeared as source code in February 2021 on Github[1] and has not been subject to peer

---

[1] https://github.com/IlyaGrebnov/libsais, last accessed: August 22, 2022.

D. Arroyuelo and B. Poblete (Eds.): SPIRE 2022, LNCS 13617, pp. 99–113, 2022.
https://doi.org/10.1007/978-3-031-20643-6_8

review in any academic context. The author claims that `libsais` is an improved implementation of the SA-IS algorithm and hence has linear time complexity [11].

The only non-recursive linear-time suffix sorting algorithm `GSACA` was introduced in 2015 by Baier and is not competitive, neither in terms of speed nor in the amount of memory consumed [1,2]. Despite the new algorithm's entirely novel approach and interesting theoretical properties [6], there has been little effort in optimising it. In 2021, Bertram et al. [3] provided a faster SACA `DSH` using the same sorting principle as `GSACA`. Their algorithm beats `DivSufSort` in terms of speed, but also has super-linear time complexity.

Our Contributions. We provide a linear-time SACA that relies on the same *grouping* principle that is employed by `DSH` and `GSACA`, but is faster than both. This is done by exploiting certain properties of Lyndon words that are not used in the other algorithms. As a result, our algorithm is more than 11% faster than `DSH` on real-world texts and at least 46% faster than Baier's `GSACA` implementation. Although our algorithm is not on par with `libsais` on real-world data, it significantly improves Baier's sorting principle and positively answers the question whether the precomputed Lyndon array can be used to accelerate `GSACA` (posed in [4]).

The rest of this paper is structured as follows: Sect. 2 introduces the definitions and notations used throughout this paper. In Sect. 3, the grouping principle is investigated and a description of our algorithm is provided. Finally, in Sect. 4 our algorithm is evaluated experimentally and compared to other relevant SACAs.

This is an abridged version of a longer paper available on arXiv [13].

2   Preliminaries

For $i, j \in \mathbb{N}_0$ we denote the set $\{k \in \mathbb{N}_0 : i \leq k \leq j\}$ by the interval notations $[i..j] = [i..j+1) = (i-1..j] = (i-1..j+1)$. For an array A we analogously denote the *subarray* from i to j by $A[i..j] = A[i..j+1) = A(i-1..j] = A(i-1..j+1) = A[i]A[i+1]\ldots A[j]$. We use zero-based indexing, i.e. the first entry of the array A is $A[0]$. A *string* S of *length* n over an *alphabet* Σ is a sequence of n characters from Σ. We denote the length n of S by $|S|$ and the i'th symbol of S by $S[i-1]$, i.e. strings are zero-indexed. Analogous to arrays we denote the *substring* from i to j by $S[i..j] = S[i..j+1) = S(i-1..j] = S(i-1..j+1) = S[i]S[i+1]\ldots S[j]$. For $j > i$ we let $S[i..j]$ be the *empty string* ε. The *suffix* i of a string S of length n is the substring $S[i..n)$ and is denoted by S_i. Similarly, the substring $S[0..i]$ is a *prefix* of S. A suffix (prefix) is *proper* if $i > 0$ ($i+1 < n$). For two strings u and v and an integer $k \geq 0$ we let uv be the concatenation of u and v and denote the k-times concatenation of u by u^k. We assume totally ordered alphabets. This induces a total order on strings. Specifically, we say a string S of length n is *lexicographically smaller*

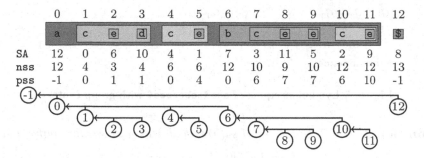

Fig. 1. Shown are the Lyndon prefixes of all suffixes of $S =$ acedcebceece\$ and the corresponding suffix array, nss-array, pss-array and pss-tree. Each box indicates a Lyndon prefix. For instance, the Lyndon prefix of $S_7 =$ ceece\$ is $\mathcal{L}_7 =$ cee. Note that $\mathcal{L}_i$ is exactly $S[i]$ concatenated with the Lyndon prefixes of i's children in the pss-tree (see Lemma 4), e.g. $\mathcal{L}_6 = S[6]\mathcal{L}_7\mathcal{L}_{10} =$ bceece.

than another string S' of length m if and only if there is some $\ell \le \min\{n, m\}$ such that $S[0..\ell) = S'[0..\ell)$ and either $n = \ell < m$ or $S[\ell] < S'[\ell]$. If S is lexicographically smaller than S' we write $S <_{lex} S'$.

A non-empty string S is a *Lyndon word* if and only if S is lexicographically smaller than all its proper suffixes [14]. The *Lyndon prefix* of S is the longest prefix of S that is a Lyndon word. We let $\mathcal{L}_i$ denote the Lyndon prefix of S_i.

In the remainder of this paper, we assume an arbitrary but fixed string S of length $n > 1$ over a totally ordered alphabet Σ with $|\Sigma| \in \mathcal{O}(n)$. Furthermore, we assume w.l.o.g. that S is *null-terminated*, that is $S[n-1] = \$$ and $S[i] > \$$ for all $i \in [0..n-1)$.

The *suffix array* SA of S is an array of length n that contains the indices of the suffixes of S in increasing lexicographical order. That is, SA forms a permutation of $[0..n)$ and $S_{SA[0]} <_{lex} S_{SA[1]} <_{lex} \cdots <_{lex} S_{SA[n-1]}$.

Definition 1 (pss-tree [4]). *Let pss be the array such that pss[i] is the index of the previous smaller suffix for each $i \in [0..n)$ (or -1 if none exists). Formally, $pss[i] := \max(\{j \in [0..i) : S_j <_{lex} S_i\} \cup \{-1\})$. Note that pss forms a tree with -1 as the root, in which each $i \in [-1..n)$ is represented by a node and pss[i] is the parent of node i. We call this tree the pss-tree. Further, we impose an order on the nodes that corresponds to the order of the indices represented by the nodes. In particular, if $c_1 < c_2 < \cdots < c_k$ are the children of i (i.e. $pss[c_1] = \cdots = pss[c_k] = i$), we say c_k is the last child of i.*

Analogous to pss[i], we define $nss[i] := \min\{j \in (i..n] : S_j <_{lex} S_i\}$ as the next smaller suffix of i. Note that $S_n = \varepsilon$ is smaller than any non-empty suffix of S, hence nss is well-defined.

In the rest of this paper, we use $S =$ acedcebceece\$ as our running example. Figure 1 shows its Lyndon prefixes and the corresponding pss-tree.

$$\mathcal{G}_1 \quad \mathcal{G}_2 \; \mathcal{G}_3 \qquad \mathcal{G}_4 \qquad \mathcal{G}_5 \; \mathcal{G}_6 \; \mathcal{G}_7 \qquad\qquad \mathcal{G}_8$$

```
G1   G2 G3      G4       G5  G6  G7            G8
 $    bceece    ce       ced cee  d             e
 12   0    6    4   10   1   7   3   2   5   8   9   11
 acedcebceece
```

Fig. 2. A Lyndon grouping of `acedcebceece$` with group contexts.

Definition 2. *Let* $\mathcal{P}_i$ *be the set of suffixes with* i *as next smaller suffix, that is*

$$\mathcal{P}_i = \{j \in [0\,..\,i) : nss[j] = i\}$$

For instance, in the example we have $\mathcal{P}_4 = \{1,3\}$ because $\mathbf{nss}[1] = \mathbf{nss}[3] = 4$.

3 GSACA

We start by giving a high level description of the sorting principle based on grouping by Baier [1,2]. Very basically, the suffixes are first assigned to lexicographically ordered groups, which are then refined until the suffix array emerges. The algorithm consists of the following steps.

- *Initialisation:* Group the suffixes according to their first character.
- *Phase I:* Refine the groups until the elements in each group have the same Lyndon prefix.
- *Phase II:* Sort elements within groups lexicographically.

Definition 3 (Suffix Grouping, adapted from [3]). *Let* S *be a string of length* n *and* SA *the corresponding suffix array. A group* $\mathcal{G}$ *with group context* α *is a tuple* $\langle g_s, g_e, |\alpha| \rangle$ *with group start* $g_s \in [0\,..\,n)$ *and group end* $g_e \in [g_s\,..\,n)$ *such that the following properties hold:*

1. *All suffixes in* $SA\,[g_s\,..\,g_e]$ *share the prefix* α, *i.e. for all* $i \in SA\,[g_s\,..\,g_e]$ *it holds* $S_i = \alpha S_{i+|\alpha|}.$
2. α *is a Lyndon word.*

We say i *is in* $\mathcal{G}$ *or* i *is an element of* $\mathcal{G}$ *and write* $i \in \mathcal{G}$ *if and only if* $i \in SA\,[g_s\,..\,g_e]$. *A suffix grouping for* S *is a set of groups* $\mathcal{G}_1, \ldots, \mathcal{G}_m$, *where the groups are pairwise disjoint and cover the entire suffix array. Formally, if* $\mathcal{G}_i = \langle g_{s,i}, g_{e,i}, |\alpha_i| \rangle$ *for all* i, *then* $g_{s,1} = 0, g_{e,m} = n - 1$ *and* $g_{s,j} = 1 + g_{e,j-1}$ *for all* $j \in [2\,..\,m]$. *For* $i, j \in [1\,..\,m]$, $\mathcal{G}_i$ *is a* lower (higher) *group than* $\mathcal{G}_j$ *if and only if* $i < j$ $(i > j)$. *If all elements in a group* $\mathcal{G}$ *have* α *as their Lyndon prefix then* $\mathcal{G}$ *is a* Lyndon group. *If* $\mathcal{G}$ *is not a Lyndon group, it is called* preliminary. *Furthermore, a suffix grouping is* Lyndon *if all its groups are Lyndon groups, and* preliminary *otherwise.*

With these notions, a suffix grouping is created in the initialisation, which is then refined in Phase I until it is Lyndon, and further refined in Phase II until the suffix array emerges. Figure 2 shows a Lyndon grouping of our running example.

In Subsects. 3.1 and 3.2 we explain Phases II and I, respectively, of our suffix array construction algorithm. Phase II is described first because it is much simpler.

```
A[0] ← n − 1;
for i = 0 → n − 1 do
    for j ∈ P_{A[i]} do
        Let k be the start of the group containing j;
        remove j from its current group and put it in a new group ⟨k, k, |L_j|⟩ immediately
            preceding j's old group;
        A[k] ← j;
    end
end
```

Algorithm 1: Phase II of GSACA [1, 2]

3.1   Phase II

In Phase II we need to refine the Lyndon grouping obtained in Phase I into the suffix array. Let $\mathcal{G}$ be a Lyndon group with context α and let $i, j \in \mathcal{G}$. Since $S_i = \alpha S_{i+|\alpha|}$ and $S_j = \alpha S_{j+|\alpha|}$, we have $S_i <_{lex} S_j$ if and only if $S_{i+|\alpha|} <_{lex} S_{j+|\alpha|}$. Hence, in order to find the lexicographically smallest suffix in $\mathcal{G}$, it suffices to find the lexicographically smallest suffix p in $\{i + |\alpha| : i \in \mathcal{G}\}$. Note that removing $p - |\alpha|$ from $\mathcal{G}$ and inserting it into a new group immediately preceding $\mathcal{G}$ yields a valid Lyndon grouping. We can repeat this process until each element in $\mathcal{G}$ is in its own singleton group. As $\mathcal{G}$ is Lyndon, we have $S_{k+|\alpha|} <_{lex} S_k$ for each $k \in \mathcal{G}$. Therefore, if all groups lower than $\mathcal{G}$ are singletons, p can be determined by a simple scan over $\mathcal{G}$ (by determining which member of $\{i + |\alpha| : i \in \mathcal{G}\}$ is in the lowest group). Consider for instance $\mathcal{G}_4 = \langle 3, 4, |\mathsf{ce}|\rangle$ from Fig. 2. We consider $4 + |\mathsf{ce}| = 6$ and $10 + |\mathsf{ce}| = 12$. Among them, 12 belongs to the lowest group, hence S_{10} is lexicographically smaller than S_4. Thus, we know $\mathsf{SA}[3] = 10$ and remove 10 from $\mathcal{G}_4$ and repeat the process with the emerging group $\mathcal{G}'_4 = \langle 4, 4, |\mathsf{ce}|\rangle$. As $\mathcal{G}'_4$ only contains 4 we know $\mathsf{SA}[4] = 4$.

If the groups are refined from lower to higher as just described, each time a group $\mathcal{G}$ is processed, all groups lower than $\mathcal{G}$ are singletons. However, sorting groups in such a way leads to a superlinear time complexity. Bertram et al. [3] provide a fast-in-practice $\mathcal{O}(n \log n)$ algorithm for this, broadly following the described approach.

In order to get a linear time complexity, we turn this approach on its head like Baier does [1, 2]: Instead of repeatedly finding the next smaller suffix in a group, we consider the suffixes in lexicographically increasing order and for each encountered suffix i, we move all suffixes that have i as the next smaller suffix (i.e. those in P_i) to new singleton groups immediately preceding their respective old groups. Corollary 1 implies that this procedure is well-defined.

Lemma 1. *For any $j, j' \in P_i$ we have $L_j \neq L_{j'}$ if and only if $j \neq j'$.*

Corollary 1. *In a Lyndon grouping, the elements of P_i are in different groups.*

Accordingly, Algorithm 1 correctly computes the suffix array from a Lyndon grouping. A formal proof of correctness is given in [1, 2]. Figure 3 shows Algorithm 1 applied to our running example.

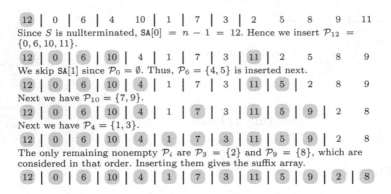

| 12 | 0 | 6 | 4 | 10 | 1 | 7 | 3 | 2 | 5 | 8 | 9 | 11 |

Since S is nullterminated, $\text{SA}[0] = n - 1 = 12$. Hence we insert $\mathcal{P}_{12} = \{0, 6, 10, 11\}$.

| 12 | 0 | 6 | 10 | 4 | 1 | 7 | 3 | 11 | 2 | 5 | 8 | 9 |

We skip $\text{SA}[1]$ since $\mathcal{P}_0 = \emptyset$. Thus, $\mathcal{P}_6 = \{4, 5\}$ is inserted next.

| 12 | 0 | 6 | 10 | 4 | 1 | 7 | 3 | 11 | 5 | 2 | 8 | 9 |

Next we have $\mathcal{P}_{10} = \{7, 9\}$.

| 12 | 0 | 6 | 10 | 4 | 1 | 7 | 3 | 11 | 5 | 9 | 2 | 8 |

Next we have $\mathcal{P}_4 = \{1, 3\}$.

| 12 | 0 | 6 | 10 | 4 | 1 | 7 | 3 | 11 | 5 | 9 | 2 | 8 |

The only remaining nonempty $\mathcal{P}_i$ are $\mathcal{P}_3 = \{2\}$ and $\mathcal{P}_9 = \{8\}$, which are considered in that order. Inserting them gives the suffix array.

| 12 | 0 | 6 | 10 | 4 | 1 | 7 | 3 | 11 | 5 | 9 | 2 | 8 |

Fig. 3. Refining a Lyndon grouping for $S = \texttt{acedcebceece\$}$ (see Fig. 2) into the suffix array, as done in Algorithm 1. Inserted elements are colored green. (Color figure online)

Note that each element $i \in [0 .. n - 1)$ has exactly one next smaller suffix, hence there is exactly one j with $i \in \mathcal{P}_j$ and thus i is inserted exactly once into a new singleton group in Algorithm 1. Therefore, it suffices to map each group from the Lyndon grouping obtained from Phase I to its current start; we use an array C that contains the current group starts.

There are two major differences between our Phase II and Baier's, both are concerned with the iteration over the $\mathcal{P}_i$-sets.

The first difference is the way in which we determine the elements of $\mathcal{P}_i$ for some i. The following observations enable us to iterate over $\mathcal{P}_i$.

Lemma 2. $\mathcal{P}_i$ is empty if and only if $i = 0$ or $S_{i-1} <_{lex} S_i$. Furthermore, if $\mathcal{P}_i \neq \emptyset$ then $i - 1 \in \mathcal{P}_i$.

Lemma 3. For some $j \in [0 .. i)$, we have $j \in \mathcal{P}_i$ if and only if j's last child is in $\mathcal{P}_i$, or $j = i - 1$ and $S_j >_{lex} S_i$.

Specifically, (if $\mathcal{P}_i$ is not empty) we can iterate over $\mathcal{P}_i$ by walking up the pss-tree starting from $i - 1$ and halting when we encounter a node that is not the last child of its parent.[2] Baier [1,2] tests whether $i - 1$ (pss$[j]$) is in $\mathcal{P}_i$ by explicitly checking whether $i - 1$ (pss$[j]$) has already been written to A using an explicit marker for each suffix. Reading and writing those markers leads to bad cache performance because the accessed memory locations are unpredictable (for the CPU/compiler). Lemmata 2 and 3 enable us to avoid reading and writing those markers. In fact, in our implementation of Phase II, the array A is the only memory written to that is not always in the cache. Lemma 2 tells us whether we need to follow the pss-chain starting at $i - 1$ or not. Namely, this is the case if and only if $S_{i-1} >_{lex} S_i$, i.e. $i - 1$ is a leaf in the pss-tree. This information is required when we encounter i in A during the outer for-loop in Algorithm 1, thus

---

[2] Note that $n - 1$ is the last child of the artificial root -1. This ensures that we always halt before we actually reach the root of the pss-tree. Moreover, Corollary 1 implies that the order in which we process the elements in $\mathcal{P}_i$ is not important.

```
A ← (n − 1)⊥ⁿ⁻¹ ; // set A[0] = n − 1, fill the rest with "undefined"
Q ← queue containing only n − 1;
i ← 1; // current index in A
while Q is not empty do
    s ← Q.size();
    repeat s times // insert elements that are currently in the queue
        v ← Q.pop();
        if pss[v] is marked then // v is last child of pss[v]
        |   Q.push(pss[v]);
        end
        A[C[G[v]]] ← v; // insert v
        if pss[v] + 1 < v then mark A[C[G[v]]]; // v − 1 is leaf
        C[G[v]] ← C[G[v]] + 1; // increment current start of v's old group
    end
    while Q.size() < w ∧ i < n ∧ A[i] ≠ ⊥ do // refill the queue
        if A[i] is marked then // A[i] − 1 is leaf
        |   Q.push(A[i] − 1);
        end
        i ← i + 1;
    end
end
```

Algorithm 2: Breadth-first approach to Phase II. The constant w is the maximum queue size and $G[i]$ is the index of the group start pointer of i's group in C.

we *mark* such an entry i in A if and only if $\mathcal{P}_i \neq \emptyset$. Implementation-wise, we use the most significant bit (MSB) of an entry to indicate whether it is marked or not. By definition, we have $S_{i-1} >_{lex} S_i$ if and only if $\text{pss}[i] + 1 < i$. Since $\text{pss}[i]$ must be accessed anyway when i is inserted into A (for traversing the pss-chain), we can insert i marked or unmarked into A. Further, Lemma 3 implies that we must stop traversing a pss-chain when the current element is not the last child of its parent. We mark the entries in pss accordingly, also using the MSB of each entry. In the rest of this paper, we assume pss to be marked in this way.

Consider for instance $i = 6$ in our running example. As $6 - 1 = 5$ is a leaf (cf. Fig. 1), we have $5 \in \mathcal{P}_6$. We can deduce the fact that 5 is indeed a leaf from $\text{pss}[6] = 0 < 5$ alone. Further, 5 is the last child of $\text{pss}[5] = 4$, so $4 \in \mathcal{P}_6$. Since 4 is not the last child of $\text{pss}[4] = 0$, we have $\mathcal{P}_6 = \{4, 5\}$.

The second major change concerns the cache-unfriendliness of traversing the $\mathcal{P}_i$-sets. This bad cache performance results from the fact that the next pss-value (and the group start pointer) cannot be fetched until the current one is in memory. Instead of traversing the $\mathcal{P}_i$-sets one after another, we opt to traversing multiple such sets in a sort of breadth-first-search manner simultaneously. Specifically, we maintain a small ($\leq 2^{10}$ elements) queue Q of elements (nodes in the pss-tree) that can currently be processed. Then we iterate over Q and process the entries one after another. Parents of last children are inserted into Q in the same order as the respective children. After each iteration, we continue to scan over A and for each encountered marked entry i insert $i - 1$ into Q until we either encounter an empty entry in A or Q reaches its maximum capacity. This is repeated until the suffix array emerges. The queue size could be unlimited, but limiting it ensures that it fits into the CPU's cache. Figure 4 shows our Phase II on the running example and Algorithm 2 describes it formally in pseudo code.

The first step is the same as in Fig. 3. Note that $\mathcal{P}_0 = \emptyset$, hence 0 is not marked for further processing.

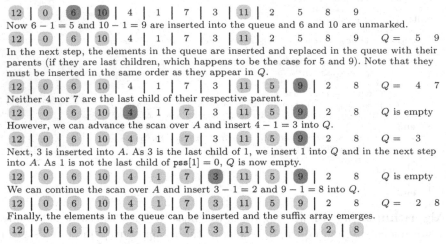

Now $6 - 1 = 5$ and $10 - 1 = 9$ are inserted into the queue and 6 and 10 are unmarked.

In the next step, the elements in the queue are inserted and replaced in the queue with their parents (if they are last children, which happens to be the case for 5 and 9). Note that they must be inserted in the same order as they appear in Q.

Neither 4 nor 7 are the last child of their respective parent.

However, we can advance the scan over A and insert $4 - 1 = 3$ into Q.

Next, 3 is inserted into A. As 3 is the last child of 1, we insert 1 into Q and in the next step into A. As 1 is not the last child of $\mathsf{pss}[1] = 0$, Q is now empty.

We can continue the scan over A and insert $3 - 1 = 2$ and $9 - 1 = 8$ into Q.

Finally, the elements in the queue can be inserted and the suffix array emerges.

Fig. 4. Refining a Lyndon grouping for $S = acedcebceece\$$ (see Fig. 2) into the suffix array using Algorithm 2. Marked entries are coloured blue while inserted but unmarked elements are coloured green. Note that the uncoloured entries are not actually present in the array A but only serve to indicate the current Lyndon grouping. (Color figure online)

Theorem 1. *Algorithm 2 correctly computes the suffix array from a Lyndon grouping.*

3.2   Phase I

In Phase I, a Lyndon grouping is derived from a suffix grouping in which the group contexts have length (at least) one. That is, the suffixes are sorted and grouped by their Lyndon prefixes. Lemma 4 describes the relationship between the Lyndon prefixes and the pss-tree that is essential to Phase I.

Lemma 4. *Let $c_1 < \cdots < c_k$ be the children of $i \in [0 \mathinner{..} n)$ in the pss-tree. $\mathcal{L}_i$ is $S[i]$ concatenated with the Lyndon prefixes of $c_1, \ldots, c_k$. More formally:*

$$\mathcal{L}_i = S[i \mathinner{..} \mathit{nss}[i]) = S[i]S[c_1 \mathinner{..} c_2) \ldots S[c_k \mathinner{..} \mathit{nss}[i]) = S[i]\mathcal{L}_{c_1} \ldots \mathcal{L}_{c_k}$$

We start from the *initial suffix grouping* in which the suffixes are grouped according to their first characters. From the relationship between the Lyndon prefixes and the pss-tree in Lemma 4 one can get the general idea of extending the context of a node's group with the Lyndon prefixes of its children (in correct order) while maintaining the sorting [1]. Note that any node is by definition in a higher group than its parent. Also, by Lemma 4 the leaves of the pss-tree are already in Lyndon groups in the initial suffix grouping. Therefore, if we consider the groups in lexicographically decreasing order (i.e. higher to lower) and append the context of the current group to each parent (and insert

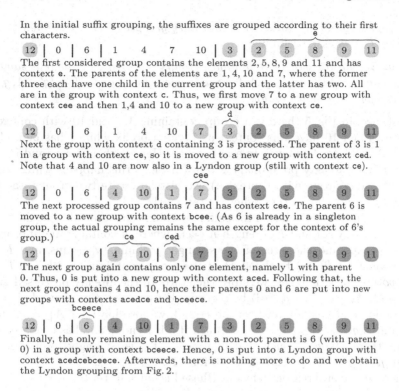

In the initial suffix grouping, the suffixes are grouped according to their first characters.

The first considered group contains the elements $2, 5, 8, 9$ and 11 and has context e. The parents of the elements are $1, 4, 10$ and 7, where the former three each have one child in the current group and the latter has two. All are in the group with context c. Thus, we first move 7 to a new group with context cee and then $1, 4$ and 10 to a new group with context ce.

Next the group with context d containing 3 is processed. The parent of 3 is 1 in a group with context ce, so it is moved to a new group with context ced. Note that 4 and 10 are now also in a Lyndon group (still with context ce).

The next processed group contains 7 and has context cee. The parent 6 is moved to a new group with context bcee. (As 6 is already in a singleton group, the actual grouping remains the same except for the context of 6's group.)

The next group again contains only one element, namely 1 with parent 0. Thus, 0 is put into a new group with context aced. Following that, the next group contains 4 and 10, hence their parents 0 and 6 are put into new groups with contexts acedce and bceece.

Finally, the only remaining element with a non-root parent is 6 (with parent 0) in a group with context bceece. Hence, 0 is put into a Lyndon group with context acedcebceece. Afterwards, there is nothing more to do and we obtain the Lyndon grouping from Fig. 2.

Fig. 5. Refining the initial suffix grouping for $S = abccabccbcc\$$ (see Fig. 2) into the Lyndon grouping. Elements in Lyndon groups are marked gray or green, depending on whether they have been processed already. Note that the applied procedure does not entirely correspond to our algorithm for Phase I; it only serves to illustrate the sorting principle. (Color figure online)

them into new groups accordingly), each encountered group is guaranteed to be Lyndon [1]. Consequently, we obtain a Lyndon grouping. Figure 5 shows this principle applied to our running example. Formally, the suffix grouping satisfies the following property during Phase I before and after processing a group:

Property 1. For any $i \in [0..n)$ with children $c_1 < \cdots < c_k$ there is $j \in [0..k]$ such that (a) $c_1, \ldots, c_j$ are in groups that have already been processed, (b) $c_{j+1}, \ldots, c_k$ are in groups that have not yet been processed, and (c) the context of the group containing i is $S[i]\mathcal{L}_{c_1} \ldots \mathcal{L}_{c_j}$. Furthermore, each processed group is Lyndon.

Additionally and unlike in Baier's original approach, all groups created during our Phase I are either Lyndon or only contain elements whose Lyndon prefix is different from the group's context.

Definition 4 (Strongly preliminary group). *We call a preliminary group* $\mathcal{G} = \langle g_s, g_e, |\alpha| \rangle$ *strongly preliminary if and only if* $\mathcal{G}$ *contains only elements* *whose Lyndon prefix is not α. A preliminary group that is not strongly preliminary is called* weakly preliminary.

Lemma 5. *For any weakly preliminary group $\mathcal{G} = \langle g_s, g_e, |\alpha| \rangle$ there is some $g' \in [g_s .. g_e)$ such that $\mathcal{G}' = \langle g_s, g', |\alpha| \rangle$ is a Lyndon group and $\mathcal{G}'' = \langle g' + 1, g_e, |\alpha| \rangle$ is a strongly preliminary group.*

For instance, in Fig. 5 there is a group containing 1,4 and 10 with context ce. However, 4 and 10 have this context as Lyndon prefix while 1 has ced. Consequently, 1 will later be moved to a new group. Hence, when Baier (and Bertram et al.) create a weakly preliminary group (in Fig. 5 this happens while processing the Lyndon group with context e), we instead create two groups, the lower containing 4 and 10 and the higher containing 1.

During Phase I we maintain the suffix grouping using the following data structures. Two arrays A and I of length n each, where A contains the unprocessed Lyndon groups and the sizes of the strongly preliminary groups, and I maps each element $s \in [0 .. n)$ to the start of the group containing s. We call $I[s]$ the *group pointer* of s. Further, we store the starts of the already processed groups in a list C . Let $\mathcal{G} = \langle g_s, g_e, |\alpha| \rangle$ be a group. For each $s \in \mathcal{G}$ we have $I[s] = g_s$. If $\mathcal{G}$ is Lyndon and has not yet been processed, we also have $s \in A[g_s .. g_e]$ for all $s \in \mathcal{G}$ and $A[g_s] < A[g_s + 1] < \cdots < A[g_e]$. If $\mathcal{G}$ is Lyndon and has been processed already, there is some j such that $C[j] = g_s$. If $\mathcal{G}$ is (strongly) preliminary we have $A[g_s] = g_e + 1 - g_s$ and $A[k] = 0$ for all $k \in (g_s .. g_e]$.

There are several reasons why our Phase I is much faster than Baier's. Firstly, we do not write preliminary groups to A. Secondly, we compute pss beforehand using an algorithm by Bille et al. [4] instead of on the fly as Baier does [1,2]. Furthermore, we have the Lyndon groups in A sorted and store the sizes of the strictly preliminary groups in A as well. The former makes finding the number of children a parent has in the currently processed group easier and faster. The latter makes the separate array of length n used by Baier [1,2] for the group sizes obsolete and is made possible by the fact that we only write Lyndon groups to A. For reasons why these changes lead to a faster algorithm see [13].

As alluded above, we follow Baier's general approach and consider the Lyndon groups in lexicographically decreasing order while updating the groups containing the parents of elements in the current group. Since children are in higher groups than their parents by definition, when we encounter some group $\mathcal{G} = \langle g_s, g_e, |\alpha| \rangle$, the children of any element in $\mathcal{G}$ are in already processed groups. Hence, by Property 1 $\mathcal{G}$ must be Lyndon. For a formal proof see [1].

In the rest of this section we explain how to actually process a Lyndon group.

Let $\mathcal{G} = \langle g_s, g_e, |\alpha| \rangle$ be the currently processed group and w.l.o.g. assume that no element in $\mathcal{G}$ has the root -1 as parent (we do not have the root in the suffix grouping, thus nodes with the root as parent can be ignored here). Furthermore, let $\mathcal{A}$ be the set of parents of elements in $\mathcal{G}$ (i.e. $\mathcal{A} = \{\text{pss}[i] : i \in \mathcal{G}, \text{pss}[i] \geq 0\}$) and let $\mathcal{G}_1 < \cdots < \mathcal{G}_k$ be those (necessarily preliminary) groups containing elements from $\mathcal{A}$. For each $g \in [1 .. k]$ let α_g be the context of $\mathcal{G}_g$.

As noted in Fig. 5, we have to consider the number of children an element in $\mathcal{A}$ has in $\mathcal{G}$. Specifically, we need to move two parents in $\mathcal{A}$ which are currently in the same group to different new groups if they have differing numbers of children in $\mathcal{G}$. Let $\mathcal{A}_\ell$ contain those elements from $\mathcal{A}$ with exactly ℓ children in $\mathcal{G}$. Maintaining Property 1 requires that, after processing $\mathcal{G}$, for some $g \in [1 .. k]$ the elements in $\mathcal{G}_g \cap \mathcal{A}_\ell$ are in groups with context $\alpha_g \alpha^\ell$. For any $\ell < \ell'$, we have $\alpha_g \alpha^\ell <_{lex} \alpha_g \alpha^{\ell'}$, thus the elements in $\mathcal{G}_g \cap \mathcal{A}_\ell$ must form a lower group than those in $\mathcal{G}_g \cap \mathcal{A}_{\ell'}$ after $\mathcal{G}$ has been processed [1,2]. To achieve this, first the parents in $\mathcal{A}_{|\mathcal{G}|}$ are moved to new groups immediately following their respective old groups, then those in $\mathcal{A}_{|\mathcal{G}|-1}$ and so on [1,2].

We proceed as follows. First, determine $\mathcal{A}$ and count how many children each parent has in $\mathcal{G}$. Then, sort the parents according to these counts using a bucket sort.[3] Further, partition the elements in each bucket into two sub-buckets depending on whether they should be inserted into Lyndon groups or strongly preliminary groups. Then, for the sub-buckets (in the order of decreasing count; for equal counts: first strongly preliminary then Lyndon sub-buckets) move the parents into new groups.[4] Because of space constraints, we do not describe the rather technical details. These can be found in the extended paper [13].

4   Experiments

Our implementation FGSACA of the optimised GSACA is publicly available.[5]

We compare our algorithm with the GSACA implementation by Baier [1,2], and the **double sort** algorithms DS1 and DSH by Bertram et al. [3]. The latter two also use the grouping principle but employ integer sorting and have super-linear time complexity. DSH differs from DS1 only in the initialisation: in DS1 the suffixes are sorted by their first character while in DSH up to 8 characters are considered. We further include DivSufSort 2.0.2 and libsais 2.7.1 since the former is used by Bertram et al. as a reference [3] and the latter is the currently fastest suffix sorter known to us.

The algorithms were evaluated on real texts (in the following PC-Real), real repetitive texts (PC-Rep-Real) and artificial repetitive texts (PC-Rep-Art) from the Pizza & Chili corpus. To test the algorithms on texts for which a 32-bit suffix array is not sufficient, we also included larger texts (Large), namely the first 10^{10} bytes from the English Wikipedia dump from 01.06.2022 and the human DNA concatenated with itself. For more detail on the data and our testing methodology see the longer version of this paper [13].

All algorithms were faster on the more repetitive datasets, on which the differences between the algorithms were also smaller. On all datasets, our algorithm

---

[3] Note that the sum of the counts is $|\mathcal{G}|$, hence the time complexity of the bucket sort is linear in the size of the group.

[4] Note that Baier broadly follows the same steps (determine parents, sort them, move them to new groups accordingly) [1,2]. However, each individual step is different because of our distinction between strongly preliminary, weakly preliminary and Lyndon groups.

[5] https://gitlab.com/qwerzuiop/lfgsaca.

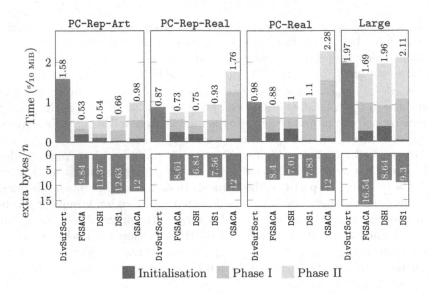

Fig. 6. Normalised running time and working memory averaged for each category. The horizontal red line indicates the time for libsais. For Large we did not test GSACA because Baier's reference implementation only supports 32-bit words. (Color figure online)

is between 46% and 60% faster than GSACA and compared to DSH about 2% faster on repetitive data, over 11% faster on PC-Real and over 13% faster on Large.

Especially notable is the difference in the time required for Phase II: Our Phase II is between 33% and 50% faster than Phase II of DSH. Our Phase I is also faster than Phase I of DS1 by a similar margin. Conversely, Phase I of DSH is much faster than our Phase I. However, this is only due to the more elaborate construction of the initial suffix grouping as demonstrated by the much slower Phase I of DS1. Compared to FGSACA, libsais is between 46% and 3% faster.

Memory-wise, for 32-bit words, FGSACA uses about 8.83 bytes per input character, while DS1 and DSH use 8.94 and 8.05 bytes/character, respectively. GSACA always uses 12 bytes/character. On Large, FGSACA expectedly requires about twice as much memory. For DS1 and DSH this is not the case, mostly because they use 40-bit integers for the additional array of length n that they require (while we use 64-bit integers). DivSufSort requires only a small constant amount of working memory and libsais never exceeded 21kiB of working memory on our test data.

Acknowledgements. This work was supported by the Deutsche Forschungsgemeinschaft (DFG - German Research Foundation) (OH 53/7-1).

A   Proofs

Lemma 1. *For any $j, j' \in \mathcal{P}_i$ we have $\mathcal{L}_j \neq \mathcal{L}_{j'}$ if and only if $j \neq j'$.*

Proof. Let $j, j' \in \mathcal{P}_i$ and $j \neq j'$. By definition of $\mathcal{P}_i$ we have $\mathbf{nss}[j] = \mathbf{nss}[j'] = i$. Since $\mathcal{L}_j = S[j \mathinner{..} \mathbf{nss}[j])$ and $\mathcal{L}_{j'} = S[j' \mathinner{..} \mathbf{nss}[j'])$, $\mathcal{L}_j$ and $\mathcal{L}_{j'}$ have different lengths, implying the claim.

Lemma 2. *$\mathcal{P}_i$ is empty if and only if $i = 0$ or $S_{i-1} <_{lex} S_i$. Furthermore, if $\mathcal{P}_i \neq \emptyset$ then $i - 1 \in \mathcal{P}_i$.*

Proof. $\mathcal{P}_0 = \emptyset$ by definition. Let $i \in [1 \mathinner{..} n)$. If $S_{i-1} >_{lex} S_i$ we have $\mathbf{nss}[i-1] = i$ and thus $i - 1 \in \mathcal{P}_i$. Otherwise ($S_{i-1} <_{lex} S_i$), assume there is some $j < i - 1$ such that $\mathbf{nss}[j] = i$. By definition, $S_j >_{lex} S_i$ and $S_j <_{lex} S_k$ for each $k \in (j \mathinner{..} i)$. But by transitivity we also have $S_j >_{lex} S_{i-1}$, which is a contradiction, hence $\mathcal{P}_i$ must be empty.

Lemma 3. *For some $j \in [0 \mathinner{..} i)$, we have $j \in \mathcal{P}_i$ if and only if j's last child is in $\mathcal{P}_i$, or $j = i - 1$ and $S_j >_{lex} S_i$.*

Proof. By Lemma 2 we may assume $\mathcal{P}_i \neq \emptyset$ and $j + 1 < i$, otherwise the claim is trivially true. If j is a leaf we have $\mathbf{nss}[j] = j + 1 < i$ and thus $j \notin \mathcal{P}_i$ by definition. Hence assume j is not a leaf and has $j' > j$ as last child, i.e. $\mathbf{pss}[j'] = j$ and there is no $k > j'$ with $\mathbf{pss}[k] = j$. It suffices to show that $j' \in \mathcal{P}_i$ if and only if $j \in \mathcal{P}_i$. Note that $\mathbf{pss}[j'] = j$ implies $\mathbf{nss}[j] > j'$.

$\implies$: From $\mathbf{nss}[j'] = i$ and thus $S_k >_{lex} S_{j'} >_{lex} S_j$ (for all $k \in (j' \mathinner{..} i)$) we have $\mathbf{nss}[j] \geq i$. Assume $\mathbf{nss}[j] > i$. Then $S_i >_{lex} S_j$ and thus $\mathbf{pss}[i] = j$, which is a contradiction.

$\impliedby$: From $S_i <_{lex} S_j <_{lex} S_{j'}$ we have $\mathbf{nss}[j'] \leq i$. Assume $\mathbf{nss}[j'] < i$ for a contradiction. For all $k \in (j \mathinner{..} j')$, $\mathbf{pss}[j'] = j$ implies $S_k >_{lex} S_{j'}$. Furthermore, for all $k \in [j' \mathinner{..} \mathbf{nss}[j'])$ we have $S_k >_{lex} S_{\mathbf{nss}[j']}$ by definition. In combination this implies $S_k >_{lex} S_{\mathbf{nss}[j']}$ for all $k \in (j \mathinner{..} \mathbf{nss}[j'])$. As $\mathbf{nss}[j] = i > \mathbf{nss}[j']$ we hence have $\mathbf{pss}[\mathbf{nss}[j']] = j$, which is a contradiction.

Theorem 1. *Algorithm 2 correctly computes the suffix array from a Lyndon grouping.*

Proof. By Lemmata 2 and 3, Algorithms 1 and 2 are equivalent for a maximum queue size of 1. Therefore it suffices to show that the result of Algorithm 2 is independent of the queue size. Assume for a contradiction that the algorithm inserts two elements i and j with $S_i <_{lex} S_j$ belonging to the same Lyndon group with context α, but in a different order as Algorithm 1 would. This can only happen if j is inserted earlier than i. Note that, since i and j have the same Lyndon prefix α, the $\mathbf{pss}$-subtrees T_i and T_j rooted at i and j, respectively, are isomorphic (see [4]). In particular, the path from the rightmost leaf in T_i to i has the same length as the path from the rightmost leaf in T_j to j. Thus, i and j are inserted in the same order as $S_{i+|\alpha|}$ and $S_{j+|\alpha|}$ occur in the suffix array. Now the claim follows inductively.

Lemma 4. *Let $c_1 < \cdots < c_k$ be the children of $i \in [0\,..\,n)$ in the pss-tree. $\mathcal{L}_i$ is $S[i]$ concatenated with the Lyndon prefixes of $c_1, \ldots, c_k$. More formally:*

$$\mathcal{L}_i = S\,[i\,..\,\mathrm{nss}[i]) = S[i]S\,[c_1\,..\,c_2)\ldots S\,[c_k\,..\,\mathrm{nss}[i]) = S[i]\mathcal{L}_{c_1}\ldots\mathcal{L}_{c_k}$$

Proof. By definition we have $\mathcal{L}_i = S\,[i\,..\,\mathrm{nss}[i])$. Assume i has $k \geq 1$ children $c_1 < \cdots < c_k$ in the pss-tree (otherwise $\mathrm{nss}[i] = i + 1$ and the claim is trivial). For the last child c_k we have $\mathrm{nss}[c_k] = \mathrm{nss}[i]$ from Lemma 3. Let $j \in [1\,..\,k)$ and assume $\mathrm{nss}[c_j] \neq c_{j+1}$. Then we have $\mathrm{nss}[c_j] < c_{j+1}$, otherwise c_{j+1} would be a child of c_j. As we have $S_{\mathrm{nss}[c_j]} <_{lex} S_{c_j}$ and $S_{c_j} <_{lex} S_{c_{j'}}$ for each $j' \in [1\,..\,j)$ (by induction), we also have $S_{\mathrm{nss}[c_j]} <_{lex} S_{i'}$ for each $i' \in (i\,..\,\mathrm{nss}[c_j])$. Since $\mathrm{nss}[i] > \mathrm{nss}[c_j]$, $\mathrm{nss}[c_j]$ must be a child of i in the pss-tree, which is a contradiction.

Lemma 5. *For any weakly preliminary group $\mathcal{G} = \langle g_s, g_e, |\alpha| \rangle$ there is some $g' \in [g_s\,..\,g_e)$ such that $\mathcal{G}' = \langle g_s, g', |\alpha| \rangle$ is a Lyndon group and $\mathcal{G}'' = \langle g'+1, g_e, |\alpha| \rangle$ is a strongly preliminary group.*

Proof. Let $\mathcal{G} = \langle g_s, g_e, |\alpha| \rangle$ be a weakly preliminary group. Let $F \subset \mathcal{G}$ be the set of elements from $\mathcal{G}$ whose Lyndon prefix is α. By Lemma 6 (below) we have $S_i <_{lex} S_j$ for any $i \in F, j \in \mathcal{G} \setminus F$. Hence, splitting $\mathcal{G}$ into two groups $\mathcal{G}' = \langle g_s, g_s + |F| - 1, |\alpha| \rangle$ and $\mathcal{G}'' = \langle g_s + |F|, g_e, |\alpha| \rangle$ results in a valid suffix grouping. Note that, by construction, the former is a Lyndon group and the latter is strongly preliminary.

Lemma 6. *For strings wu and wv over Σ with $u <_{lex} wu$ and $v >_{lex} wv$ we have $wu <_{lex} wv$.*

Proof. Note that there is no j such that $wv = w^j$ since otherwise v would be a prefix of wv and $v <_{lex} wv$ would hold. Hence, there are $k \in \mathbb{N}, \ell \in [0\,..\,|w|), b \in \Sigma$ and $m \in \Sigma^*$ such that $wv = w^k w\,[0\,..\,\ell)\,bm$ and $b > w[\ell]$. There are two cases:

- $wu = w^j$ for some $j \geq 1$
    - If $j\,|w| \leq k\,|w| + \ell$, then wu is a prefix of wv.
    - Otherwise, the first different symbol in wu and wv is at index $p = k\,|w| + \ell$ and we have $(wu)[p] = w^j[p] = w[\ell] < b = (wv)[p]$.
- There are $i \in \mathbb{N}, j \in [0\,..\,|w|), a \in \Sigma$ and $q \in \Sigma^*$ such that $wu = w^i w\,[0\,..\,j)\,aq$ and $a < w[j]$.
    - If $|w^i w\,[0\,..\,j)| \leq |w^k w\,[0\,..\,\ell)|$, the first different symbol is at position $p = |w^i w\,[0\,..\,j)|$ with $wu[p] = a < w[j] \leq wv[p]$.
    - Otherwise, the first different symbol is at position $p = |w^k w\,[0\,..\,\ell)|$ with $wv[p] = b > w[\ell] = wu[p]$.

In all cases, the claim follows.

References

1. Baier, U.: Linear-time suffix sorting. Master's thesis, Ulm University (2015)
2. Baier, U.: Linear-time suffix sorting - a new approach for suffix array construction. In: Grossi, R., Lewenstein, M. (eds.) 27th Annual Symposium on Combinatorial Pattern Matching. Leibniz International Proceedings in Informatics, vol. 54. Schloss Dagstuhl - Leibniz-Zentrum für Informatik (2016)
3. Bertram, N., Ellert, J., Fischer, J.: Lyndon words accelerate suffix sorting. In: Mutzel, P., Pagh, R., Herman, G. (eds.) 29th Annual European Symposium on Algorithms. Leibniz International Proceedings in Informatics, vol. 204, pp. 15:1–15:13. Schloss Dagstuhl - Leibniz-Zentrum für Informatik (2021)
4. Bille, P., et al.: space efficient construction of Lyndon arrays in linear time. In: Czumaj, A., Dawar, A., Merelli, E. (eds.) 47th International Colloquium on Automata, Languages, and Programming. Leibniz International Proceedings in Informatics, vol. 168, pp. 14:1–14:18. Schloss Dagstuhl-Leibniz-Zentrum für Informatik (2020)
5. Fischer, J., Kurpicz, F.: Dismantling DivSufSort. In: Holub, J., Žd'árek, J. (eds.) Proceedings of the Prague Stringology Conference 2017, pp. 62–76 (2017)
6. Franek, F., Paracha, A., Smyth, W.F.: The linear equivalence of the suffix array and the partially sorted Lyndon array. In: Holub, J., Žd'árek, J. (eds.) Proceedings of the Prague Stringology Conference 2017, pp. 77–84 (2017)
7. Goto, K.: Optimal time and space construction of suffix arrays and LCP arrays for integer alphabets. In: Holub, J., Žd'árek, J. (eds.) Proceedings of the 23rd Prague Stringology Conference, pp. 111–125 (2017)
8. Li, Z., Li, J., Huo, H.: Optimal in-place suffix sorting. Inf. Comput. **285**, 104818 (2022). https://doi.org/10.1016/j.ic.2021.104818. ISSN 0890-5401
9. Manber, U., Myers, G.: Suffix arrays: a new method for on-line string searches. In: Proceedings of the First Annual ACM-SIAM Symposium on Discrete Algorithms, pp. 319–327. Society for Industrial and Applied Mathematics (1990)
10. Nong, G.: Practical linear-time O(1)-workspace suffix sorting for constant alphabets. ACM Trans. Inf. Syst. **31**(3), 1–15 (2013)
11. Nong, G., Zhang, S., Chan, W.H.: Linear suffix array construction by almost pure induced-sorting. In: 2009 Data Compression Conference, pp. 193–202 (2009)
12. Ohlebusch, E.: Bioinformatics algorithms: sequence analysis, genome rearrangements, and phylogenetic reconstruction. Oldenbusch Verlag (2013)
13. Olbrich, J., Ohlebusch, E., Büchler, T.: On the optimisation of the GSACA suffix array construction algorithm (2022). https://doi.org/10.48550/ARXIV.2206.12222
14. Pierre Duval, J.: Factorizing words over an ordered alphabet. J. Algorithms **4**(4), 363–381 (1983)

String Compression

Balancing Run-Length Straight-Line Programs

Gonzalo Navarro, Francisco Olivares, and Cristian Urbina[✉]

CeBiB — Center for Biotechnology and Bioengineering, Department of Computer
Science, University of Chile, Santiago, Chile
crurbina@dcc.uchile.cl

Abstract. It was recently proved that any SLP generating a given string
w can be transformed in linear time into an equivalent balanced SLP
of the same asymptotic size. We show that this result also holds for
RLSLPs, which are SLPs extended with run-length rules of the form
$A \rightarrow B^t$ for $t > 2$, deriving $\exp(A) = \exp(B)^t$. An immediate consequence is the simplification of the algorithm for extracting substrings of
an RLSLP-compressed string. We also show that several problems like
answering RMQs and computing Karp-Rabin fingerprints on substrings
can be solved in $\mathcal{O}(g_{rl})$ space and $\mathcal{O}(\log n)$ time, g_{rl} being the size of the
smallest RLSLP generating the string, of length n. We extend the result
to solving more general operations on string ranges, in $\mathcal{O}(g_{rl})$ space and
$\mathcal{O}(\log n)$ applications of the operation. In general, the smallest RLSLP
can be asymptotically smaller than the smallest SLP by up to an $\mathcal{O}(\log n)$
factor, so our results can make a difference in terms of the space needed
for computing these operations efficiently for some string families.

Keywords: Run-length straight-line programs · Substring range
problems · Repetitive strings

1 Introduction

Enormous collections of data are being generated at every second nowadays.
Already storing this data is becoming a relevant and practical challenge. Compression serves to represent the data within reduced space. Still, just storing
the data in compressed form is not sufficient in many cases; one also requires
to construct data structures that support various queries within the compressed
space. For example, *index* data structures support the search for short patterns
in compressed strings. In areas like Bioinformatics, these collections of strings
are often very repetitive [22], which makes traditional compressors and indexes
based on Shannon's entropy unsuitable for this task [19].

Over the years, several compressors and data structures exploiting repetitiveness have been devised. Examples of this are the Lempel-Ziv family [16,18]

Funded in part by Basal Funds FB0001, Fondecyt Grant 1-200038, and two Conicyt
Doctoral Scholarships, ANID, Chile.

and the run-length Burrows-Wheeler transform (BWT) [3,8]. While compressors based on Lempel-Ziv achieve the best compression ratios, indexes based on them are not very fast and provide limited functionality. On the other hand, indexes based on the BWT can efficiently solve a variety of queries over strings, but their compression ratio is far from optimal for repetitive sequences [13].

Somewhere in between of Lempel-Ziv and BWT compression is *grammar compression*. This approach consists in constructing a deterministic context-free grammar generating only the string to be compressed; such grammars are called straight-line programs (SLPs). Although finding the smallest SLP generating a string is NP-complete [4], there exist several heuristics [17,20] and approximations [11,23] producing SLPs of small size. The popularity of SLPs probably comes from their simplicity to expose repetitive patterns on strings, which is useful to avoid redundant computation in compressed space [15,25]. This makes SLPs ideal for indexing and answering queries in compressed space [2,10].

A problem that complicates such computations is that the parse tree of the grammars can be arbitrarily tall. While tasks like accessing a symbol of the string in time proportional to the parse tree height is almost trivial, achieving $\mathcal{O}(\log n)$ time on general grammars requires much more sophistication [2]. Recently, Ganardi et al. [10] showed that any SLP can be balanced without paying an (asymptotic) increase in its size. This simplified several problems that were difficult for general SLPs, but easy if the depth of their parse tree is $\mathcal{O}(\log n)$. Accessing a symbol in time $\mathcal{O}(\log n)$ is nearly optimal, actually [24].

An extended grammar compression mechanism are the run-length SLPs introduced by Nishimoto et al. [21]. An RLSLPs is an SLP extended with run-length rules of the form $A \to B^t$ for some $t > 2$, which derive $\mathsf{exp}(A) = \mathsf{exp}(B)^t$. While the size of the smallest SLP generating a string of length n is always $\Omega(\log n)$, the smallest RLSLP can be of size $\mathcal{O}(1)$ for some string families, which exhibit a logarithmic gap between the compression power of SLPs and RLSLPs. RLSLP have recently gained popularity for indexing. For example, all known *locally consistent grammars* are RLSLPs, and they have been a key component in the most recent indices for repetitive text collections. A locally consistent grammar is built through consecutive applications of a locally consistent parsing, which is a method to partition a string into non-overlapping blocks, such that equal substrings are equally parsed with the possible exception of their margins. Gagie et al. [9] built an index based on locally consistent grammars using $\mathcal{O}(r \log \log n)$ space, with which they were able to count the occ occurrences of a length-m pattern in optimal time $\mathcal{O}(m)$ and locate them in optimal time $\mathcal{O}(m + occ)$, where r is the number of runs in the BWT [3] of the string. Kociumaka et al. [5] also built a locally consistent grammar to index a string. Their grammar can count and locate the pattern in optimal time using $\mathcal{O}(\gamma \log \frac{n}{\gamma} \log^\epsilon n)$ space, where γ is the size of the smallest string attractor of the string [14].

In this paper we extend the results of Ganardi et al. to RLSLPs, that is, we show that one can always balance an RLSLP in linear time without increasing its asymptotic size. This result yields a considerable simplification to the algorithm for accessing any symbol of the string in logarithmic time [5, Appendix A]. It has other implications, like computing range minimum queries (RMQs) [7] or Karp-Rabin fingerprints [12], in $\mathcal{O}(\log n)$ time and within $\mathcal{O}(g_{rl})$ space. We generalize those concepts and show how to compute a wide class of semiring-like functions over substrings of an RLSLP-compressed string within $\mathcal{O}(g_{rl})$ space and $\mathcal{O}(\log n)$ applications of the function.

2   Terminology

2.1   Strings

Let Σ be any finite set of *symbols* (an *alphabet*). A *string* w is any finite tuple of elements in Σ. The *length* of a string is the length of the tuple, and the *empty string* of length 0 is denoted by ε. The set Σ^* is formed by all the strings that can be defined over Σ. For any string $w = w_1 \ldots w_n$, its i-th symbol is denoted by $w[i] = w_i$. Similarly, $w[i : j] = w_i \ldots w_j$ with $1 \leq i \leq j \leq n$, or ε if $j < i$. We also define $w[: i] = w_1 \ldots w_i$ and $w[i :] = w_i \ldots w_n$. If $x[1 : n]$ and $y[1 : m]$ are strings, the concatenation operation xy is defined as $xy = x_1 \ldots x_n y_1 \ldots y_m$. If $w = xyz$, then y (resp. x, z) is a *substring* (resp. *prefix*, *suffix*) of w.

2.2   Straight-Line Programs

A *straight-line program* (SLP) is a deterministic context-free grammar generating a unique string w. More formally, an SLP is a context free grammar $G = (V, \Sigma, R, S)$ where V is the set of variables (or non-terminals), Σ is the set of terminal symbols (disjoint from V), $R \subseteq V \times (V \cup \Sigma)^*$ is the set of rules and S is the initial variable; satisfying that each variable has only one rule associated, and that the variables are ordered in such a way that the starting variable is the greater of them, and any variable can only refer to other variables strictly lesser than itself or terminals, in the right-hand side of its rule. Any variable A derives a unique string $\mathtt{exp}(A)$, and the string generated by the SLP, is the string generated by its starting variable. The size of an SLP is defined as the sum of the lengths of the right-hand side of its rules. The size of the smallest SLP generating a string is denoted by g, and is a relevant measure of repetitiveness. An SLP generating a non-empty string is often given in so-called Chomsky Normal Form, that is, with all its rules being of the form $A \to BC$ or $A \to a$ for A, B, C variables, and a a terminal symbol.

While computing the smallest grammar is an NP-hard problem [4], there exist several heuristic providing log-approximations of the smallest SLP [11,23]. SLPs are popular as compression devices because several problems over strings can be solved efficiently using their SLP representation, without ever decompressing them. Examples of this are accessing to arbitrary positions of w, extracting substrings, and many other kind of queries [2]. For several queries, it is convenient to have a balanced SLP, that is, an SLP whose parse tree has $\mathcal{O}(\log n)$ depth. Recently, Ganardi et al. showed that any SLP can be balanced [10].

2.3   Directed Acyclic Graph of an SLP

A *directed acyclic graph* (DAG) is a directed multigraph D without cycles (nor loops). We denote by $|D|$ the number of edges in this DAG. For our purposes, we assume that any DAG has a distinguished node r, satisfying that any other node can be reached from r, and has no incoming edges. We also assume that if a node has k outgoing edges, they are numbered from 1 to k. The *sink nodes* of a DAG are the nodes without outgoing edges. The set of sink nodes of D is denoted by W. We denote the number of paths from u to v as $\pi(u, v)$, and $\pi(u, V) = \sum_{v \in V} \pi(u, v)$ for a set V of nodes. The number of paths from the root to the sink nodes is $n(D) = \pi(r, W)$.

One can interpret an SLP generating a string w as a DAG D: There is a node for each variable in the SLP, the root node is the initial variable, terminal rules of the form $A \to a$ are the sink nodes, and a variable with rule $A \to B_1 B_2 \ldots B_k$ has outgoing edges (A, i, B_i) for $i \in [1..k]$. Note that if D is a DAG representing G, then $n(D) = |\exp(G)| = |w|$.

2.4   Run-Length Straight-Line Programs

A *run-length straight-line program* (RLSLP) is an SLP extended with *run-length* rules [21]. An RLSLP can have rules of the form:

- $A \to a$, for some terminal symbol a.
- $A \to A_1 A_2 \ldots A_k$, for some variables $A_1, \ldots, A_k$ and $k > 1$.
- $A \to B^t$, for some $t > 2$.

The string generated by a variable A with rule $A \to B^t$ is $\exp(B)^t$. A run-length rule is considered to have size 2 (one word is needed to store the exponent). We denote by g_{rl} to the size of the smallest RLSLP generating the string. The depth of the RLSLP is the depth of its associated equivalent SLP, obtained by *unfolding* its run-length rules $A \to B^t$ into rules of the form $A \to BB \ldots B$ of length t. Observe that a rule of the form $A \to A_1 A_2 \ldots A_k$ can always be transformed into $\mathcal{O}(k)$ rules of size 2, with one of them derivating the same string as A. Doing this for all rules can increase the depth of the RLSLP, but if k is bounded by a constant, then this increase is only by a constant factor.

3   Balancing Run-Length Straight-Line Programs

The idea utilized by Ganardi et al. to transform an SLP G into an equivalent balanced SLP of size $\mathcal{O}(|G|)$ [10, Theorem 1.2], can be adapted to work with RLSLPs. First, we state some definitions and results proved in their work, which we need to obtain our result.

Definition 1. *(Ganardi et al. [10, page 5]) Let D be a DAG, and define the pairs $\lambda(v) = (\lfloor \log_2 \pi(r, v) \rfloor, \lfloor \log_2 \pi(v, W) \rfloor)$. The symmetric centroid decomposition (SC-decomposition) of a DAG D produces a set of edges between nodes with the same λ pairs defined as $E_{scd}(D) = \{(u, i, v) \in E \mid \lambda(u) = \lambda(v)\}$, partitioning D into disjoint paths called SC-paths (some of them possibly empty).*

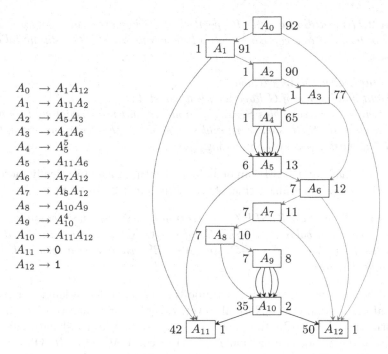

$$A_0 \rightarrow A_1 A_{12}$$
$$A_1 \rightarrow A_{11} A_2$$
$$A_2 \rightarrow A_5 A_3$$
$$A_3 \rightarrow A_4 A_6$$
$$A_4 \rightarrow A_5^5$$
$$A_5 \rightarrow A_{11} A_6$$
$$A_6 \rightarrow A_7 A_{12}$$
$$A_7 \rightarrow A_8 A_{12}$$
$$A_8 \rightarrow A_{10} A_9$$
$$A_9 \rightarrow A_{10}^4$$
$$A_{10} \rightarrow A_{11} A_{12}$$
$$A_{11} \rightarrow 0$$
$$A_{12} \rightarrow 1$$

Fig. 1. The DAG and SC-decomposition of an unfolded RLSLP generating the string $0(0(01)^5 1^2)^6 (01)^5 1^3$. The value to the left of a node is the number of paths from the root to that node, and the value to the right is the number of paths from the node to sink nodes. Red edges belong to the SC-decomposition of the DAG. Blue (resp. green) edges branch from an SC-path to the left (resp. to the right). (Color figure online)

The set E_{scd} can be computed in $\mathcal{O}(|D|)$ time. If D is the DAG of an SLP G this becomes $\mathcal{O}(|G|)$. The following lemma justifies the name "SC-paths".

Lemma 1. *(Ganardi et al. [10, Lemma 2.1]) Let $D = (V, E)$ be a DAG. Then every node has at most one outgoing and at most one incoming edge from $E_{scd}(D)$. Furthermore, every path from the root r to a sink node contains at most $2 \log_2 n(D)$ edges that do not belong to $E_{scd}(D)$.*

Note that the sum of the lengths of all SC-paths is at most the number of nodes of the DAG, or the number of variables of the SLP. An example of the SC-decomposition of a DAG can be seen in Fig. 1.

The following definition and technical lemma are needed to construct the building blocks of our balanced RLSLPs.

Definition 2. *(Ganardi et al. [10, page 7]) A weighted string is a string $w \in \Sigma^*$ equipped with a weight function $|| \cdot || : \Sigma \rightarrow \mathbb{N} \backslash \{0\}$, which is extended homomorphically. If A is a variable in an SLP G, then we also write $||A||$ for the weight of the string $exp(A)$ derived from A.*

Lemma 2. *(Ganardi et al. [10, Proposition 2.2]) For every non-empty weighted string w of length n one can construct in linear time an SLP G with the following properties:*

- *G contains at most $3n$ variables*
- *All right-hand sides of G have length at most 4*
- *G contains suffix variables $S_1, ..., S_n$ producing all non-trivial suffixes of w*
- *every path from S_i to some terminal symbol a in the derivation tree of G has length at most $3 + 2(\log_2 ||S_i|| - \log_2 ||a||)$*

We prove that any RLSLP can be balanced without asymptotically increasing its size. Our proof generalizes that of [10, Theorem 1.2] for SLPs.

Theorem 1. *Given an RLSLP G generating a string w, it is possible to construct an equivalent balanced RLSLP G' of size $\mathcal{O}(|G|)$, in linear time, with only rules of the form $A \to a, A \to BC$, and $A \to B^t$, where a is a terminal, B and C are variables, and $t > 2$.*

Proof. Without loss of generality, assume that G has rules of length at most 2, so it is almost in Chomsky Normal Form, except that it has run-length rules. Transform the RLSLP G into an SLP H by unfolding its run-length rules, and then obtain the SC-decomposition $E_{scd}(D)$ of the DAG D of H. Observe that the SC-paths of H use the same variables of G, so it holds that the sum of the lengths of all the SC-paths of H is less than the number of variables of G. Also, note that any variable A of G having a rule of the form $A \to B^t$ for some $t > 2$ is necessarily an endpoint of an SC-path in D, otherwise A would have t outgoing edges in $E_{scd}(D)$, which cannot happen.[1] This implies that the balancing procedure of Ganardi et al. over H, which transforms the rules of variables that are not the endpoint of an SC-path in the DAG D, will not touch variables that originally were run-length in G.

Let $\rho = (A_0, d_0, A_1), (A_1, d_1, A_2), \ldots, (A_{p-1}, d_{p-1}, A_p)$ be an SC-path of D. It holds that for each A_i with $i \in [0..p-1]$, in the SLP H, its rule goes to two distinct variables, one to the left and one to the right. For each variable A_i, with $i \in [0..p-1]$, there is a variable A'_{i+1} that is not part of the path. Let $A'_1 A'_2 \ldots A'_p$ be the sequence of those variables. Let $L = L_1 L_2 \ldots L_s$ be the subsequence of left variables of the previous sequence. Then construct an SLP of size $\mathcal{O}(s) \le \mathcal{O}(p)$ for the sequence L (seen as a string) as in Lemma 2, using $|\exp(L_i)|$ in H as the weight function. In this SLP, any path from the suffix nonterminal S_i to a variable L_j has length at most $3 + 2(\log_2 ||S_i|| - \log_2 ||L_j||)$. Similarly, construct an SLP of size $\mathcal{O}(t) \le \mathcal{O}(p)$ for the sequence $R = R_1 R_2 \ldots R_t$ of right symbols in reverse order, as in Lemma 2, but with prefix variables P_i instead of suffix variables. Each variable A_i, with $i \in [0..p-1]$, derives the same string as $w_\ell A_p w_r$, for some suffix w_ℓ of L and some prefix w_r of R. We can find rules deriving these prefixes and suffixes in the SLPs produced in the previous

---

[1] Seen another way, $\lambda(A) \neq \lambda(B)$ because $\log_2 \pi(A, W) = \log_2(t \cdot \pi(B, W)) > 1 + \log_2 \pi(B, W)$.

step, so for any variable A_i, we construct an equivalent rule of length at most 3. Add these equivalent rules, and the left and right SLP rules to a new RLSLP G'. Do this for all SC-paths. Finally, we add the original terminal variables and run-length variables of the RLSLP G, so G' is an RLSLP equivalent to G.

The SLP constructed for L has all its rules of length at most 4, and $3s \leq 3p$ variables. The same happens with R. The other constructed rules also have length at most 3, and there are p of them. Summing over all SC-paths we have $\mathcal{O}(|G|)$ size. The original terminal variables and run-length variables of G have rules of size at most 4, and we keep them. Thus, the RLSLP G' has size $\mathcal{O}(|G|)$.

Any path in the derivation tree of G' is of length $\mathcal{O}(\log n)$. To see why, let $A_0, \ldots, A_p$ be an SC-path. Consider a path from a variable A_i to an occurrence of a variable that is in the right-hand side of A_p in G'. Clearly this path has length at most 2. Now consider a path from A_i to a variable A'_j in L with $i < j \leq p$. By construction this path is of the form $A_i \to S_k \to^* A'_j$ for some suffix variable S_k (if the occurrence of A'_j is a left symbol), and its length is at most $1 + 3 + 2(\log_2 ||S_k|| - \log_2 ||A'_j||) \leq 4 + 2\log_2 ||A_i|| - 2\log_2 ||A'_j||$. Analogously, if A'_j is a right variable, the length of the path is bounded by $1 + 3 + 2(\log_2 ||P_k|| - \log_2 ||A'_j||) \leq 4 + 2\log_2 ||A_i|| - 2\log_2 ||A'_j||$. Finally, consider a maximal path to a leaf in the parse tree of G'. Factorize it as

$$A_0 \to^* A_1 \to^* \cdots \to^* A_k$$

where each A_i is a variable of H (and also of G). Paths $A_i \to^* A_{i+1}$ are like those defined in the paragraph above, satisfying that their length is bounded by $4 + 2\log_2 ||A_i|| - 2\log_2 ||A_{i+1}||$. Observe that between each A_i and A_{i+1}, in the DAG D there is almost an SC-path, except that the last edge is not in E_{scd}. The length of this path is at most

$$\sum_{i=0}^{k-1} (4 + 2\log_2 ||A_i|| - 2\log_2 ||A_{i+1}||) \leq k + 2\log_2 ||A_0|| - 2\log_2 ||A_k||$$

By Lemma 1, $k \leq 2\log_2 n$, which yields the $\mathcal{O}(\log n)$ upper bound. The construction time is linear, because the SLPs of Lemma 2 are constructed in linear time in the lengths of the SC-paths (summing to $\mathcal{O}(|G|)$), and $E_{scd}(D)$ can be obtained in time $\mathcal{O}(|G|)$ (instead of $\mathcal{O}(H)$) if we represent in the DAG D the edges of a variable A with rule $A \to B^t$ as a single edge extended with the power t. This way, when traversing the DAG from root to sinks and sinks to root to compute λ values, it holds that $\pi(A, W) = t \cdot \pi(B, W)$, and that $\pi(r, B) = t \cdot \pi(r, A) + c$, where c are the paths from root incoming from other variables. Thus, each run-length edge must be traversed only once, not t times.

To have rules of size at most two, delete rules in G' of the form $A \to B$ (replacing all A's by B's), and note that rules of the form $A \to BCDE$ or $A \to BCD$ can be decomposed into rules of length 2, with only a constant increase in size and depth.   □

4 Substring Range Operations in $\mathcal{O}(g_{rl})$ Space

4.1 Karp-Rabin Fingerprints

To answer signature $\kappa(w[p:q]) = (\sum_{i=p}^{q} w[i] \cdot c^{i-p}) \bmod \mu$, for a suitable integer c and prime number μ, we use the following identity for any $p' \in [p..q-1]$:

$$\kappa(w[p:q]) = \left(\kappa(w[p:p']) + \kappa(w[p'+1:q]) \cdot c^{p'-p+1} \right) \bmod \mu \qquad (1)$$

and then it holds

$$\kappa(w[p:p']) = \left(\kappa(w[p:q]) - \kappa(w[p'+1:q]) \cdot c^{p'-p+1} \right) \bmod \mu$$

$$\kappa(w[p'+1:q]) \cdot = \left(\frac{\kappa(w[p:q]) - \kappa(w[p:p'])}{c^{p'-p+1}} \right) \bmod \mu,$$

which implies that, to answer $\kappa(w[p:q])$, we can compute $\kappa(w[1:p-1])$ and $\kappa(w[1:q])$ and then subtract one to another. For that reason, we only consider computing fingerprints of text prefixes. Then, the recursive calls to our algorithm just need to know the right boundary of a prefix, namely computing signature on the substring $\exp(A)[1:j]$ of the string expanded by a symbol A of our grammar can be expressed as $\kappa(A, j)$.

Suppose that we want to compute the signature of a prefix $w[1:j]$ and that there is a rule $A \to BC$ such that $\exp(A) = w[1:q]$, with $j \leq q$. If $j = |\exp(B)|$ or $j = q$, we can have stored $\kappa(\exp(A))$ and $\kappa(\exp(B))$ and answer directly the query. On the other hand, if $j < |\exp(B)|$, we can answer $\kappa(\exp(B)[1:j])$ by descending in the derivation tree of B. Otherwise, $|\exp(B)| < j < |\exp(A)|$, then we can use Eq. 1 and answer $(\kappa(\exp(B)) + \kappa(\exp(C)[1:j-|\exp(B)|] \cdot c^{|\exp(B)|}) \bmod \mu$, where $\kappa(\exp(C)[1:j-|\exp(B)|]$ is obtained by descending in the derivation tree of C. Then, in addition to storing $\kappa(\exp(A))$ for every nonterminal A, we also need to store $c^{|\exp(A)|} \bmod \mu$ and $|\exp(A)|$. Therefore, the cost of computing fingerprints is just the depth of the derivation tree of A.

The same does not apply for run-length rules $A \to B^t$, because we cannot afford the space consumption of storing $c^{t' \cdot |\exp(B)|} \bmod \mu$ for every $1 \leq t' \leq t$, as this could give us a structure bigger than $\mathcal{O}(g_{rl})$. Instead, we can treat run-length rules as regular rules $A \to B \ldots B$. Then, we can use the following identity

$$\kappa(\exp(B^{t'})) = \left(\kappa(\exp(B)) \cdot \frac{c^{|\exp(B)| \cdot t'} - 1}{c^{|\exp(B)|} - 1} \right) \bmod \mu.$$

Namely, to compute $\kappa(\exp(B^{t'}))$ we can have previously stored $c^{|\exp(B)|} \bmod \mu$ and $(c^{|\exp(B)|} - 1)^{-1} \bmod \mu$ and then compute the exponentiation in time $\mathcal{O}(\log t)$. With this, if $j \in [t' \cdot |\exp(B)| + 1..(t'+1) \cdot |\exp(B)|]$ we can handle run-length rules signatures $\kappa(\exp(B^t)[1:j])$ as

$$\left(\kappa(\exp(B^{t'})) + \kappa(\exp(B)[1:j-t' \cdot |\exp(B)|]) \cdot c^{t' \cdot |\exp(B)|} \right) \bmod \mu,$$

where $\kappa(\exp(B)[1 : j - t' \cdot |\exp(B)|])$ is obtained by descending in the derivation tree of B. We are saving space by storing our structure at the cost of increasing computation time. As we show later, this time is in fact logarithmic.

A Structure for Karp-Rabin Signatures. We construct a structure over a balanced RLSLP from Theorem 1, using some auxiliary arrays. We define an array $L[A] = |\exp(A)|$ consisting of the length of the expansion of each nonterminal A. For terminals a, we assume $L[a] = 1$. Also, we define arrays K_1 and K_2 such that, for each nonterminal A,

$$K_1[A] = \kappa(\exp(A)),$$
$$K_2[A] = c^{L[A]} \bmod \mu,$$

with the Karp-Rabin fingerprint of the string expanded by A and the last power of c used in the signature multiplied by c, namely the first power needed for signing the second part of the string expanded by A. For terminals a we assume $K_1[a] = a \bmod \mu$ and $K_2[a] = c \bmod \mu$. In addition, for rules $A \to B^t$ we store

$$E[A] = (K_2[B] - 1)^{-1} \bmod \mu.$$

The arrays L, K_j, and E add only $\mathcal{O}(g_{rl})$ extra space. With these auxiliary structures, we can compute fingerprints in $\mathcal{O}(\log n)$ time.

Theorem 2 (cf. [1,5]). *It is possible to construct an index of size $\mathcal{O}(g_{rl})$ supporting Karp-Rabin fingerprints for prefixes of $w[1 : n]$ in $\mathcal{O}(\log n)$ time.*

Proof. Let G be a balanced RLSLP of size $\mathcal{O}(g_{rl})$ constructed as in Theorem 1. We construct arrays L, K_i, and E as shown above. To compute $\kappa(A, j)$, we do as follows:

1. If $j = L[A]$, return $K_1[A]$.
2. If $A \to BC$, then:
   (a) If $j \le L[B]$, return $\kappa(B, j)$.
   (b) If $L[B] < j$, return $\big(K_1[B] + \kappa(C, j - L[B]) \cdot K_2[B]\big) \bmod \mu$.
3. If $A \to B^t$ for $t > 2$, then:
   (a) If $j \le L[B]$, return $\kappa(B, j)$.
   (b) If $j \in [t'L[B] + 1..(t' + 1)L[B]]$ with $1 \le t' < t$, let $e = K_2[B]^{t'}$ and $f = (e - 1) \cdot E[A] \bmod \mu$, then return

$$(K_1[B] \cdot f + \kappa(B, j - t'L[B]) \cdot e) \bmod \mu.$$

Every step of the algorithm takes $\mathcal{O}(1)$ time, so the cost is the depth of the derivation tree of G. The only exception is case 3(b), in which we have an exponentiation. For a non-terminal $A \to B^t$, this exponentiation takes $\mathcal{O}(\log t)$ time, which is $\mathcal{O}(\log(|\exp(A)|/|\exp(B)|))$ time for managing every run-length rule. We show next that $\mathcal{O}(\log(|\exp(A)|/|\exp(B)|))$ telescopes to $\mathcal{O}(\log |\exp(A)|)$, thus we obtain $\mathcal{O}(\log n)$ time for the overall algorithm time.

$$A_0 \rightarrow A_1 A_2$$
$$A_1 \rightarrow A_3 A_4$$
$$A_2 \rightarrow A_4 A_5$$
$$A_3 \rightarrow A_7^3$$
$$A_4 \rightarrow A_7 A_6$$
$$A_5 \rightarrow A_6^3$$
$$A_6 \rightarrow 1$$
$$A_7 \rightarrow 0$$

$$\kappa(A_0, 9) = (K_1[A_1] + \kappa(A_2, 4) \cdot K_2[A_1]) \bmod 3 = 2$$
$$\downarrow$$
$$\kappa(A_2, 4) = (K_1[A_4] + \kappa(A_5, 2) \cdot K_2[A_4]) \bmod 3 = 2$$
$$\downarrow$$
$$\kappa(A_5, 2) = (K_1[A_6] \cdot f + \kappa(A_6, 1) \cdot e) \bmod 3 = 0$$
$$\downarrow$$
$$\kappa(A_6, 1) = K_1[A_6] = 1$$

Fig. 2. Example of a balanced RLSLP for the string $0^4 1 0 1^4$ (left) and fingerprint computation over a length-8 prefix of the string generated by this RLSLP (right), with $c = 2$, $\mu = 3$, $K_1[A_1, A_4, A_6] = [1, 2, 1]$, $K_2[A_1, A_4, A_6] = [2, 1, 2]$, $f = 1$, and $e = 2$.

The telescoping argument is as follows. We prove by induction that the cost $k(A)$ to compute $\kappa(A, j)$ is at most $h(A) + \log |\exp(A)|$, where $h(A)$ is the height of the parse tree of A and j is arbitrary. Then in case 2 we have $k(A) \leq 1 + \max(k(B), k(C))$, which by induction is $\leq 1 + \max(h(B), h(C)) + \log |\exp(A)| = h(A) + \log |\exp(A)|$. In case 3 we have $k(A) \leq 1 + \log(|\exp(A)|/|\exp(B)|) + k(B)$, and since by induction $k(B) \leq h(B) + \log |\exp(B)|$, we obtain $k(A) \leq h(A) + \log |\exp(A)|$. Since G is balanced, this implies $k(A) = \mathcal{O}(\log n)$ when A is the root symbol. □

Figure 2 shows an example of this procedure.

4.2   Range Minimum Queries

A *range minimum query* (RMQ) over a string returns the position of the leftmost occurrence of the minimum within a range. For these type of queries, we can provide an $\mathcal{O}(g_{rl})$ space and $\mathcal{O}(\log n)$ time solution. In the Appendix A we also show how to efficiently compute the related PSV/NSV queries.

Theorem 3. *It is possible to construct an index of size $\mathcal{O}(g_{rl})$ supporting RMQs in $\mathcal{O}(\log n)$ time.*

Proof. Let G be a balanced RLSLP of size $\mathcal{O}(g_{rl})$ constructed as in Theorem 1. We define $\mathrm{rmq}(A, i, j)$ as the pair (a, k) where a is the least symbol in $\exp(A)[i : j]$, and k is the absolute position within $\exp(A)$ of the leftmost occurrence of a in $\exp(A)[i : j]$. Store the values $L[A] = |\exp(A)|$, and $M[A] = \mathrm{rmq}(A, 1, L[A])$, for every variable A, as arrays. These arrays add only $\mathcal{O}(g_{rl})$ extra space. To compute $\mathrm{rmq}(A, i, j)$, do as follows:

1. If $i = 1$ and $j = L[A]$, return $M[A]$.
2. If $A \rightarrow BC$, then:
   (a) If $i, j \leq L[B]$, return $\mathrm{rmq}(B, i, j)$.
   (b) If $i, j > L[B]$, let $(a, k) = \mathrm{rmq}(C, i - L[B], j - L[B])$. Return $(a, L[B] + k)$.

(c) If $i \leq L[B]$ and $L[B] < j$ with $j - i + 1 < L[A]$, let $(a_1, k_1) = \text{rmq}(B, i, L[B])$ and $(a_2, k_2) = \text{rmq}(C, 1, j - L[B])$. Return (a_1, k_1) if $a_1 \leq a_2$, or $(a_2, L[B] + k_2)$ if $a_2 < a_1$.

3. If $A \to B^t$ for $t > 2$, then:

(a) If $i, j \in [t'L[B]+1..(t'+1)L[B]]$, let $(a, k) = \text{rmq}(B, i - t'L[B], j - t'L[B])$. Return $(a, t'L[B] + k)$.

(b) If $i \in [t'L[B]+1..(t'+1)L[B]]$ and $j \in [t''L[B]+1..(t''+1)L[B]]$ for some $t' < t''$. Let $(a_l, k_l) = \text{rmq}(B, i - t'L[B], L[B])$, $(a_r, k_r) = \text{rmq}(B, 1, j - t''L[B])$ and $(a_c, k_c) = M[B]$ (only if $t'' - t' > 1$). Return (a, k), where $a = \min(a_l, a_r, a_c)$, and k is either $t'L[B] + k_l$, $t''L[B] + k_r$, or $(t'+1)L[B] + k_c$ (only if $t'' - t' > 1$), depending on which of these positions correspond to an absolute position of a in $\exp(A)$, and is the leftmost of them.

We analyze the number of recursive calls of the algorithm above. For cases 2(a), 2(b) and 3(a) there is only one recursive call, over a variable which is deeper in the derivation tree of G. In cases 2(c) and 3(b), it could be that two recursive calls occur, but overall, this can happen only one time in the whole run of the algorithm. The reason is that when two recursive calls occur at the same depth, from that point onward, the algorithm will be computing $\text{rmq}(\cdot)$ over suffixes or prefixes of expansions of variables. If we try to compute for example $\text{rmq}(A, i, L[A])$, and A is of the form $A \to BC$, if $i < L[B]$, the call over B is again a suffix call. If $A \to B^t$ for some $t > 2$, and we want to compute $\text{rmq}(A, i, L[A])$, we end with a recursive call over a suffix of B too. Hence, there are only $\mathcal{O}(\log n)$ recursive calls to $\text{rmq}(\cdot)$. The non-recursive step takes constant time, even for run-length rules, so we obtain $\mathcal{O}(\log n)$ time. $\square$

4.3   More General Functions

More generally, we can compute a wide class of functions in $\mathcal{O}(g_{rl})$ space and $\mathcal{O}(\log n)$ applications of the function.

Theorem 4. *Let f be a function from strings to a set of size $n^{\mathcal{O}(1)}$, such that $f(xy) = h(f(x), f(y), |x|, |y|)$ for any strings x and y, where h is a function computable in time $\mathcal{O}(\text{time}(h))$. Let $w[1 : n]$ be a string. It is possible to construct an index to compute $f(w[i : j])$ in $\mathcal{O}(g_{rl})$ space and $\mathcal{O}(\text{time}(h) \cdot \log n)$ time.*

Proof. Let G be a balanced RLSLP of size g_{rl} constructed as in Theorem 1. Store the values $L[A] = |\exp(A)|$ and $F[A] = f(\exp(A))$, for every variable A, as arrays. These arrays add only $\mathcal{O}(g_{rl})$ extra space because the values in F fit in $\mathcal{O}(\log n)$-bit words. To compute $f(A, i, j) = f(\exp(A)[i : j])$, we do as follows:

1. If $i = 1$ and $j = L[A]$, return $F[A]$.
2. If $A \to BC$, then:
   (a) If $i, j \leq L[B]$, return $f(B, i, j)$.
   (b) If $i, j > L[B]$, return $f(C, i - L[B], j - L[B])$.
   (c) If $i \leq L[B] < j$, return

$$h(f(B, i, L[B]), f(C, 1, j - L[B]), L[B] - i + 1, j - L[B]).$$

3. If $A \to B^t$ for $t > 2$, then:
   (a) If $i, j \in [t'L[B]+1..(t'+1)L[B]]$, return $f(B, i - t'L[B], j - t'L[B])$.
   (b) If $i \in [t'L[B]+1..(t'+1)L[B]]$ and $j \in [t''L[B]+1..(t''+1)L[B]]$ for some $t' < t''$, let

$$f_l = f(B, i - t'L[B], L[B])$$
$$f_r = f(B, 1, j - t''L[B])$$
$$f_c(0) = f(\varepsilon)$$
$$f_c(1) = F[B]$$
$$f_c(i) = h(f_c(i/2), f_c(i/2), L[B]^{i/2}, L[B]^{i/2}) \text{ for even } i$$
$$f_c(i) = h(f_c(1), f_c(i-1), L[B], L[B]^{i-1}) \text{ for odd } i$$
$$h_l = h(f_l, f_c(t'' - t' - 1), (t'+1)L[B] - i + 1, (t'' - t' - 1)L[B])$$

then return

$$h(h_l, f_r, (t'+1)L[B] - i + 1 + (t'' - t' - 1)L[B], j - t''L[B] + 1)$$

Just like when computing RMQs in Theorem 3, there is at most one call in the whole algorithm invoking two non-trivial recursive calls. To estimate the cost of each recursive call, the same analysis as for Theorem 2 works, because the expansion of whole nonterminals is handled in constant time as well, and the $\mathcal{O}(\log t)$ cost of the run-length rules telescopes in the same way.

The precise telescoping argument is as follows. We prove by induction that the cost $c(A)$ to compute $f(A, i, L[A])$ or $f(A, 1, j)$ (i.e., the cost of suffix or prefix calls) is at most $\texttt{time}(h) \cdot (h(A) + \log |\exp(A)|)$, where $h(A)$ is the height of the parse tree of A and i, j are arbitrary. Then in case 2 we have $c(A) \le \texttt{time}(h) + \max(c(B), c(C))$, which by induction is at most $\texttt{time}(h) \cdot (1 + \max(h(B), h(C)) + \log |\exp(A)|) = \texttt{time}(h) \cdot (h(A) + \log |\exp(A)|)$. In case 3 we have that the cost is $c(A) \le \texttt{time}(h) \cdot \log(|\exp(A)|/|\exp(B)|) + c(B)$, which by induction yields

$$c(A) \le \texttt{time}(h) \cdot (\log(|\exp(A)|/|\exp(B)|) + h(B) + \log |\exp(B)|))$$
$$\le \texttt{time}(h) \cdot (h(A) + \log |\exp(A)|)$$

In the case that two non-trivial recursive calls are made at some point when computing $f(A_k, i, j)$, this is the unique point in the algorithm where it happens, so we charge only $\texttt{time}(h) \cdot (h(A_k) + \log |\exp(A_k)|)$ to the cost of A_k. Then the total cost of the algorithm starting from A_0 is at most $\texttt{time}(h) \cdot (h(A_0) + \log |\exp(A_0)|)$ plus the cost $\texttt{time}(h) \cdot (h(A_k) + \log |\exp(A_k)|)$ that we did not charge to A_k. This at most doubles the cost, maintaining it within the same order. Because the grammar is balanced, we obtain $\mathcal{O}(\texttt{time}(h) \cdot \log n)$ time. $\square$

5   Conclusion

In this work, we have shown that any RLSLP can be balanced in linear time without increasing it asymptotic size. This allows us to compute several substring range queries like RMQ, PSV/NSV (in the Appendix A), and

Karp-Rabin fingerprints $\mathcal{O}(\log n)$ time within $\mathcal{O}(g_{rl})$ space. More generally, in $\mathcal{O}(g_{rl})$ space we can compute the wide class of substring functions that satisfy $f(xy) = h(f(x), f(y), |x|, |y|)$, in $\mathcal{O}(\log n)$ times the cost of computing h. Our work also simplifies some previously established results like retrieving substrings in $\mathcal{O}(\log n)$ space and within $\mathcal{O}(g_{rl})$ space.

An open challenge is to efficiently count the number of occurrences of a pattern in the string, within $\mathcal{O}(g_{rl})$ space [5, Appendix A].

A   PSV and NSV Queries

Other relevant queries are *previous smaller value* (PSV) and *next smaller value* (NSV) [6,9], defined as follows:

- $\mathtt{psv}(i) = \max(\{j \mid j < i, w[j] < w[i]\} \cup \{0\})$
- $\mathtt{nsv}(i) = \min(\{j \mid j > i, w[j] < w[i]\} \cup \{n+1\})$
- $\mathtt{psv}'(i, d) = \max(\{j \mid j < i, w[j] < d\} \cup \{0\})$
- $\mathtt{nsv}'(i, d) = \min(\{j \mid j > i, w[j] < d\} \cup \{n+1\})$

Note that the first two queries can be computed by accessing $w[i]$ in $\mathcal{O}(\log n)$ time, and then calling one of the latter two queries, respectively. We show that the latter queries can be answered in $\mathcal{O}(g_{rl})$ space and $\mathcal{O}(\log n)$ time.

Theorem 5. *It is possible to construct an index of size $\mathcal{O}(g_{rl})$ supporting PSV and NSV queries in $\mathcal{O}(\log n)$ time.*

Proof. Let G be a balanced RLSLP of size $\mathcal{O}(g_{rl})$ constructed as in Theorem 1. Store the values $L[A] = |\exp(A)|$ and $M[A] = \min(\{\exp(A)[i] \mid i \in [1..L[A]]\})$, for every variable A, as arrays. These arrays add only $\mathcal{O}(g_{rl})$ extra space. To compute $\mathtt{psv}'(A, i, d)$, do as follows:

1. If $i = 1$ or $M[A] \geq d$, return 0.
2. If $A \to a$, return 1.
3. If $A \to BC$, then:
   (a) If $i \leq L[B] + 1$, return $\mathtt{psv}'(B, i, d)$.
   (b) If $L[B] + 1 < i$, let $k = \mathtt{psv}'(C, i - L[B], d)$. If $k > 0$, return $L[B] + k$, otherwise, return $\mathtt{psv}'(B, i, d)$.
4. If $A \to B^t$ for $t > 2$, then:
   (a) If $i \leq L[B] + 1$, return $\mathtt{psv}'(B, i, d)$.
   (b) If $i \in [t'L[B] + 1..(t' + 1)L[B]]$, let $k = \mathtt{psv}'(B, i - t'L[B], d)$. If $k > 0$, return $t'L[B] + k$. Otherwise, return $(t' - 1)L[B] + \mathtt{psv}'(B, i, d)$.
   (c) If $L[A] < i$, return $(t - 1)L[B] + \mathtt{psv}'(B, i, d)$.

The guard in point 1 guarantees that, in the simple case where i is beyond $|\exp(A)|$, at most one recursive call needs more than $\mathcal{O}(1)$ time. In general, we can make two calls in case 3(b), but then the second call (inside B) is of the simple type from there on. The case of run-length rules is similar. Thus, we obtain $\mathcal{O}(\log n)$ time. The query $\mathtt{nsv}'$ is handled similarly.   □

References

1. Bille, P., Gørtz, I.L., Cording, P.H., Sach, B., Vildhøj, H.W., Vind, S.: Fingerprints in compressed strings. J. Comput. Syst. Sci. **86**, 171–180 (2017). https://doi.org/10.1016/j.jcss.2017.01.002, https://www.sciencedirect.com/science/article/pii/S0022000017300028
2. Bille, P., Landau, G.M., Raman, R., Sadakane, K., Satti, S.R., Weimann, O.: Random access to grammar-compressed strings and trees. SIAM J. Comput. **44**(3), 513–539 (2015). https://doi.org/10.1137/130936889
3. Burrows, M., Wheeler, D.: A block-sorting lossless data compression algorithm. Tech. report, DIGITAL SRC RESEARCH REPORT (1994)
4. Charikar, M., et al.: The smallest grammar problem. IEEE Trans. Inf. Theory **51**(7), 2554–2576 (2005)
5. Christiansen, A., Ettienne, M., Kociumaka, T., Navarro, G., Prezza, N.: Optimal-time dictionary-compressed indexes. ACM Trans. Algorithms **17**, 1–39 (2020). https://doi.org/10.1145/3426473
6. Fischer, J., Mäkinen, V., Navarro, G.: Faster entropy-bounded compressed suffix trees. Theoret. Comput. Sci. **410**(51), 5354–5364 (2009)
7. Fischer, J., Heun, V.: Space-efficient preprocessing schemes for range minimum queries on static arrays. SIAM J. Comput. **40**(2), 465–492 (2011). https://doi.org/10.1137/090779759
8. Gagie, T., Navarro, G., Prezza, N.: Optimal-time text indexing in BWT-runs bounded space. In: Proceedings 29th Annual ACM-SIAM Symposium on Discrete Algorithms (SODA), pp. 1459–1477 (2018)
9. Gagie, T., Navarro, G., Prezza, N.: Fully functional suffix trees and optimal text searching in BWT-runs bounded space. J. ACM **67**(1), 1–54 (2020). https://doi.org/10.1145/3375890
10. Ganardi, M., Jeż, A., Lohrey, M.: Balancing straight-line programs. J. ACM **68**(4), 1–40 (2021). https://doi.org/10.1145/3457389
11. Jeż, A.: Approximation of grammar-based compression via recompression. Theoret. Comput. Sci. **592**, 115–134 (2015)
12. Karp, R.M., Rabin, M.O.: Efficient randomized pattern-matching algorithms. IBM J. Res. Dev. **31**(2), 249–260 (1987). https://doi.org/10.1147/rd.312.0249
13. Kempa, D., Kociumaka, T.: Resolution of the burrows-wheeler transform conjecture. Commun. ACM **65**(6), 91–98 (2022). https://doi.org/10.1145/3531445
14. Kempa, D., Prezza, N.: At the roots of dictionary compression: string attractors. In: Proceedings of the 50th Annual ACM SIGACT Symposium on Theory of Computing (2018). https://doi.org/10.1145/3188745.3188814
15. Kini, D., Mathur, U., Viswanathan, M.: Data race detection on compressed traces. In: Proceedings of the 2018 26th ACM Joint Meeting on European Software Engineering Conference and Symposium on the Foundations of Software Engineering, pp. 26–37. ESEC/FSE 2018, Association for Computing Machinery, New York, NY, USA (2018). https://doi.org/10.1145/3236024.3236025
16. Kreft, S., Navarro, G.: Lz77-like compression with fast random access. In: 2010 Data Compression Conference, pp. 239–248 (2010)
17. Larsson, N., Moffat, A.: Offline dictionary-based compression. In: Proceedings DCC 1999 Data Compression Conference (Cat. No. PR00096), pp. 296–305 (1999)
18. Lempel, A., Ziv, J.: On the complexity of finite sequences. IEEE Trans. Inf. Theory **22**(1), 75–81 (1976)

19. Navarro, G.: Indexing highly repetitive string collections, part I: repetitiveness measures. ACM Comput. Surv. **54**(2), article 29 (2021)
20. Nevill-Manning, C.G., Witten, I.H.: Identifying hierarchical structure in sequences: a linear-time algorithm. J. Artif. Intell. Res. **7**(1), 67–82 (1997)
21. Nishimoto, T., Inenaga, S., Bannai, H., Takeda, M.: Fully dynamic data structure for LCE queries in compressed space. In: 41st International Symposium on Mathematical Foundations of Computer Science (MFCS 2016). Leibniz International Proceedings in Informatics (LIPIcs), vol. 58, pp. 72:1–72:15 (2016)
22. Przeworski, M., Hudson, R., Di Rienzo, A.: Adjusting the focus on human variation. Trends Genetics: TIG **16**(7), 296–302 (2000)
23. Rytter, W.: Application of Lempel-Ziv factorization to the approximation of grammar-based compression. Theoret. Comput. Sci. **302**(1), 211–222 (2003)
24. Verbin, E., Yu, W.: Data structure lower bounds on random access to grammar-compressed strings. In: Proceedings 24th Annual Symposium on Combinatorial Pattern Matching (CPM), pp. 247–258 (2013)
25. Zhang, M., Mathur, U., Viswanathan, M.: Checking LTL[F, G, X] on compressed traces in polynomial time. In: Proceedings of the 29th ACM Joint Meeting on European Software Engineering Conference and Symposium on the Foundations of Software Engineering, pp. 131–143. ESEC/FSE 2021, Association for Computing Machinery, New York, NY, USA (2021). https://doi.org/10.1145/3468264.3468557

Substring Complexities on Run-Length Compressed Strings

Akiyoshi Kawamoto[✉] and Tomohiro I[ID]

Kyushu Institute of Technology, 680-4 Kawazu, Iizuka, Fukuoka 820-8502, Japan
kawamoto.akiyoshi256@mail.kyutech.jp, tomohiro@ai.kyutech.ac.jp

Abstract. Let $S_T(k)$ denote the set of distinct substrings of length k in a string T, then its cardinality $|S_T(k)|$ is called the k-th substring complexity of T. Recently, $\delta = \max\{|S_T(k)|/k : k \geq 1\}$ has been shown to be a good compressibility measure of highly-repetitive strings. In this paper, given T of length n in the run-length compressed form of size ρ, we show that δ can be computed in $C_{\mathsf{sort}}(\rho, n)$ time and $O(\rho)$ space, where $C_{\mathsf{sort}}(\rho, n) = O(\min(\rho \lg \lg \rho, \rho \lg_\rho n))$ is the time complexity for sorting ρ integers with $O(\lg n)$ bits each in $O(\rho)$ space in the Word-RAM model with word size $\Omega(\lg n)$.

Keywords: Substring complexity · Compressibility measure · Run-length compression

1 Introduction

Data compression has been one of the central topics in computer science and recently has become even more important, as data continues growing faster than ever before. One data category that is rapidly increasing is *highly-repetitive strings*, which have many long common substrings. Typical examples of highly-repetitive strings are genomic sequences collected from similar species and versioned documents. It is known that highly-repetitive strings can be compressed much smaller than entropy-based compressors by utilizing repetitive-aware data compressions such as LZ76 [14], bidirectional macro scheme [21], grammar compression [12], collage system [11], and run-length compression of Burrows-Wheeler Transform [4]. Moreover, much effort has been devoted to designing efficient algorithms to conduct useful operations on compressed data (without explicitly decompressing it) such as random access and string pattern matching. We refer readers to the comprehensive surveys [16,17] for recent developments in this area.

Since the compressibility of highly-repetitive strings is not captured by the information entropy, how to measure compressibility for highly-repetitive strings is a long standing question. Addressing this problem, Kempa and Prezza [10] proposed a new concept they call *string attractors*; a set of positions Γ is called an attractor of a string T if and only if every substring of T has at least one occurrence containing a position in Γ. They showed that there is a string attractor

D. Arroyuelo and B. Poblete (Eds.): SPIRE 2022, LNCS 13617, pp. 132–143, 2022.
https://doi.org/10.1007/978-3-031-20643-6_10

behind existing repetitive-aware data compressions and the size γ of the smallest attractor lower bounds and well approximates their size. In addition, this and subsequent studies revealed that it is possible to design "universal" compressed data structures built upon any string attractor Γ to support operations such as random access and string pattern matching [10,18,19]. A drawback of string attractors is that computing the size γ of the smallest string attractors is NP-hard [10]. As a substitution of γ, Christiansen et al. [5] proposed another repetitive-aware compressibility measure δ defined as the maximum of normalized substring complexities $\max\{|S_T(k)|/k : k \geq 1\}$, where $S_T(k)$ denotes the set of substrings of length k in T. They showed that $\delta \leq \gamma$ always holds and there is an $O(\delta \lg(n/\delta))$-size data structure for supporting efficient random access and string pattern matching.

Although substring complexities were used in [20] to approximate the number of LZ76 phrases in sublinear time, only recently δ gained an attention as a gold standard of repetitive-aware compressibility measures [5]. Recent studies have shown that δ possesses desirable properties as a compressibility measure. One of the properties is the robustness. For example, we would hope for a compressibility measure to monotonically increase while appending/prepending characters to a string. It was shown that δ has this monotonicity [13] while γ does not [15]. Akagi et al. [1] studied the sensitivity of repetitive measures to edit operations and the results confirmed the robustness of δ.

From an algorithmic perspective, for a string of length n, δ can be computed in $O(n)$ time and space [5]. This contrasts with $\hat{\gamma}$ and the smallest compression size in some compression schemes like bidirectional macro scheme, grammar compression and collage system, which are known to be NP-hard to compute. Exploring time-space tradeoffs, Bernardini et al. [3] presented an algorithm to compute δ with sublinear additional working space. Efficient computation of δ has a practical importance for the $O(\delta \lg(n/\delta))$-size data structure of [5] as once we know δ, we can reduce the main-memory space needed to build the data structure (see Conclusions of [5]).

In this paper, we show that δ can be computed in $C_{\mathsf{sort}}(\rho, n)$ time and $O(\rho)$ space, where $C_{\mathsf{sort}}(\rho, n)$ is the time complexity for sorting ρ integers with $O(\lg n)$ bits each in $O(\rho)$ space in the Word-RAM model with word size $\Omega(\lg n)$. We can easily obtain $C_{\mathsf{sort}}(\rho, n) = O(\rho \lg \rho)$ by comparison sort and $C_{\mathsf{sort}}(\rho, n) = O(\rho \lg_\rho n)$ by radix sort that uses $\Theta(\rho)$-size buckets. Plugging in more advanced sorting algorithms for word-RAM model, $C_{\mathsf{sort}}(\rho, n) = O(\rho \lg \lg \rho)$ [8]. In randomized setting, $C_{\mathsf{sort}}(\rho, n) = O(\rho)$ if $\lg n = \Omega(\lg^{2+\epsilon} \rho)$ for some fixed $\epsilon > 0$ [2], and otherwise $C_{\mathsf{sort}}(\rho, n) = O(\rho\sqrt{\lg \frac{\lg n}{\lg \rho}})$ [9].

2   Preliminaries

Let Σ be a finite *alphabet*. An element of Σ^* is called a *string* over Σ. The length of a string w is denoted by $|w|$. The empty string ε is the string of length 0, that is, $|\varepsilon| = 0$. Let $\Sigma^+ = \Sigma^* - \{\varepsilon\}$. The concatenation of two strings x and y is denoted by $x \cdot y$ or simply xy. When a string w is represented by the concatenation of strings x, y and z (i.e. $w = xyz$), then x, y and z are called a *prefix, substring,* and *suffix* of w, respectively. A substring x of w is called *proper* if $|x| < |w|$.

The i-th character of a string w is denoted by $w[i]$ for $1 \leq i \leq |w|$, and the substring of a string w that begins at position i and ends at position j is denoted by $w[i..j]$ for $1 \leq i \leq j \leq |w|$, i.e., $w[i..j] = w[i]w[i+1]\ldots w[j]$. For convenience, let $w[i..j] = \varepsilon$ if $j < i$. For two strings x and y, let $\mathsf{lcp}(x,y)$ denote the length of the longest common prefix between x and y. For a character c and integer $e \geq 0$, let c^e denote the string consisting of a single character c repeated e times.

A substring $w[i..j]$ is called a *run* of w if and only if it is a maximal repeat of a single character, i.e., $w[i-1] \neq w[i] = w[i+1] = \cdots = w[j-1] = w[j] \neq w[j+1]$ (the border cases $w[i-1] \neq w[i]$ and $w[j] \neq w[j+1]$ are simply ignored if $i = 1$ and $j = |w|$, respectively). We obtain the run-length encoding of a string by representing each run by a pair of character and the number of repeats in $O(1)$ words of space. For example, a string **aabbbaabb** has four runs **aa**, **bbb**, **aa** and **bb**, and its run-length encoding is $(\mathsf{a}, 2), (\mathsf{b}, 3), (\mathsf{a}, 2), (\mathsf{b}, 2)$ or we just write as $\mathsf{a}^2\mathsf{b}^3\mathsf{a}^2\mathsf{b}^2$.

Throughout this paper, we refer to T as a string over Σ of length n with ρ runs and $\delta = \max\{|S_T(k)|/k : k \geq 1\}$. Our assumption on computational model is the Word-RAM model with word size $\Omega(\lg n)$. We assume that every character in Σ is interpreted as an integer with $O(1)$ words or $O(\lg n)$ bits, and the order of two characters in Σ can be determined in constant time.

Let $\mathcal{T}$ denote the trie representing the suffixes of T that start with the run's boundaries in T, which we call the *runs-suffix-trie* of T. Note that $\mathcal{T}$ has at most ρ leaves and at most ρ branching internal nodes, and thus, $\mathcal{T}$ can be represented in $O(\rho)$ space by compacting non-branching internal nodes and representing edge labels by the pointers to the corresponding substrings in the run-length encoded string of T. We call the compacted runs-suffix-trie the *runs-suffix-tree*. In order to work in $O(\rho)$ space, our algorithm actually works on the runs-suffix-tree of T, but our conceptual description will be made on the runs-suffix-trie considering that non-branching internal nodes remain. A node is called *explicit* when we want to emphasize that the node is present in the runs-suffix-tree, too.

Let V denote the set of nodes of $\mathcal{T}$. For any $v \in V$, let $\mathsf{str}(v)$ be the string obtained by concatenating the edge labels from the root to v, and $\mathsf{d}(v) = |\mathsf{str}(v)|$. We say that v represents the string $\mathsf{str}(v)$ and sometimes identify v with $\mathsf{str}(v)$ if it is clear from the context. For any $v \in V$ and $1 \leq k \leq \mathsf{d}(v)$, let $\mathsf{str}_k(v)$ be the suffix of $\mathsf{str}(v)$ of length k. Let $\hat{V}$ denote the set of nodes that do not have a child v with $\mathsf{str}(v) = c^e$ for some character c and integer $e > 0$. Figure 1 shows an example of $\mathcal{T}$ for $T = $ **aabbbaabbaaa**.

3   Connection Between $S_T(k)$ and the Runs-Suffix-Trie

We first recall a basic connection between $S_T(k)$ and the nodes of the *suffix trie* of T that is the trie representing "all" the suffixes of T [7,22]. Since a k-length substring w in T is a k-length prefix of some suffix of T, w can be uniquely

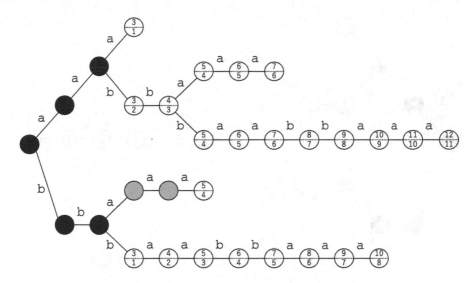

Fig. 1. An example of runs-suffix-trie $\mathcal{T}$ for $T = \mathsf{aabbbaabbaaa}$. The nodes in black $\{\varepsilon, \mathsf{a}, \mathsf{aa}, \mathsf{b}, \mathsf{bb}\}$ are the nodes in $V - \hat{V}$. With the notations introduced in Sect. 3, the nodes in gray bba and bbaa cannot be deepest matching nodes due to bbba and bbbaa, respectively. The other nodes belong to D and two integers in node $v \in D$ represent $\mathsf{d}(v)$ (upper integer) and $|\mathsf{t}(v)|$ (lower integer), where $\mathsf{t}(v)$ is defined in Sect. 3.

associated with the node v representing w by the path from the root to v. Thus, the set $S_T(k)$ of substrings of length k is captured by the nodes of depth k. This connection is the basis of the algorithm presented in [5, Lemma 5.7] to compute δ in $O(n)$ time and space.

We want to establish a similar connection between $S_T(k)$ and the nodes of the runs-suffix-trie $\mathcal{T}$ of T. Since only the suffixes that start with run's boundaries are present in $\mathcal{T}$, there could be a substring w of T that is not represented by a path from the root to some node. Still, we can find an occurrence of any non-empty substring w in a path starting from a node in $V - \hat{V}$: Suppose that $w = T[i..j]$ is an occurrence of w in T and $i' \leq i$ is the starting position of the run containing i, then there is a node v such that $\mathsf{str}(v) = T[i'..j]$ and $\mathsf{str}_{|w|}(v) = w$. Formally, we say that node v is a *matching node* for a string w if and only if $\mathsf{str}(v) = w[1]^e w$ for some integer $e \geq 0$. Note that there could be more than one matching node for a string w, but the *deepest matching node* for w is unique because two distinct matching nodes must have different values of e. The following lemma summarizes the above discussion.

Lemma 1. *For any substring w of T, there is a unique deepest matching node in $\mathcal{T}$.*

The next lemma implies that a deepest matching node for some string is in $\hat{V}$.

Lemma 2. *A node $v \notin \hat{V}$ cannot be a deepest matching node for any string (Fig. 2).*

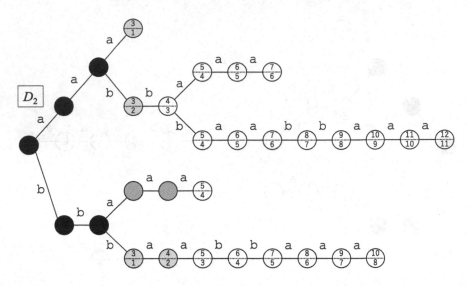

Fig. 2. Displaying the nodes in D_2 in orange filled color. The nodes aaa, aab, bbb and bbba are respectively the deepest matching nodes for aa, ab, bb and ba. (Color figure online)

Proof. Since $v \notin \hat{V}$, there exists a child v' of v such that $\mathsf{str}(v') = c^e$ for some character c and integer $e > 0$. It also implies that $\mathsf{str}(v) = c^{e-1}$. Thus, v' is always a deeper matching node for any suffix of $\mathsf{str}(v)$ and the claim of the lemma follows.

Let D be the set of nodes v such that there is a string for which v is the deepest matching node. Also, let $D_k \subseteq D$ denote the set of nodes v such that there is a string of length k for which v is the deepest matching node. For fixed k, it is obvious that a node v in D_k is the deepest matching node for a unique substring of length k, which is $\mathsf{str}_k(v)$. Together with Lemma 1, there is a bijection between $S_T(k)$ and D_k, which leads to the following lemma.

Lemma 3. *For any $1 \le k \le n$, $|S_T(k)| = |D_k|$.*

By Lemma 3, δ can be computed by $\max\{|D_k|/k : k \ge 1\}$.

Next we study some properties on D, which is used to compute δ efficiently. For any $v \in \hat{V}$, let $\mathsf{s}(v)$ denote the longest proper prefix of the first run of $\mathsf{str}(v)$, and let $\mathsf{t}(v)$ be the string such that $\mathsf{str}(v) = \mathsf{s}(v) \cdot \mathsf{t}(v)$. In other words, $\mathsf{t}(v)$ is the shortest string for which v is a matching node. For example, if $\mathsf{str}(v) = $ aabba then $\mathsf{s}(v) = $ a and $\mathsf{t}(v) = $ abba.

Lemma 4. *For any $v \in \hat{V}$, $\{k : v \in D_k\}$ is $[|\mathsf{t}(v)|..\mathsf{d}(v)]$ or $\emptyset$.*

Proof. First of all, by definition it is clear that v cannot be a matching node for a string of length shorter than $|\mathsf{t}(v)|$ or longer than $\mathsf{d}(v)$. Hence, only the integers in $[|\mathsf{t}(v)|..\mathsf{d}(v)]$ can be in $\{k : v \in D_k\}$, and the claim of the lemma says that $\{k : v \in D_k\}$ either takes them all or nothing.

We prove the lemma by showing that if v is not the deepest matching node for some suffix w of $\mathsf{str}(v)$ with $|\mathsf{t}(v)| \leq |w| \leq \mathsf{d}(v)$, then v is not the deepest matching node for any string. The assumption implies that there is a deeper matching node v' for w. Let $c = \mathsf{t}(v)[1]$, then $\mathsf{str}(v) = c^e \mathsf{t}(v)$ with $e = |\mathsf{s}(v)|$. Note that w is obtained by prepending zero or more c's to $\mathsf{t}(v)$, and hence, $\mathsf{str}(v')$ is written as $c^{e'} w$ for some $e' > e$. Therefore, v and v' are matching nodes for any suffix of $\mathsf{str}(v)$ of length in $[|\mathsf{t}(v)|..\mathsf{d}(v)]$. Since v' is deeper than v, the claim holds.

Lemma 5. *For any $u \in D$, every child v of u is in D.*

Proof. We show a contraposition, i.e., $u \notin D$ if $v \notin D$. Let $c = \mathsf{t}(v)[1]$, then $\mathsf{str}(v) = c^e \mathsf{t}(v)$ with $e = |\mathsf{s}(v)|$. Note that, by Lemma 4, $v \in D$ if and only if v is the deepest matching node for $\mathsf{t}(v)$. We assume that $u \in \hat{V}$ since otherwise $u \notin D$ is clear. Then, it holds that $u \in D$ if and only if u is the deepest matching node for $\mathsf{t}(v)[1..|\mathsf{t}(v)| - 1]$. The assumption of $v \notin D$ then implies that there is a deeper matching node v' for $\mathsf{t}(v)$ such that $\mathsf{str}(v') = c^{e'} \mathsf{t}(v)$ with $e' > e$. Since the parent u' of v' is deeper than u and a matching node for $\mathsf{t}(v)[1..|\mathsf{t}(v)| - 1]$, we conclude that $u \notin D$.

By Lemma 5, we can identify some deepest matching node v such that all the ancestors of v are not in D and all the descendants of v are in D. We call such a node v a *DMN-root*, and let D_{root} denote the set of DMN-roots. In Sect. 4 we will show how to compute D_{root} in $C_{\mathsf{sort}}(\rho, n)$ time and $O(\rho)$ space.

Figure 3 shows how D_k changes when we increase k.

Now we focus on the difference $\alpha_k = |D_k| - |D_{k-1}|$ between $|D_k|$ and $|D_{k-1}|$, and show that we can partition $[1..n]$ into $O(\rho)$ intervals so that α_k remains the same within each interval.

Lemma 6. $|\{k : \alpha_{k+1} \neq \alpha_k\}| = O(\rho)$.

Proof. We prove the lemma by showing that $|\{k : |D_{k+1} - D_k| \neq |D_k - D_{k-1}|\}| = O(\rho)$ and $|\{k : |D_k - D_{k+1}| \neq |D_{k-1} - D_k|\}| = O(\rho)$.

It follows from Lemma 4 that a node v in $(D_{k+1} - D_k)$ satisfies $k+1 = |\mathsf{t}(v)|$. If the parent u of v is in D, then $u \in (D_k - D_{k-1})$ since $k = |\mathsf{t}(u)|$. In the opposite direction, every child of $u \in (D_k - D_{k-1})$ is in $(D_{k+1} - D_k)$ by Lemma 5. Therefore $|D_{k+1} - D_k|$ and $|D_k - D_{k-1}|$ can differ only if one of the following conditions holds:

1. there is a node $v \in (D_{k+1} - D_k)$ whose parent is not in D, which means that $v \in D_{\mathsf{root}}$;
2. there is a node $u \in (D_k - D_{k-1})$ that has either no children or more than one child, which means that u is an explicit node.

Note that each node v in D_{root} contributes to the first case when $k + 1 = |\mathsf{t}(v)|$ and each explicit node u can contribute to the second case when $k = |\mathsf{t}(u)|$. Hence $|\{k : |D_{k+1} - D_k| \neq |D_k - D_{k-1}|\}| = O(\rho)$.

It follows from Lemma 4 that a node v in $(D_k - D_{k+1})$ satisfies $k = \mathsf{d}(v)$. If the parent u of v is in D, then $u \in (D_{k-1} - D_k)$ since $k-1 = \mathsf{d}(u)$. In the opposite

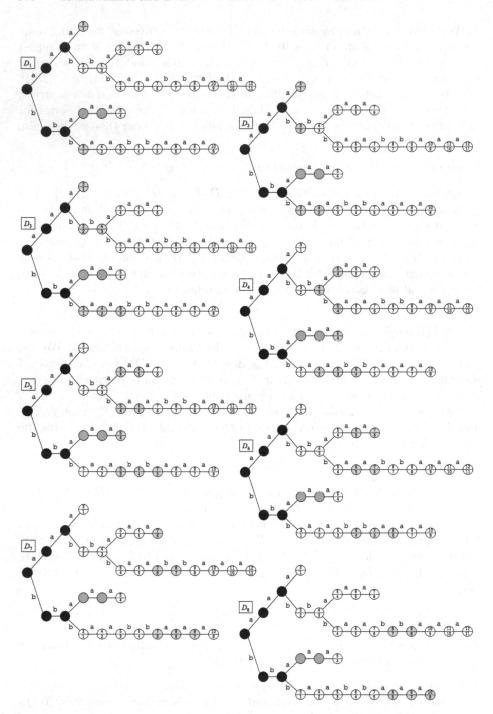

Fig. 3. Displaying the nodes in D_k for $k = 1, 2, \ldots, 8$ in orange filled color. Two integers in node $v \in D$ represent $\mathsf{d}(v)$ (upper integer) and $|\mathsf{t}(v)|$ (lower integer). By Lemma 4, node $v \in D$ belongs to D_k for any $k \in [|\mathsf{t}(v)|..\mathsf{d}(v)]$. (Color figure online)

direction, every child of $u \in (D_{k-1} - D_k)$ is in $(D_k - D_{k+1})$ by Lemmas 5 and 4. Therefore $|D_k - D_{k+1}|$ and $|D_{k-1} - D_k|$ can differ only if one of the following conditions holds:

1. there is a node $v \in (D_k - D_{k+1})$ whose parent is not in D, which means that $v \in D_{\text{root}}$;
2. there is a node $u \in (D_{k-1} - D_k)$ that has either no children or more than one child, which means that u is an explicit node.

Note that each node v in D_{root} contributes to the first case when $k = \mathsf{d}(v)$ and each explicit node u can contribute to the second case when $k - 1 = \mathsf{d}(u)$. Hence $|\{k : |D_k - D_{k+1}| \neq |D_{k-1} - D_k|\}| = O(\rho)$.

Putting all together, $|\{k : \alpha_{k+1} \neq \alpha_k\}| = O(\rho)$.

The next lemma will be used in our algorithm presented in Sect. 4.

Lemma 7. *Assume that $\alpha_k = \alpha$ for any $k \in (k'..k'']$. Then, $|S_T(k)|/k$, in the range $k \in [k'..k'']$, is maximized at k' or k''.*

Proof. For any $k \in [k'..k'']$ with $k = k' + x$, $|S_T(k)|/k$ can be represented by

$$f(x) = \frac{|S_T(k')| + \alpha x}{(k' + x)}.$$

By differentiating $f(x)$ with respect to x, we get

$$f'(x) = \frac{\alpha(k' + x) - (|S_T(k')| + \alpha x)}{(k' + x)^2} = \frac{\alpha k' - |S_T(k')|}{(k' + x)^2}.$$

Since $f(x)$ is monotonically increasing if $\alpha k' - |S_T(k')| > 0$, and otherwise monotonically non-increasing, $|S_T(k)|/k$, in the range $[k'..k'']$, is maximized at k' or k''.

4 Algorithm

Based on the properties of D established in Sect. 3, we present an algorithm, given ρ-size run-length compressed string T, to compute δ in $C_{\text{sort}}(\rho, n)$ time and $O(\rho)$ space.

We first build the runs-suffix-tree of T.

Lemma 8. *Given a string T in run-length compressed form of size ρ, the runs-suffix-tree of T of size $O(\rho)$ can be computed in $C_{\text{sort}}(\rho, n)$ time and $O(\rho)$ space.*

Proof. Let w be a string of length ρ that is obtained by replacing each run of T with a meta-character in $[1..\rho]$, where the meta-character of a run c^e is determined by the rank of the run sorted over all runs in T using the sorting key of the pair (c, e) represented in $O(\lg n)$ bits. Since (c, e) is represented in $O(\lg n)$ bits, we can compute w in $C_{\text{sort}}(\rho, n)$ time and $O(\rho)$ space. Then we build the suffix tree of w in $O(\rho)$ time and space using any existing linear-time algorithm for building suffix trees over integer alphabets (use e.g. [6]). Since runs with the same character but different exponents have different meta-characters, we may need to merge some prefixes of edges outgoing from a node. Fixing this in $O(\rho)$ time, we get the runs-suffix-tree of T.

Note that we can reduce $O(\lg n)$-bit characters to $O(\lg \rho)$-bit characters during the process of sorting in the proof of Lemma 8. From now on we assume that a character is represented by an integer in $[1..\rho]$. In particular, pointers to some data structures associated with each character can be easily maintained in $O(\rho)$ space.

We augment the runs-suffix-tree in $O(\rho)$ time and space so that we can support *longest common prefix queries* that compute the lcp value for any pair of suffixes starting with run's boundary in constant time. We can implement this with a standard technique that employs lowest common ancestor queries over suffix trees (see e.g. [7]), namely, we just compute the string depth of the lowest common ancestor of two leaves corresponding to the suffixes.

Using this augmented runs-suffix-tree, we can compute the set D_{root} of DMN-roots in $O(\rho)$ time.

Lemma 9. D_{root} *can be computed in* $O(\rho)$ *time.*

Proof. For each leaf l, we compute, in the root-to-leaf path, the deepest node $\mathsf{b}(l)$ satisfying that there exists v such that $\mathsf{t}(\mathsf{b}(l)) = \mathsf{t}(v)$ and $\mathsf{s}(\mathsf{b}(l))$ is a proper prefix of $\mathsf{s}(v)$. If $\mathsf{b}(l)$ is not a leaf, the child of $\mathsf{b}(l)$ (along the path) is the DMN-root on the path. Since $\mathsf{str}(\mathsf{b}(l))$ and $\mathsf{str}(v)$ are the same if we remove their first runs, we can compute the longest one by using lcp queries on the pair of leaves having the same character in the previous run. Let $\mathsf{rem}(l)$ denote the string that can be obtained by removing the first run from $\mathsf{str}(l)$.

For any character c, let L_c be the doubly linked list of the leaves starting with the same character c and sorted in the lexicographic order of $\mathsf{rem}(\cdot)$ (we remark that it is not the lex. order of $\mathsf{str}(\cdot)$). Such lists for "all" characters c can be computed in a batch in $O(\rho)$ time by scanning all leaves in the lexicographic order. Suppose that we arrive at a leaf l' and there is a leaf l with $\mathsf{str}(l) = c^e \mathsf{str}(l')$ for some character c and integer $e > 0$, then we append l to L_c.

Now we focus on the leaves that start with the same character c. Given L_c, Algorithm 1 computes $|\mathsf{b}(\cdot)|$'s in increasing order of the exponents of their first runs. When we process a leaf l, every leaf with a shorter first run than l is removed from the list so that we can efficiently find two lexicographically closest leaves (in terms of $\mathsf{rem}(\cdot)$ lex. order) with longer first runs. Let e be the exponent of the first run of $\mathsf{str}(l)$. By a linear search to lex. smaller (resp. larger) direction from l in the current list, we can find the lex. predecessor p (resp. the lex. successor s) that have longer first run than e (just ignore it if such a leaf does not exist or set sentinels at both ends of the list). Note that the exponent of the first run of l' is e for any leaf l' in between p and s. Then, we compute $|\mathsf{b}(l')| = e + \max(\mathsf{lcp}(\mathsf{rem}(l'), \mathsf{rem}(p)), \mathsf{lcp}(\mathsf{rem}(l'), \mathsf{rem}(s)))$ and remove l' from the list. We can process all the leaves in L_c in linear time since any leaf l in L_c is visited once and removed from the list after we compute $|\mathsf{b}(l)|$ by two lcp queries.

Algorithm 1. How to compute $|\mathsf{b}(\cdot)|$ for the leaves in L_c.

Input. Doubly linked list L_c of the leaves starting with character c and sorted in the lexicographic order of $\mathsf{rem}(\cdot)$.

Output. $|\mathsf{b}(\cdot)|$ for the leaves in L_c.

1 $L \leftarrow L_c$;   /* initialize tentative list L by L_c */
2 **foreach** l in L_c *in increasing order of the exponents of their first runs* **do**
3   **if** l *is removed from* L **then** continue ;
4      $e \leftarrow$ the exponent of the first run of l;
5   compute predecessor p of l (in $\mathsf{rem}(\cdot)$ lex. order) that has longer first run than e;
6   compute successor s of l (in $\mathsf{rem}(\cdot)$ lex. order) that has longer first run than e;
7   **foreach** l' *existing in between* p *and* s *in* L **do**
8   output $|\mathsf{b}(l')| = e + \max(\mathsf{lcp}(\mathsf{rem}(l'), \mathsf{rem}(p)), \mathsf{lcp}(\mathsf{rem}(l'), \mathsf{rem}(s)))$ for l';
9   remove l' from L;

After computing $|\mathsf{b}(\cdot)|$ for all leaves in $O(\rho)$ total time, it is easy to locate the nodes of D_{root} by traversing the runs-suffix-tree in $O(\rho)$ time.

We finally come to our main contribution of this paper.

Theorem 1. *Given ρ-size run-length compressed string T of length n, we can compute δ in $C_{\mathsf{sort}}(\rho, n)$ time and $O(\rho)$ space.*

Proof. Thanks to Lemma 7, in order to compute δ it suffices to take the maximum of $|S_T(k)|/k$ for k at which α_k and α_{k+1} differ. According to the proof of Lemma 6, we obtain $\{k : \alpha_{k+1} \neq \alpha_k\}$ by computing the nodes contributing the changes of α_k, which are the DMN-roots and explicit nodes. As shown in Lemmas 8 and 9, DMN-roots can be computed in $C_{\mathsf{sort}}(\rho, n)$ time and $O(\rho)$ space. Note that the contributions of $v \in D_{\mathsf{root}}$ to $\alpha_{k+1} - \alpha_k$ are 1 at $k = |\mathsf{t}(v)| - 1$ and -1 at $k = \mathsf{d}(v)$, and the contributions of an explicit node u in D to $\alpha_{k+1} - \alpha_k$ are $h-1$ at $k = |\mathsf{t}(u)|$ and $1-h$ at $k = \mathsf{d}(u)+1$, where h is the number of children of u. We list the information by the pairs $(|\mathsf{t}(v)| - 1, 1), (\mathsf{d}(v), -1)\}, (|\mathsf{t}(u)|, h - 1)$ and $(\mathsf{d}(u)+1, 1-h)$ and sort them in increasing order of the first element in $C_{\mathsf{sort}}(\rho, n)$ time. We obtain $\{k : \alpha_{k+1} \neq \alpha_k\}$ by the set of the first elements, and we can compute $\alpha_{k+1} - \alpha_k$ by summing up the second elements for fixed k. By going through the sorted list, we can keep track of $|S_T(k')|$ for $k' \in \{k : \alpha_{k+1} \neq \alpha_k\}$, and thus, compute $\delta = \max\{|S_T(k')|/k' : k' \in \{k : \alpha_{k+1} \neq \alpha_k\}\}$ in $O(\rho)$ time and space.

Acknowledgements. This work was supported by JSPS KAKENHI (Grant Number 22K11907).

References

1. Akagi, T., Funakoshi, M., Inenaga, S. Sensitivity of string compressors and repetitiveness measures (2021). arXiv:2107.08615
2. Andersson, A., Hagerup, T., Nilsson, S., Raman, R.: Sorting in linear time? J. Comput. Syst. Sci. **57**(1), 74–93 (1998)
3. Bernardini, G., Fici, G., Gawrychowski, P., Pissis, S.P.: Substring complexity in sublinear space (2020). arXiv:2007.08357
4. Burrows, M., Wheeler, D.: A block-sorting lossless data compression algorithm. Technical report, HP Labs (1994)
5. Christiansen, A. R., Ettienne, M. B., Kociumaka, T., Navarro, G., Prezza, N.: Optimal-time dictionary-compressed indexes. ACM Trans. Algorithms, **17**(1):8:1–8:39 (2021). https://doi.org/10.1145/3426473
6. Farach, M.: Optimal suffix tree construction with large alphabets. In: 38th Annual Symposium on Foundations of Computer Science, FOCS 1997, Miami Beach, Florida, USA, 19–22 October 1997, pp. 137–143 (1997)
7. Gusfield, D.: Algorithms on Strings, Trees, and Sequences. Cambridge University Press, New York (1997)
8. Han, Y.: Deterministic sorting in o(nloglogn) time and linear space. J. Algorithms **50**(1), 96–105 (2004)
9. Han, Y., Thorup, M.: Integer sorting in 0(n sqrt (log log n)) expected time and linear space. In Proceedings of 43rd Symposium on Foundations of Computer Science (FOCS) 2002. IEEE Computer Society, pp. 135–144 (2002)
10. Kempa, D., Prezza, N.: At the roots of dictionary compression: string attractors. In: Proceedings of the 50th Annual ACM SIGACT Symposium on Theory of Computing (STOC) 2018, pp. 827–840 (2018)
11. Kida, T., Matsumoto, T., Shibata, Y., Takeda, M., Shinohara, A., Arikawa, S.: Collage system: a unifying framework for compressed pattern matching. Theor. Comput. Sci. **1**(298), 253–272 (2003)
12. Kieffer, J.C., Yang, E.-H.: Grammar-based codes: a new class of universal lossless source codes. IEEE Trans. Inf. Theory **46**(3), 737–754 (2000)
13. Kociumaka, T., Navarro, G., Prezza, N.: Towards a definitive measure of repetitiveness. In: Kohayakawa, Y., Miyazawa, F.K. (eds.) LATIN 2021. LNCS, vol. 12118, pp. 207–219. Springer, Cham (2020). https://doi.org/10.1007/978-3-030-61792-9_17
14. Lempel, A., Ziv, J.: On the complexity of finite sequences. IEEE Trans. Inf. Theory **22**(1), 75–81 (1976)
15. Mantaci, S., Restivo, A., Romana, G., Rosone, G., Sciortino, M.: A combinatorial view on string attractors. Theor. Comput. Sci. **850**, 236–248 (2021)
16. Navarro, G.: Indexing highly repetitive string collections, part I: repetitiveness measures. ACM Comput. Surv. **54**(2):29:1–29:31 (2021)
17. Navarro, G.: Indexing highly repetitive string collections, part II: compressed indexes. ACM Comput. Surv. **54**(2):26:1–26:32 (2021)
18. Navarro, G., Prezza, N.: Universal compressed text indexing. Theor. Comput. Sci. **762**, 41–50 (2019)
19. Prezza, N.: Optimal rank and select queries on dictionary-compressed text. In: Pisanti, N., Pissis, S.P., (eds.) Proceedings of 30th Annual Symposium on Combinatorial Pattern Matching (CPM) 2019, vol. 128 of LIPIcs, pp. 4:1–4:12. Schloss Dagstuhl - Leibniz-Zentrum für Informatik (2019)

20. Raskhodnikova, S., Ron, D., Rubinfeld, R., Smith, A.D.: Sublinear algorithms for approximating string compressibility. Algorithmica **65**(3), 685–709 (2013)
21. Storer, J.A., Szymanski, T.G.: Data compression via textural substitution. J. ACM **29**(4), 928–951 (1982)
22. Weiner, P.: Linear pattern-matching algorithms. In: Proceedings of 14th IEEE Annual Symposium on Switching and Automata Theory, pp. 1–11 (1973)

Information Retrieval

How Train–Test Leakage Affects Zero-Shot Retrieval

Maik Fröbe[1](✉), Christopher Akiki[2], Martin Potthast[2], and Matthias Hagen[1]

[1] Martin-Luther-Universität Halle-Wittenberg, Wittenberg, Germany
maik.froebe@informatik.uni-halle.de
[2] Leipzig University, Leipzig, Germany

Abstract. Neural retrieval models are often trained on (subsets of) the millions of queries of the MS MARCO/ORCAS datasets and then tested on the 250 Robust04 queries or other TREC benchmarks with often only 50 queries. In such setups, many of the few test queries can be very similar to queries from the huge training data—in fact, 69% of the Robust04 queries have near-duplicates in MS MARCO/ORCAS. We investigate the impact of this unintended train–test leakage by training neural retrieval models on combinations of a fixed number of MS MARCO/ORCAS queries, which are very similar to actual test queries, and an increasing number of other queries. We find that leakage can improve effectiveness and even change the ranking of systems. However, these effects diminish the smaller, and thus more realistic, the extent of leakage is in all training instances.

Keywords: Neural information retrieval · Train–test leakage · BERT · T5

1 Introduction

Training transformer-based retrieval models requires large amounts of data unavailable in many traditional retrieval benchmarks [34]. Data-hungry training regimes became possible with the 2019 release of MS MARCO [10] and its 367,013 queries that were subsequently enriched by the ORCAS click log [8] with 10 million queries. Fine-tuning models trained on MS MARCO to other benchmarks or using them without fine-tuning in zero-shot scenarios is often very effective [34,36,47]. For example, monoT5 [36] which was trained on MS MARCO data only is the most effective model for the Robust04 document ranking task at the time of writing.[1] Furthermore, the reference implementations of monoT5 and monoBERT [37] in retrieval frameworks such as PyTerrier [32] or Pyserini/PyGaggle [26] all use models trained only on MS MARCO by default. However, when MS MARCO was officially split into train and test data, cross-benchmark use was not anticipated, so that MS MARCO's training queries may overlap with the test queries of other datasets (e.g., Robust04). We investigate the impact of

---

[1] https://paperswithcode.com/sota/ad-hoc-information-retrieval-on-trec-robust04.

D. Arroyuelo and B. Poblete (Eds.): SPIRE 2022, LNCS 13617, pp. 147–161, 2022.
https://doi.org/10.1007/978-3-031-20643-6_11

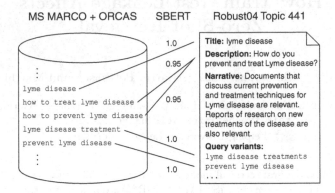

Fig. 1. MS MARCO/ORCAS queries with high Sentence-BERT (SBERT) similarity to Robust04 Topic 441.

such a train–test leakage by training neural models on MS MARCO document ranking data with different proportions of controlled leakage to Robust04 and the TREC 2017 and 2018 Common Core tracks as test datasets.

To identify probably leaking queries, we run a semantic nearest-neighbor search using Sentence-BERT [38] and compare each MS MARCO/ORCAS query to the title, description, and manual query variants [3,4] of the topics in Robust04 and the TREC 2017 and 2018 Common Core tracks. Figure 1 illustrates this procedure for Topic 441 (lyme disease) from Robust04. Our manual review of the leakage candidates shows that 69% to 76% of the topics have near-duplicates in MS MARCO/ORCAS. To analyze the effect of this potential train–test leakage on neural retrieval models, we create three types of training datasets per test corpus, in variants with 1,000 to 128,000 training instances (query + (non-)relevant document): (1) a fixed number of instances derived from test queries from the test corpora (1000 for Robust04 and 200 for each of the two Common Core tracks), augmented by other random non-leaking MS MARCO/ORCAS instances to simulate an upper bound on train–test leakage effects, (2) a fixed number of leaking MS MARCO/ORCAS instances (1000 for Robust04 and 200 each for the two Common Core tracks) supplemented by other random non-leaking instances, and (3) random MS MARCO/ORCAS instances, ensuring that no train–test leakage candidates are included.

In our experiments, we observe leakage-induced improvements in effectiveness for Robust04 and the two Common Core tracks, which can even change the ranking of systems. However, the average improvements in overall effectiveness are often not significant and decrease as the proportion of leakage in the training data becomes smaller and more representative of realistic training scenarios. However, given the swaps in system rankings as well as leakage effects on search results that we observed, we do advise caution: In any case, a rigorous experimental setup demands for maximizing its reliability, so that train–test leaks should still be avoided.[2]

---

[2] All code and data is publicly available at https://github.com/webis-de/SPIRE-22.

2 Background and Related Work

Disjoint training, validation, and test datasets are essential to properly evaluate the effectiveness of machine learning models [7]. Duplication between training and test data can lead to incorrectly high "effectiveness" by memorizing instances rather than learning the target concept. In practice, however, the training and test data often still contain redundancies. For text data, paraphrases, synonyms, etc., can be especially problematic, resulting in train–test leaks [19,24,29]. For instance, the training and test sets of the ELI5 dataset [13] for question answering were created using TF-IDF as a heuristic to eliminate redundancies between them. This proved insufficient as 81% of the test questions turned out to be paraphrases of training questions, which clearly favored models that memorized the training data [24]. Recently, Zhan et al. [46] found that 79% of the TREC 2019 Deep Learning track topics have similar or duplicated queries in the training data and proposed new data splits to evaluate the interpolation and extrapolation effectiveness of models. However, not all types of train–test leaks are unintentional. The TREC 2017 and 2018 Common Core tracks [1] intentionally reused topics from Robust04 to allow participants to use the relevance judgments for training. Indeed, approaches trained on the Robust04 judgments were more effective than others [1]. In this paper, we study whether a similar effect can be observed for unintentional leakage from the large MS MARCO and ORCAS datasets.

Training retrieval models on MS MARCO and applying them to another corpus is a form of transfer learning [34]. Transfer learning is susceptible to train–test leakage since the train and test data are often generated independently without precautions to prevent leaks [6]. Research on leakage in transfer learning focuses on membership inference [35,41] (predicting if a model has seen an instance during training) and property inference [2,17] (predicting properties of the training data). Both inferences rely on the observation that neural models may memorize some training instances to generalize through interpolation [5,7] and to similar test instances [15,16]. It is unclear whether and how neural retrieval models in a transfer learning scenario are affected by leakage. Memorized relevant instances might reduce effectiveness for different test queries while improving it for similar queries, like the examples in Fig. 1. We take the first steps to investigate the effects of such a train–test leakage.

When the target corpus contains only few training instances, transferred retrieval models are often more effective without fine-tuning, in a zero-shot setting [47]; for instance, when training on MS MARCO and testing on TREC datasets [34,36,47]. A frequently used target TREC dataset is Robust04 [42] with 250 topics and a collection of 528,155 documents published between 1989 and 1996 by the Financial Times, the Federal Register, the Foreign Broadcast Information Service, and the LA Times.[3] Later, the TREC Common Core track 2017 [1] reused 50 of the 250 Robust04 topics on the New York Times Anno-

---

[3] https://trec.nist.gov/data/cd45/index.html.

tated Corpus [39][4] (1,864,661 documents published between 1987 and 2007) and the Common Core track 2018 reused another 25 Robust04 topics (and introduced 25 new topics) on the Washington Post Corpus[5] (595,037 documents published between 2012 and 2017). At the Robust04 track, 311,410 relevance judgments were collected, 30,030 at the TREC 2017 Common Core track, and 26,233 at the TREC 2018 Common Core track. Interestingly, every Robust04 topic and every topic from the Common Core tracks 2017 and 2018 was augmented with at least eight query variants compiled by expert searchers, and made available as an additional resource [3,4].

Research on paraphrase detection [12,43] and semantic question matching [40] is of great relevance to the identification of potentially leaking queries between training and test data. Reimers and Gurevych [38] and Lin et al. [28] showed that pooling or averaging the output of contextual word embeddings of pre-trained transformer encoders like BERT [11] is not suited for paraphrase detection—both, with respect to efficiency and accuracy. Sentence-BERT [38] solves this issue by adopting a BERT-based triplet network structure and a contrastive loss that attempts to learn a global and a local structure suited for detecting semantically related sentences. We therefore use Sentence-BERT in a version specifically trained for paraphrase detection to identify leaking queries.

3   Identifying Leaking Queries

To examine the impact of possible leaks from MS MARCO/ORCAS to the TREC datasets Robust04 and Common Core 2017 and 2018, we compare the former's queries (367,013 plus 10 million) to the 275 topics of the latter three. Since lexical similarity may not be sufficient, as indicated by the ELI5 issue [24] mentioned above, we compute semantic similarity scores using Sentence-BERT [38].[6] We store the Sentence-BERT embeddings of all MS MARCO and ORCAS queries in two Faiss indexes [23] and query them for the 100 nearest neighbors (exact; cosine similarity) of each topic's title, description, and query variants.

To determine the threshold for the Sentence-BERT similarity score beyond which we consider a query a source of leakage for a topic, one human annotator assessed whether a query is leaking for a TREC topic (title, description, query variants) for a stratified sample of 100 pairs of queries and topics with a similarity above 0.8. Against these manual judgments, a similarity threshold of 0.91 is the lowest that yields a 0.9 precision for deciding that a query is leaking for a topic. Table 1 shows the number of topics for which queries above this threshold can be found. From MS MARCO and ORCAS combined, 3,960 queries are leakage candidates for one of 181 Robust04 topics (72% of the 250 topics). From the two Common Core tracks, 37 and 38 topics have leakage candidates (76% of the 50 topics, respectively)—high similarities mostly against the query variants.

---

[4] https://catalog.ldc.upenn.edu/LDC2008T19.

[5] https://trec.nist.gov/data/wapost/.

[6] Of the available pre-trained Sentence-BERT models, we use the paraphrase detection model: https://huggingface.co/sentence-transformers/paraphrase-MiniLM-L6-v2.

Table 1. Number of topics (T) in Robust04 and the TREC 2017 and 2018 Common Core tracks for which similar queries (number as Q) in MS MARCO (MSM) and the union of MSM and ORCAS (+ORC) exist in terms of the query having a Sentence-BERT score > 0.91 against the topic's title, description, or a query variant.

Candidates	Robust04				Core 2017				Core 2018			
	MSM		+ ORC		MSM		+ ORC		MSM		+ ORC	
	T	Q	T	Q	T	Q	T	Q	T	Q	T	Q
Title	33	83	140	1,775	2	12	23	176	2	2	21	110
Description	2	3	8	50	0	0	0	0	0	0	1	2
Variants	45	116	167	3,356	6	16	38	602	9	26	38	973
Union	53	151	181	3,960	7	18	38	645	9	26	38	973

Some of these leakage candidates still are false positives (threshold precision of 0.9). To only use actual leaking queries in our train–test leakage experiments, two annotators manually reviewed the 5 most similar candidates per topic above the 0.91 threshold (a total of 827 candidates; not all topics had 5 candidates). Given the title, description, and narrative of a topic, the annotators labeled the similarity of a query to the topic title according to Jansen et al.'s reformulation types [22]: a query can be *identical* to the topic title (differences only in inflection or word order), be a *generalization* (subset of words), a *specialization* (superset of words), a *reformulation* (some synonymous terms), or it can be on a *different topic*. An initial kappa test on 103 random of the 827 candidates showed moderate agreement (Cohen's kappa of 0.59; 103 queries: we aimed for 100 but included all queries for a topic when one was selected). After discussing the 103 cases with the annotators, they agreed on all cases and we had them each independently label half of the remaining 724 candidates. Table 2 shows the annotation results: 172 topics of Robust04 (i.e., 69%) have manually verified leaking queries (648 total), as well as 37 topics of Common Core 2017 (74%) and 38 of Common Core 2018 (76%). A large portion of the true-positive leaking queries are identical to or specializations of a topic's title (57.5% of 721). In our below train–test leakage experiments, we only use manually verified true-positive leaking queries as the source of leakage from MS MARCO/ORCAS.

4   Experimental Analysis

Focusing on zero-shot settings, we train neural retrieval models on specifically designed datasets to assess the effect of train–test leakage from MS MARCO/ORCAS to Robust04 and TREC 2017 and 2018 Common Core. We analyze the models' effectiveness in five-fold cross-validation experiments, report detailed results for varying training set sizes for monoT5 (which has the highest effectiveness in our experiments), and study the effects of leaked instances on the retrieval scores and the resulting rankings.

Table 2. Statistics of the 827 manually annotated leakage candidate queries. (a) Number of true and false candidates. (b) Annotated query reformulation types.

(a) Manually annotated candidates.

Corpus	Candidates	Queries	Topics
Robust04	true	648	172
	false	93	53
Core 2017	true	138	37
	false	21	11
Core 2018	true	157	38
	false	19	7

(b) Reformulation types.

Type	Queries
Identical	187
Generalization	124
Specialization	228
Reformulation	182
Different Topic	106

Training Datasets. For each of the three test scenarios (Robust04 and the two Common Core scenarios), we construct three types of training datasets: (1) 'No Leakage' with random non-leaking queries (balanced between MS MARCO and ORCAS as in previous experiments [8]), (2) 'MSM Leakage' with a fixed number of random manually verified leaking queries from MS MARCO/ORCAS (500 queries for Robust04, 100 queries for Common Core) supplemented by no-leakage queries until a desired size is reached, and (3) 'Test Leakage' with a fixed number of queries from the actual test data (500 for Robust04, 100 for Common Core; oversampling: each topic twice (but different documents) to match 'MSM Leakage') supplemented by no-leakage queries until a desired size is reached. 'Test Leakage' is meant as an "upper bound" for any train–test leakage effect.

For each type, we construct datasets with 1,000 to 128,000 instances (500 to 64,000 queries; each with one relevant and one non-relevant document). Since MS MARCO/ORCAS queries only have annotated relevant documents, we follow Nogueira et al. [36] and sample "non-relevant" instances from the top-100 BM25 results for such queries. For the 'Test Leakage' scenario, we use the actual TREC judgments to sample the non-/relevant instances. In our 72 training datasets (3 test scenarios, 3 types, 8 sizes), the number of leaked instances is held constant to analyze the effect of a decreasing (and thus more realistic) ratio of leakage. Larger training data would result in more costly training, but our chosen sizes already suffice to observe a diminishing effect of train–test leakage.

Trained Models. For our experimental analyses, we use models based on mono-BERT [37] and monoT5 [36] as implemented in PyGaggle [26], and models based on Duet [33], KNRM [44], and PACRR [21] as implemented in Capreolus [45] (default configurations each). In pilot experiments with 32,000 'No Leakage' instances, these models had higher nDCG@10 scores on Robust04 than CEDR [30], HINT [14], PARADE [25], and TK [20]. Following Nogueira et al. [36], we use the base versions of BERT and T5 to spend the computational

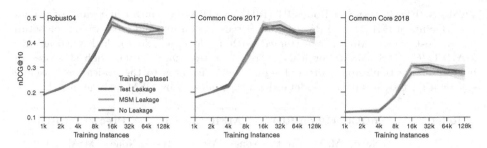

Fig. 2. Effectiveness of monoT5 measured as nDCG@10 on the topics with leakage (172 topics for Robust04, 37 and 38 for the 2017 and 2018 editions of the Common Core track). Models trained on datasets of varying size with no leakage (No), leakage from MS MARCO/ORCAS (MSM), or leakage from the test data (Test).

budget on training many models instead of a single large one. Since the training is not deterministic, each model is trained on each of the 72 training sets five times for one epoch with varying seeds (used to shuffle the training queries; configured in PyTorch). We use ir_datasets [31] for data wrangling and, following previously suggested training regimes [36,37,45], pass the relevant and the non-relevant document of a query consecutively to a model in the same batch during training. During inference, all models re-rank the top-100 BM25 results (Capreolus, default configuration) and we break potential score ties in rankings via alphanumeric ordering by document ID (with random IDs, this leads to a random distribution for other document properties such as text length [27]).

Leakage-Induced nDCG Improvements for MonoT5. Figure 2 shows the average nDCG@10 of monoT5, the most effective model in our experiments, for different training set sizes, tested only on topics with leaked queries. For small training sets, monoT5 achieves rather low nDCG@10 values and cannot exploit the leakage. The nDCG@10 increases with more training data on all benchmarks, peaking at 16,000 or 32,000 instances. At the peaks, monoT5 trained with leakage is more effective than without, and training on test leakage leads to a slightly higher nDCG@10 than leakage from MS MARCO/ORCAS (MSM). However, the difference between test and MSM leakage is larger for Robust04 (with some documents published as early as 1989) compared to the newer Common Core tracks (with documents published closer to the publication date of MS MARCO). On the Common Core data, MSM leakage is almost as effective as test leakage.

Leakage-Induced Effectiveness Improvements for Other Models. We employ a five-fold cross-validation setup for Duet, KNRM, monoBERT, monoT5, and PACRR to study whether leakage-induced effectiveness improvements can also be observed for other models when a grid search in the cross-validation setup can choose the training set size with the highest leakage effect for each model. We report the effectiveness of the models as nDCG@10, Precision@1, and the

Table 3. Effectiveness on Robust04 (R04) as nDCG@10, mean first rank of a relevant document (MFR), and Precision@1 (Prec@1) in a five-fold cross-validation setup on all test topics. Models are trained with no leakage (None), leakage from MS MARCO/ORCAS (MSM), or leakage from the test data (Test). Highest scores in bold; † marks Bonferroni-corrected significant differences to the no-leakage baseline (Student's t-test, $p = 0.05$). Model order swaps induced by MSM leakage in red.

Model	nDCG@10 on R04			MFR on R04			Prec@1 on R04		
	None	MSM	Test	None	MSM	Test	None	MSM	Test
Duet [33]	0.201	0.198	**0.224**†	2.420	2.682	**2.340**	0.297	0.261	**0.304**
KNRM [44]	0.194	0.214†	**0.309**†	2.348	2.309	**1.976**†	0.293	0.313	**0.329**
monoBERT [37]	0.394	0.373†	**0.396**	1.688	1.725	**1.639**	0.434	**0.454**	0.414
monoT5 [36]	0.461	0.457	**0.478**†	1.443	**1.416**	1.417	0.562	0.578	**0.590**
PACRR [21]	0.382	0.364†	**0.391**	1.663	1.604	**1.579**†	0.458	0.462	**0.502**

mean first rank of a relevant document (MFR) [18].[7] While effectiveness scores measured via nDCG@10 and Precision@1 have the property that higher values are better (a score of 1 indicates "best" effectiveness), for MFR, lower scores are better—but still a score of 1 is the best case indicating that the document on rank 1 always is relevant. In all effectiveness evaluations, we conduct significance tests via Student's t-test ($p = 0.05$) with Bonferroni correction for multiple testing as implemented in PyTerrier [32].

Table 3 shows the five-fold cross-validated effectiveness on Robust04 for the five models when optimizing each fold for nDCG@10, MFR, or Precision@1 in a grid search. Models trained on test leakage almost always are more effective than their no-leakage counterparts (exception: Precision@1 of monoBERT) and actual test leakage usually helps more than leakage from MS MARCO/ORCAS (MSM; exceptions: MFR of monoT5 and Precision@1 of monoBERT). Overall, on Robust04, models trained with MSM leakage are often less effective than their no-leakage counterparts (e.g., the nDCG@10 of monoBERT). A possible explanation might be the large time gap between the Robust04 document publication dates (documents published between 1987 and 2007) and the MS MARCO data (crawled in 2018). A similar observation was made during the TREC 2021 Deep Learning track [9]. The transition from Version 1 of MS MARCO to Version 2 (crawled three years later) caused some models to prefer older documents since they saw old documents as positive instances and newer ones as negative instances during training. Still, MSM leakage can lead to swaps in model ranking on Robust04. For instance, KNRM trained with MSM leakage achieves a higher nDCG@10 and Precision@1 than Duet without leakage, while KNRM trained without leakage is less effective than Duet.

---

[7] We use MFR instead of the mean reciprocal rank (MRR) as suggested by Fuhr [18]. His criticism of MRR was recently supported by further empirical evidence [48].

Table 4. Effectiveness on Common Core 2017 (CC17) as nDCG@10, mean first rank of a relevant document (MFR), and Precision@1 (Prec@1) in a five-fold cross-validation setup on all test topics. Models are trained with no leakage (None), leakage from MS MARCO/ORCAS (MSM), or leakage from the test data (Test). Highest scores in bold; † marks Bonferroni-corrected significant differences to the no-leakage baseline (Student's t-test, $p = 0.05$). Model order swaps induced by MSM leakage in red.

Model	nDCG@10 on R04			MFR on R04			Prec@1 on R04		
	None	MSM	Test	None	MSM	Test	None	MSM	Test
Duet [33]	0.374	0.373	**0.376**	1.620	1.512	**1.485**	0.500	0.480	**0.540**
KNRM [44]	0.316	**0.343**†	0.330	1.587	**1.512**	1.568	0.480	0.520	0.480
monoBERT [37]	0.402	0.407	**0.419**	1.625	**1.605**	1.634	**0.480**	0.460	0.460
monoT5 [36]	0.445	0.464	**0.490**†	1.363	1.384	**1.359**	0.660	0.620	**0.680**
PACRR [21]	0.406	0.403	**0.413**	**1.390**	1.515	1.546	0.540	0.520	**0.580**

Table 4 shows the five-fold cross-validated effectiveness on the TREC 2017 Common Core track for the five models when optimizing each fold for nDCG@10, MFR, or Precision@1 in a grid search. In contrast to Robust04, more models improve their effectiveness when trained with MSM leakage as the time gap between the New York Times Annotated Corpus and MS MARCO is smaller than for Robust04. MonoT5 with actual test leakage is the most effective model for all three measures, and monoT5 trained on MSM leakage is more effective than the no-leakage counterpart in nDCG@10 and MFR. MSM leakage also may cause model order swaps at higher positions in the systems' nDCG@10 ordering: monoBERT with MSM leakage would slightly overtake PACRR. Still, most of the effectiveness improvements on this dataset caused by MSM leakage or test leakage are not significant (exception: the nDCG@10 differences for monoT5 with test leakage and KNRM with MSM leakage to the no-leakage counterparts).

Table 5 shows the five-fold cross-validated effectiveness on the TREC 2018 Common Core track for the five models when optimizing each fold for nDCG@10, MFR, or Precision@1 in a grid search. In contrast to Robust04 and the 2017 edition of the Common Core track, training with MSM leakage improves the effectiveness in all cases for all three measures. While most of the leakage-induced effectiveness improvements are not statistically significant, the model order even changes on the top MFR position, where PACRR with MSM leakage would overtake monoT5 without leakage.

Discussion.
The results in Tables 3, 4 and 5 show that leakage from MS MARCO/ ORCAS (MSM) can have an impact on the retrieval effectiveness, even when only a small number of instances are leaked, as in our experiments. While the changes on Robust04 are rather negligible, the impact is larger for the Common Core tracks with document publication dates closer to the ones from MS MARCO. Interestingly, MSM leakage-induced nDCG@10 improvements sometimes can

Table 5. Effectiveness on Common Core 2018 (CC18) as nDCG@10, mean first rank of a relevant document (MFR), and Precision@1 (Prec@1) in a five-fold cross-validation setup on all test topics. Models are trained with no leakage (None), leakage from MS MARCO/ORCAS (MSM), or leakage from the test data (Test). Highest scores in bold; † marks Bonferroni-corrected significant differences to the no-leakage baseline (Student's t-test, $p = 0.05$). Model order swaps induced by MSM leakage in red.

Model	nDCG@10 on R04			MFR on R04			Prec@1 on R04		
	None	MSM	Test	None	MSM	Test	None	MSM	Test
Duet [33]	0.285	**0.301**	0.295	1.993	**1.812**	2.231	0.320	**0.380**	0.260
KNRM [44]	0.201	**0.256**†	0.238†	3.099	**2.768**	3.125	0.100	**0.160**	0.140
monoBERT [37]	0.364	**0.380**	0.366	1.810	**1.683**	1.719	0.460	**0.560**	0.460
monoT5 [36]	0.417	**0.448**	0.445	1.577	**1.503**	1.512	0.480	**0.540**	**0.540**
PACRR [21]	0.376	**0.406**	0.393	1.649	**1.383**†	1.485	0.520	**0.560**	0.540

lead to swaps in model ordering despite the improvements not being significant in most cases. This exemplifies that experimental effectiveness comparisons might be invalid when some models had access to leaked instances during training.

Memorization of Leaked Instances. To analyze whether the models memorize leaked instances, we compare the retrieval scores and resulting ranks of leaked documents in the test rankings when training includes or does not include leakage. For leaked documents not returned in the top-100 BM25 results—the models only re-rank these—, we determine a hypothetical rank by calculating the score of this document for the query and inserting the document at the corresponding rank in the to-be-re-ranked 100 documents (including breaking score-ties by document ID). Each leaked document thus has a maximal rank of 101.

Table 6 shows the mean rank of relevant documents when they were included during training (with leakage) or not (without leakage). Models perfectly memorizing their positive training instances (i.e., relevant documents for test queries) would rank these documents at substantially higher positions than models that did not see the same instance during training. However, while the mean rank of leaked relevant documents improves for most cases, the improvement is mostly negligible. For instance, the mean rank of leaked relevant documents for the very effective monoT5 and monoBERT models improves only slightly compared to their no-leakage counterparts on all three corpora. But the difference increases (still rather negligibly, though) on the corpora on which leakage was more effective. In combination with the high standard deviations, one can hardly see memorization effects for the positions of leaked relevant documents in the final rankings. We thus also inspect the retrieval scores of the leaked documents.

Table 7 shows the mean retrieval score of the relevant documents when they were included during training (with leakage) or not (without leakage). Models that memorize the leaked relevant training documents should increase their

Table 6. Mean rank of the (leaked) relevant training documents (± standard deviation) for models trained with and without leakage from MS MARCO/ORCAS (MSM leakage) or from the test data (test leakage). Ranks macro-averaged over all topics for test leakage and over all topics with leaking queries for MSM leakage.

Model	Robust04		Common Core 17		Common Core 18	
	With	Without	With	Without	With	Without
MSM leak.						
Duet	41.70 ±45.88	46.79 ±46.51	34.52 ±32.67	35.98 ±32.93	43.39 ±33.52	45.67 ±32.99
KNRM	82.36 ±31.88	84.74 ±30.15	43.24 ±31.74	43.68 ±31.50	53.12 ±32.14	53.45 ±32.14
monoBERT	23.08 ±28.71	23.58 ±28.22	46.97 ±34.95	47.11 ±35.49	41.79 ±36.16	42.48 ±36.39
monoT5	20.13 ±26.77	20.15 ±26.64	35.68 ±31.69	36.46 ±31.88	29.86 ±28.24	30.31 ±28.27
PACRR	42.41 ±44.86	42.43 ±44.67	35.79 ±33.71	36.28 ±33.83	34.76 ±36.45	35.70 ±36.87
Test leak.						
Duet	90.04 ±26.98	90.65 ±26.41	45.78 ±30.03	46.55 ±30.34	46.31 ±29.85	46.35 ±30.13
KNRM	89.95 ±26.43	91.20 ±25.24	47.37 ±32.80	47.49 ±32.81	50.53 ±32.40	50.13 ±32.26
monoBERT	47.01 ±31.84	47.39 ±31.80	46.64 ±31.51	47.12 ±31.51	43.19 ±31.51	44.04 ±31.66
monoT5	45.28 ±32.09	45.37 ±31.96	46.35 ±31.47	47.45 ±31.83	40.16 ±31.18	40.95 ±31.24
PACRR	80.89 ±34.05	82.60 ±33.07	53.59 ±31.49	52.91 ±31.20	52.25 ±32.69	52.28 ±32.32

Table 7. Mean retrieval score of the (leaked) relevant training documents (± standard deviation; higher scores = "more relevant") for models trained with/without leakage from MS MARCO/ORCAS (MSM) or the test data (Test). Scores macro-averaged over all topics for test leakage and over all topics with leaking queries for MSM leakage.

Model	Robust04		Common Core 17		Common Core 18	
	With	Without	With	Without	With	Without
MSM leak.						
Duet	0.89 ±1.22	0.78 ±1.18	0.52 ±1.18	0.47 ±1.17	0.16 ±0.79	0.09 ±0.75
KNRM	-2.06 ±3.64	-2.58 ±3.43	-2.53 ±3.75	-3.08 ±3.52	-2.32 ±3.24	-2.78 ±3.01
monoBERT	-0.75 ±0.44	-0.72 ±0.41	-0.88 ±0.49	-0.85 ±0.47	-0.92 ±0.50	-0.89 ±0.48
monoT5	-1.05 ±1.14	-1.19 ±1.20	-1.32 ±1.26	-1.48 ±1.31	-1.51 ±1.34	-1.65 ±1.38
PACRR	2.59 ±3.25	2.29 ±3.11	2.78 ±3.35	2.46 ±3.20	2.25 ±3.08	1.95 ±2.96
Test leak.						
Duet	0.07 ±0.61	-0.11 ±0.56	0.22 ±0.68	-0.01 ±0.66	0.30 ±0.69	0.09 ±0.68
KNRM	-2.71 ±3.79	-3.41 ±3.71	-2.78 ±3.65	-3.28 ±3.63	-3.08 ±4.14	-3.59 ±4.14
monoBERT	-0.91 ±0.45	-1.04 ±0.53	-0.85 ±0.44	-0.90 ±0.48	-0.85 ±0.46	-0.92 ±0.49
monoT5	-1.70 ±1.24	-2.37 ±1.53	-1.47 ±1.09	-1.98 ±1.37	-1.52 ±1.24	-2.01 ±1.50
PACRR	2.31 ±4.27	1.92 ±3.09	1.83 ±4.40	1.98 ±3.13	2.66 ±3.31	2.26 ±3.23

score, and we indeed observe that the retrieval score of leaked relevant documents increases in most cases compared to their no-leakage counterpart (exception: monoBERT for MSM leakage and PACRR for test leakage from Common Core 2017). The difference between the score differences of leakage mod-

Table 8. Macro-averaged increase of the rank-offset between the leaked relevant and non-relevant documents ($\pm$ standard deviation) for models trained on MSM leakage (Δ on MSM) or on test leakage (Δ on Test) over the no-leakage variants.

Model	Δ on MSM			Δ on Test		
	R04	C17	C18	R04	C17	C18
Duet	$6.378_{\pm32.15}$	$3.119_{\pm19.17}$	$2.647_{\pm19.23}$	$0.809_{\pm17.69}$	$1.430_{\pm19.33}$	$1.023_{\pm20.10}$
KNRM	$0.640_{\pm19.22}$	$0.979_{\pm15.23}$	$0.398_{\pm14.55}$	$1.335_{\pm11.75}$	$0.012_{\pm14.92}$	$0.140_{\pm15.18}$
monoBERT	$0.692_{\pm17.97}$	$0.076_{\pm17.19}$	$0.369_{\pm20.04}$	$3.886_{\pm20.39}$	$0.980_{\pm17.44}$	$3.497_{\pm25.98}$
monoT5	$0.443_{\pm8.60}$	$0.390_{\pm9.28}$	$0.789_{\pm9.91}$	$3.443_{\pm19.96}$	$2.242_{\pm9.84}$	$1.819_{\pm10.98}$
PACRR	$0.043_{\pm19.30}$	$0.764_{\pm10.93}$	$0.452_{\pm12.38}$	$1.952_{\pm17.71}$	$0.271_{\pm16.96}$	$0.753_{\pm14.16}$

els and non-leakage models is larger for leakage from the test data than for MSM leakage in 13 of the 15 cases (with a maximum difference for monoT5 from a test leakage difference of $0.67 = 2.37 - 1.70$ to an MSM leakage difference of $0.14 = 1.19 - 1.05$). To investigate the "full picture" with respect to also negative leaked instances (i.e., non-relevant documents), we next also study the rank offsets between the positive and the negative leaked instances.

Table 8 shows the macro-averaged increase in the rank difference of the leaked relevant and non-relevant documents between models trained with and without leakage. The leakage increases the rank offset for all five analyzed models (e.g., 6.4 ranks for Duet on Robust04 with MSM leakage). Interestingly, an in-depth inspection showed that most of the increased differences are caused by lowered ranks of the leaked non-relevant documents (e.g., 2 ranks lower for monoT5) while the leaked relevant documents improve their ranks only slightly (e.g., 0.3 ranks higher for monoT5).

Discussion. Overall, our results in Tables 6, 7 and 8 indicate that memorization happens but has little impact. Larger memorization effects might be desirable in practical scenarios where a retrieval system that memorizes good results can simply present them when the same query is submitted again. However, for empirical evaluations in scientific publications or at shared tasks, (unintended) leakage memorization at a larger scale might still lead to unwanted outcomes.

5   Conclusion

Our study of train–test leakage effects for neural retrieval models was inspired by the observation that 69% of the Robust04 topics, a dataset often used to test neural models, have very similar queries in the MS MARCO/ORCAS datasets, that are often used to train neural models. At first glance, this overlap might seem alarming since train–test leakage is known to invalidate experimental evaluations. We thus analyzed train–test leakage effects for five neural models (Duet, KNRM, monoBERT, monoT5, and PACRR) in scenarios with different amounts of leakage. While our experiments show leakage-induced effectiveness improvements that may even lead to swaps in model ranking, our overall results are reassuring: the effects on nDCG@10 are rather small and not significant in most cases. They also become smaller the smaller (and more realistic) the amount of leakage among all training instances is. Still, even if only a few nDCG@10 differences were significant, we noticed a memorization effect: the rank offset between leaked relevant and non-relevant documents increased on all scenarios.

Train–test leakage should thus still be avoided in academic experiments but the practical consequences for real search engines might be different. The observed increased rank offset might be a highly desirable effect when presuming that queries already seen during training are submitted again after a model has been deployed to production. An interesting direction for future research is to enlarge our experiments to investigate more of the few cases where train–test leakage slightly reduced the effectiveness.

References

1. Allan, J., Harman, D., Kanoulas, E., Li, D., Gysel, C., Voorhees, E.: TREC 2017 common core track overview. In: Proceedings of TREC 2017, vol. 500–324. NIST (2017)
2. Ateniese, G., Mancini, L., Spognardi, A., Villani, A., Vitali, D., Felici, G.: Hacking smart machines with smarter ones: how to extract meaningful data from machine learning classifiers. Int. J. Secur. Netw. 10(3), 137–150 (2015)
3. Benham, R., et al.: RMIT at the 2017 TREC CORE track. In: Proceedings of TREC 2017, NIST Special Publication, vol. 500-324. NIST (2017)
4. Benham, R., et al.: RMIT at the 2018 TREC CORE track. In: Proceedings of TREC 2018, NIST Special Publication, vol. 500-331. NIST (2018)
5. Berthelot, D., Raffel, C., Roy, A., Goodfellow, I.: Understanding and improving interpolation in autoencoders via an adversarial regularizer. In: Proceedings of ICLR 2019. OpenReview.net (2019)
6. Chen, C., Wu, B., Qiu, M., Wang, L., Zhou, J.: A comprehensive analysis of information leakage in deep transfer learning. CoRR abs/2009.01989 (2020)
7. Chollet, F.: Deep Learning with Python. Simon and Schuster (2021)
8. Craswell, N., Campos, D., Mitra, B., Yilmaz, E., Billerbeck, B.: ORCAS: 20 million clicked query-document pairs for analyzing search. In: Proceedings of CIKM 2020, pp. 2983–2989. ACM (2020)
9. Craswell, N., Mitra, B., Yilmaz, E., Campos, D.: Overview of the TREC 2021 deep learning track. In: Voorhees, E.M., Ellis, A. (eds.) Notebook. NIST (2021)
10. Craswell, N., Mitra, B., Yilmaz, E., Campos, D., Voorhees, E.: Overview of the TREC 2019 deep learning track. In: Proceedings of TREC 2019, NIST Special Publication. NIST (2019)
11. Devlin, J., Chang, M.W., Lee, K., Toutanova, K.: BERT: pre-training of deep bidirectional transformers for language understanding. In: Proceedings of NAACL 2019, Minneapolis, Minnesota, pp. 4171–4186. Association for Computational Linguistics (2019)
12. Dolan, W.B., Brockett, C.: Automatically constructing a corpus of sentential paraphrases. In: Proceedings of the Third International Workshop on Paraphrasing (IWP 2005) (2005)

13. Fan, A., Jernite, Y., Perez, E., Grangier, D., Weston, J., Auli, M.: ELI5: long form question answering. In: Proceedings of ACL 2019, pp. 3558–3567. ACL (2019)
14. Fan, Y., Guo, J., Lan, Y., Xu, J., Zhai, C., Cheng, X.: Modeling diverse relevance patterns in ad-hoc retrieval. In: Proceedings of SIGIR 2018, pp. 375–384. ACM (2018)
15. Feldman, V.: Does learning require memorization? A short tale about a long tail. In: Proceedings of STOC 2020, pp. 954–959. ACM (2020)
16. Feldman, V., Zhang, C.: What neural networks memorize and why: discovering the long tail via influence estimation. In: Proceedings of NeurIPS 2020 (2020)
17. Fredrikson, M., Jha, S., Ristenpart, T.: Model inversion attacks that exploit confidence information and basic countermeasures. In: Proceedings of CCS 2015, pp. 1322–1333. ACM (2015)
18. Fuhr, N.: Some common mistakes in IR evaluation, and how they can be avoided. SIGIR Forum **51**(3), 32–41 (2017)
19. He, H., Garcia, E.: Learning from imbalanced data. IEEE Trans. Knowl. Data Eng. **21**(9), 1263–1284 (2009)
20. Hofstätter, S., Zlabinger, M., Hanbury, A.: Interpretable & time-budget-constrained contextualization for re-ranking. In: Proceedings of ECAI 2020, Frontiers in Artificial Intelligence and Applications, vol. 325, pp. 513–520. IOS Press (2020)
21. Hui, K., Yates, A., Berberich, K., Melo, G.: PACRR: a position-aware neural IR model for relevance matching. In: Proceedings of EMNLP 2017, pp. 1049–1058. ACL (2017)
22. Jansen, B., Booth, D., Spink, A.: Patterns of query reformulation during web searching. J. Assoc. Inf. Sci. Technol. **60**(7), 1358–1371 (2009)
23. Johnson, J., Douze, M., Jégou, H.: Billion-scale similarity search with GPUs. IEEE Trans. Big Data **7**(3), 535–547 (2021)
24. Krishna, K., Roy, A., Iyyer, M.: Hurdles to progress in long-form question answering. In: Proceedings of NAACL 2021, pp. 4940–4957. ACL (2021)
25. Li, C., Yates, A., MacAvaney, S., He, B., Sun, Y.: PARADE: passage representation aggregation for document reranking. CoRR abs/2008.09093 (2020)
26. Lin, J., Ma, X., Lin, S., Yang, J., Pradeep, R., Nogueira, R.: Pyserini: a Python toolkit for reproducible information retrieval research with sparse and dense representations. In: Proceedings of SIGIR 2021, pp. 2356–2362. ACM (2021)
27. Lin, J., Yang, P.: The impact of score ties on repeatability in document ranking. In: Proceedings of SIGIR 2019, pp. 1125–1128. ACM (2019)
28. Lin, S., Yang, J., Lin, J.: Distilling dense representations for ranking using tightly-coupled teachers. CoRR abs/2010.11386 (2020)
29. Linjordet, T., Balog, K.: Sanitizing synthetic training data generation for question answering over knowledge graphs. In: Proceedings of ICTIR 2020, pp. 121–128. ACM (2020)
30. MacAvaney, S., Yates, A., Cohan, A., Goharian, N.: CEDR: contextualized embeddings for document ranking. In: Proceedings of SIGIR 2019, pp. 1101–1104. ACM (2019)
31. MacAvaney, S., Yates, A., Feldman, S., Downey, D., Cohan, A., Goharian, N.: Simplified data wrangling with ir_datasets. In: Proceedings of SIGIR 2021, pp. 2429–2436. ACM (2021)
32. Macdonald, C., Tonellotto, N., MacAvaney, S., Ounis, I.: PyTerrier: declarative experimentation in Python from BM25 to dense retrieval. In: Proceedings of CIKM 2021, pp. 4526–4533. ACM (2021)

33. Mitra, B., Diaz, F., Craswell, N.: Learning to match using local and distributed representations of text for web search. In: Proceedings of WWW 2017, pp. 1291–1299. ACM (2017)
34. Mokrii, I., Boytsov, L., Braslavski, P.: A systematic evaluation of transfer learning and pseudo-labeling with BERT-based ranking models. In: Proceedings of SIGIR 2021, pp. 2081–2085. ACM (2021)
35. Nasr, M., Shokri, R., Houmansadr, A.: Comprehensive privacy analysis of deep learning: passive and active white-box inference attacks against centralized and federated learning. In: Proceedings of SP 2019, pp. 739–753. IEEE (2019)
36. Nogueira, R., Jiang, Z., Pradeep, R., Lin, J.: Document ranking with a pretrained sequence-to-sequence model. In: Findings of EMNLP 2020, vol. EMNLP 2020, pp. 708–718. ACL (2020)
37. Nogueira, R., Yang, W., Cho, K., Lin, J.: Multi-stage document ranking with BERT. CoRR abs/1910.14424 (2019)
38. Reimers, N., Gurevych, I.: Sentence-BERT: sentence embeddings using Siamese BERT-networks. In: Proceedings of EMNLP 2019, pp. 3980–3990. ACL (2019)
39. Sandhaus, E.: The New York times annotated corpus. Linguist. Data Consortium Philadelphia 6(12), e26752 (2008)
40. Sharma, L., Graesser, L., Nangia, N., Evci, U.: Natural language understanding with the Quora question pairs dataset. CoRR abs/1907.01041 (2019)
41. Shokri, R., Stronati, M., Song, C., Shmatikov, V.: Membership inference attacks against machine learning models. In: Proceedings of SP 2017, pp. 3–18. IEEE (2017)
42. Voorhees, E.: The TREC robust retrieval track. SIGIR Forum 39(1), 11–20 (2005)
43. Wahle, J.P., Ruas, T., Meuschke, N., Gipp, B.: Are neural language models good plagiarists? A benchmark for neural paraphrase detection. In: Proceedings of JCDL 2021, pp. 226–229 (2021)
44. Xiong, C., Dai, Z., Callan, J., Liu, Z., Power, R.: End-to-end neural ad-hoc ranking with kernel pooling. In: Proceedings of SIGIR 2017, pp. 55–64. ACM (2017)
45. Yates, A., Arora, S., Zhang, X., Yang, W., Jose, K., Lin, J.: Capreolus: a toolkit for end-to-end neural ad hoc retrieval. In: Proceedings of WSDM 2020, pp. 861–864. ACM (2020)
46. Zhan, J., Xie, X., Mao, J., Liu, Y., Zhang, M., Ma, S.: Evaluating extrapolation performance of dense retrieval. CoRR abs/2204.11447 (2022)
47. Zhang, X., Yates, A., Lin, J.: A little bit is worse than none: ranking with limited training data. In: Proceedings of SustaiNLP 2020, pp. 107–112. Association for Computational Linguistics (2020)
48. Zobel, J., Rashidi, L.: Corpus bootstrapping for assessment of the properties of effectiveness measures. In: Proceedings of CIKM 2020, pp. 1933–1952. ACM (2020)

Computational Biology

Genome Comparison on Succinct Colored de Bruijn Graphs

Lucas P. Ramos[1]([⊠])[iD], Felipe A. Louza[2][iD], and Guilherme P. Telles[1][iD]

[1] Instituto de Computação, UNICAMP, Campinas, SP, Brazil
lucaspr98@gmail.com
[2] Universidade Federal de Uberlândia, Uberlândia, Brazil

Abstract. The improvements in DNA sequence technologies have increased the volume and speed at which genomic data is acquired. Nevertheless, due to the difficulties for completely assembling a genome, many genomes are left in a draft state, in which each chromosome is represented by a set of sequences with partial information on their relative order. Recently, some approaches have been proposed to compare genomes by comparing extracted paths from de Bruijn graphs and comparing such paths. The idea of using data from de Bruijn graphs is interesting because such graphs are built by many practical genome assemblers. In this article we introduce gcBB, a method for comparing genomes represented as succinct de Bruijn graphs directly, without resorting to sequence alignments, by means of the entropy and expectation measures based on the Burrows-Wheeler Similarity Distribution (BWSD). We have compared phylogenies of genomes obtained by other methods to those obtained with gcBB, achieving promising results.

Keywords: Succinct de Bruijn graphs · Genomic comparison · Phylogenetics

1 Introduction

Computing similarity measures between strings is a problem that must be solved often in many areas of Computer Science, such as bioinformatics, plagiarism detection and classification. Similarity measures based on the Burrows-Wheeler transform (BWT) [3], as the eBWT-based distances [12,13] and the Burrows-Wheeler Similarity Distribution (BWSD) [22], are particularly attractive because the BWT provides a self-index [14] and can be computed in linear time on the string length.

In the process of sequencing a genome, a large amount of short strings (reads) is first obtained. The reads cover each DNA nucleotide many times; such coverage varies across sequencing projects and may be as high as 200 times per nucleotide on average. The reads must then be assembled based on the overlaps among them. This is a hard problem in general, further complicated by the huge number of reads that may be obtained with the current DNA sequencing technologies,

D. Arroyuelo and B. Poblete (Eds.): SPIRE 2022, LNCS 13617, pp. 165–177, 2022.
https://doi.org/10.1007/978-3-031-20643-6_12

by the presence of repetitions in the target DNA, by sequencing errors and by other sources of ambiguities and technical limitations. Completely assembling a genome is thus a difficult task and many sequenced genomes are left in a draft state, that is, a set of strings (contigs) that may include information on their relative order (scaffolds) instead of a single string for each chromosome [16].

Different graphs have been used for genome assembly, such as overlap graphs [17], de Bruijn graphs [4], string graphs [20] and repeat graphs [7]. A graph is built early in the assembly process and then a series of algorithms extracts paths and connectivity information to obtain a tentative sequence or arrangement of sequences for the whole genome.

Many assemblers are based on the de Bruijn graph (*e.g.* [8,9,20]), that may be stored succinctly using the BOSS representation [2], that is based on the BWT. Colors may be added to the edges of a de Bruijn graph, enabling the representation of a set of strings from distinct genomes on colored de Bruijn graphs. Recent approaches have been proposed for the comparison of genomes by the extraction of paths from their colored de Bruijn graphs [10,15].

In this paper we introduce gcBB, a space-efficient algorithm to compare genomes using their BOSS representations and the BWSD. Given a set of genomes, a colored de Bruijn graph in the BOSS representation is built for each pair of genomes and BWSD measures are evaluated to assess the similarity between the genomes. Our method showed promising resulting in experiments that compared the phylogenies for genomes of 12 Drosophila species built with gcBB and with the methods by Lyman *et al.* [10] and by Polevikov and Kolmogorov [15].

2   Definitions and Notation

A string is the juxtaposition of symbols from an ordered alphabet Σ. Let S be a string of length n. We index its symbols from 1 to n. A substring of S is $S[i,j] = S[i]\ldots S[j]$ with $1 \leq i \leq j \leq n$. The substring $S[1,i]$ is referred to as a prefix of S and $S[i,n]$ is referred to as a suffix of S. The i-th circular rotation (or conjugate or simply rotation) of S is the string $S[i+1]\ldots S[n]S[1]\ldots S[i]$. When $i = 0$ the rotation is equal to S.

For clearer definitions it is convenient to use a special marker symbol $ at the end of S. This symbol does not occur elsewhere in S and is the smallest symbol in Σ. In this way, all rotations of S are distinct.

The suffix array [11] of a string S of length n is an array SA containing the permutation of $\{1,\ldots,n\}$ that gives the suffixes of S in lexicographic order, that is, $S[SA[1],n] < S[SA[2],n] < \ldots < S[SA[n],n]$.

By $\mathrm{lcp}(S_1, S_2)$ we denote the length of the longest common prefix of strings S_1 and S_2. The LCP array for a string S of length n is the array of integers containing the lcp of consecutive suffixes in the suffix array. Formally, $LCP[i] = \mathrm{lcp}(S[SA[i],n], S[SA[i-1],n])$ for $1 < i \leq n$ and $LCP[1] = 0$.

The Burrows-Wheeler Transform (BWT) [3] of a string S is a reversible transformation of S that permutes its symbols. The resulting string, denoted by

i	SA	BWT	LCP	$S[\text{SA}[i], n]$
1	12	a	0	$
2	11	r	0	a$
3	8	d	1	abra$
4	1	$	4	abracadabra$
5	4	r	1	acadabra$
6	6	c	1	adabra$
7	9	a	0	bra$
8	2	a	3	bracadabra$
9	5	a	0	cadabra$
10	7	a	0	dabra$
11	10	b	0	ra$
12	3	b	2	racadabra$

Fig. 1. Suffix array, BWT and LCP array for S = abracadabra$.

BWT, often allows better compression because equal symbols tend to be clustered. Moreover, the BWT is the core of many indexing structures for text [14].

The BWT is the last column of a matrix $\mathcal{M}$ having the sorted rotations of S as rows. In $\mathcal{M}$, the first column is called F and the last column is called L. Since $S[n]$ = $, sorting the rotations of S is equivalent to sorting the suffixes of S and the BWT may be defined in terms of the suffix array as $\text{BWT}[i] = S[\text{SA}[i] - 1]$ if $\text{SA}[i] \neq 1$ or $\text{BWT}[i]$ = $ otherwise. Figure 1 shows the suffix array, the LCP array and the BWT for S = abracadabra$.

Let $\mathcal{S} = \{S_1, S_2, \ldots, S_d\}$ be a collection of d strings of lengths $n_1, n_2, \ldots, n_d$. We define the concatenation of all strings in $\mathcal{S}$ as $S^{cat} = S_1[1, n_1 - 1]\$_1 S_2[1, n_2 - 1]\$_2 \cdots S_d[1, n_d - 1]\$_d$ that is, each terminal symbol $ is replaced by a (separator) symbol $\$_i$, with $\$_i < \$_j$ if and only if $i < j$. The length of S^{cat} is $N = \sum_{i=1}^{d} n_i$. The suffix array for a collection $\mathcal{S}$ is the suffix array $\text{SA}[1, N]$ computed for S^{cat}. The BWT for $\mathcal{S}$ is obtained from the SA of S^{cat} as well.

We define the context of a suffix $S^{cat}[i, N]$ as the substring $S^{cat}[i, j]$ such that $S^{cat}[j]$ is the first occurrence of some $\$_k$ in $S^{cat}[i, N]$. The document array is an array of integers DA that stores which document each suffix in SA "belongs" to. More formally, $\text{DA}[i] = j$ if $S^{cat}[\text{SA}[i], N]$ has the context that ends with $\$_j$.

2.1 Burrows-Wheeler Similarity Distribution

The BWT of two strings S_1 and S_2 can be used to compute similarity measures between the strings based on the observation that as more symbols of S_1 and S_2 are intermixed in the BWT, a larger number of substrings are shared between them [12].

The Burrows-Wheeler similarity distribution (BWSD) [22] between S_1 and S_2, denoted by $\text{BWSD}(S_1, S_2)$, is a probability mass function defined as follows. Given the BWT of $\mathcal{S} = \{S_1, S_2\}$, we define a bitvector α of size $n_1 + n_2$ such that $\alpha[p] = 0$ if $\text{BWT}[p] = \$_2$ or $\text{BWT}[p] \in S_1$ and $\alpha[p] = 1$ if $\text{BWT}[p] = \$_1$ or $\text{BWT}[p] \in S_2$. The bitvector α can be represented as a sequence of runs

$r = 0^{k_1}1^{k_2}0^{k_3}1^{k_4}\ldots0^{k_m}1^{k_{m+1}}$ where i^{k_j} means that i repeats k_j times, and only k_1 and k_{m+1} may be zero. The largest possible value for k_j is $k_{\max} = \max(n_1, n_2)$.

Let t_n be the number of occurrences of an exponent n in r. Let $s = t_1 + t_2 + \ldots + t_{k_j} + \ldots + t_{k_{\max}}$. The BWSD$(S_1, S_2)$ is the probability mass function

$$P\{k_j = k\} = t_k/s \text{ for } k = 1, 2, \ldots, k_{\max}.$$

Two similarity measures were defined on the BWSD of S_1 and S_2.

Definition 1. $D_M(S_1, S_2) = E(k_j) - 1$, *where* $E(k_j)$ *is the expectation of* $BWSD(S_1, S_2)$.

Definition 2. $D_E(S_1, S_2) = -\sum_{k\geq1, t_k\neq0}(t_k/s)\log_2(t_k/s)$ *is the Shannon entropy of* $BWSD(S_1, S_2)$.

Note that if S_1 and S_2 are equal then $k_{\max} = 1$, $P\{k_j = 1\} = \frac{n_1+n_2}{n_1+n_2} = 1$, $D_M(S_1, S_2) = 0$ and $D_E(S_1, S_2) = 0$. Also, if the α for BWT(S_1, S_2) is equal to the α for BWT(S_2, S_1), then both have the same BWSD and $D_E(S_1, S_2) = D_E(S_2, S_1)$ and $D_M(S_1, S_2) = D_M(S_2, S_1)$.

2.2   Succinct de Bruijn Graphs

Let $\mathcal{S} = \{S_1, S_2, \ldots, S_d\}$ be a collection of strings (reads of a genome). Assume that $\mathcal{S}$ is modified by concatenating k symbols \$ at the beginning of each string in $\mathcal{S}$. We will refer to a string of length k as a k-mer.

A de Bruijn graph (of order k) for $\mathcal{S}$ has one vertex for each k-mer in a string in $\mathcal{S}$. There is an edge from vertex u to vertex v labeled $v[k]$ if the substring $u[1]u[2]\ldots u[k]v[k]$ occurs in a string in $\mathcal{S}$. We say that the k-mer related to a vertex u is the vertex label and denote it by $\overrightarrow{u}$. A de Bruijn graph may represent a set of genomes by adding a color to each edge (colors and genomes are in a one-to-one mapping) and allowing parallel edges. Such graph is called colored de Bruijn graph.

Notice that (i) an edge from u to v corresponds to the existence of an overlap of length $k-1$ between the suffix of $\overrightarrow{u}$ and the prefix of $\overrightarrow{v}$ and (ii) the concatenation of edge labels along a path of length k that arrives at a vertex v whose label does not have a \$ will be $\overrightarrow{v}$.

BOSS [2] is a succinct representation of the de Bruijn graph that enables efficient navigation across vertices and edges. Let n and m be respectively the number of vertices and edges of a de Bruijn graph G. Assume that the vertices $v_1, v_2, \ldots, v_n$ in G are sorted according to the co-lexicographic order of their labels, *i.e.*, the lexicographic order of the reverse of their labels, $\overleftarrow{v_i} = \overrightarrow{v_i}[k]\ldots\overrightarrow{v_i}[1]$ for each vertex v_i.

We define *Node* as a conceptual matrix containing the co-lexicographically sorted set of k-mers in $\mathcal{S}$. For each vertex v_i, we define W_i as the sequence of symbols of the outgoing edges of v_i in lexicographic order. If v_i has no outgoing edges then $W_i = \$$.

The BOSS representation is composed by the following components:

1. The string $W[1, m] = W_1 W_2 \dots W_n$. Observe that $|W| = |Node|$ and $Node[i]$ denotes the vertex from which $W[i]$ leaves.
2. The bitvector $W^-[1, m]$ such that $W^-[i] = 0$ if there exists $j < i$ such that $W[j] = W[i]$ and the suffixes of length $k - 1$ of $Node[j]$ and of $Node[i]$ are identical, or $W^-[i] = 1$ otherwise.
3. The bitvector $last[1, m]$ such that $last[i] = 1$ if $i = n$ or $Node[i]$ is different from $Node[i + 1]$, or $last[i] = 0$ otherwise.
4. The counter array $C[1, \sigma]$ such that $C[c]$ stores the number of symbols smaller than c in the last column of the conceptual matrix $Node$.

For DNA sequences the alphabet is $\Sigma = \{\texttt{A}, \texttt{T}, \texttt{C}, \texttt{G}, \texttt{N}, \$\}$ with size $\sigma = 6$. Storing the string W requires $m\lceil \log_2 \sigma \rceil = 3m$ bits, the bitvectors W^- and $last$ require $2m$ bits and the counter array C requires $\sigma \log m = 6 \log m$ bits. Therefore, the overall space to store the BOSS structure is $5m + 6 \log m$ bits.

Egidi et al. [5] proposed an algorithm called eGap for computing the multi-string BWT and the LCP array in external memory. As an application the authors showed how to compute the BOSS representation with a sequential scan over the BWT and the LCP array built for the collection $\mathcal{S}$ with all strings reversed in $O(N)$ time.

3   gcBB – Genome Comparison Using BOSS and BWSD

Given a set of genomes (each one as a collection of reads) and a value for k, our method, called gcBB, constructs the colored BOSS and then computes the BWSD and the similarity measures for each pair. The output is a distance matrix with the expectation and entropy BWSD distances among all pairs of the genomes in the set. The intuition is that intermixed edges in the colored BOSS are related to shared nodes in their graphs and to similarities in the genomes. gcBB has three phases, as follows.

Phase 1: First, we construct the BWT and the LCP array for the collection of reads of each genome in external memory using eGap [5]. We also compute an auxiliary array that gives the lengths of each context, called CL. For each pair of genomes, we merge their arrays while generating the document array DA. Note that DA can be stored in a bitvector, since we merge only pairs of genomes. The resulting arrays are written to external memory.

For the set of genomes $\mathcal{S}_1 = \{\texttt{TACTCA}, \texttt{TACACT}\}$ and $\mathcal{S}_2 = \{\texttt{GACTCG}\}$, Fig. 2 shows the output of eGap for each set and the resulting merge.

Phase 2: From the merged BWT and LCP arrays we construct the colored BOSS representation as described in [5] and we compute the bitvector $\texttt{colors}[1, m]$ and the array $\texttt{coverage}[1, m]$. The $\texttt{colors}$ bitvector indicates from which genome each edge came and can be easily obtained from DA. The $\texttt{coverage}$ array gives the number of times a $(k + 1)$-mer represented by an edge occurs in its genome.

i	BWT	LCP	CL	DA	context
1	T	0	1	0	$\$_1$
2	T	0	1	0	$\$_2$
3	G	0	1	1	$\$_3$
4	C	0	5	0	ACAT$\$_2$
5	$\$_3$	2	7	0	ACTCAT$\$_1$
6	C	1	3	1	AG$\$_3$
7	C	1	3	0	AT$\$_1$
8	C	2	3	0	AT$\$_2$
9	T	0	6	0	CACAT$\$_2$
10	T	2	4	1	CAG$\$_3$
11	A	2	4	0	CAT$\$_1$
12	T	3	4	0	CAT$\$_2$
13	G	1	6	1	CTCAG$\$_3$
14	A	4	6	0	CTCAT$\$_1$
15	A	0	2	1	G$\$_3$
16	$\$_2$	1	7	1	GCTCAG$\$_3$
17	A	0	2	0	T$\$_1$
18	A	1	2	0	T$\$_2$
19	$\$_1$	1	7	0	TCACAT$\$_2$
20	C	3	5	1	TCAG$\$_3$
21	C	3	5	0	TCAT$\$_1$

i	BWT	LCP	CL	context
1	T	0	1	$\$_1$
2	T	0	1	$\$_2$
3	C	0	5	ACAT$\$_2$
4	$\$_2$	2	7	ACTCAT$\$_1$
5	C	1	3	AT$\$_1$
6	C	2	3	AT$\$_2$
7	T	0	6	CACAT$\$_2$
8	A	2	4	CAT$\$_2$
9	T	3	4	CAT$\$_1$
10	A	1	6	CTCAT$\$_1$
11	A	0	2	T$\$_1$
12	A	1	2	T$\$_2$
13	$\$_1$	1	7	TCACAT$\$_2$
14	C	3	5	TCAT$\$_1$

i	BWT	LCP	CL	context
1	G	0	1	$\$_1$
2	C	0	3	AG$\$_1$
3	T	0	4	CAG$\$_1$
4	G	1	6	CTCAG$\$_1$
5	A	0	2	G$\$_1$
6	$\$_1$	1	7	GCTCAG$\$_1$
7	C	0	5	TCAG$\$_1$

(a)   (b)   (c)

Fig. 2. The BWT, LCP and CL arrays output by eGap for genomes (a) S_1 and (b) S_2. (c) Merged BWT, LCP, CL arrays and DA for S_1S_2. The context column is not produced by eGap.

We also compute two extra arrays, LCS$[1, m]$ and KL$[1, m]$. The LCS array contains the longest common suffix between consecutive k-mers in *Node* and the KL array contains the size of each vertex label not including the $ symbols. These arrays are obtained from LCP and CL.

Consider the merged arrays of genomes S_1 and S_2 obtained in Phase 1 and $k = 3$. The colored BOSS representation obtained for S_1 and S_2 is shown in Fig. 3.

Phase 3: The distances between each pair of genomes are computed by evaluating the BWSD on the `colors` bitvector, obtaining the expectation and entropy distance matrices. Note that the colored BOSS representation contains the edges of every k'-mer from the merged genomes, for $1 \leq k' \leq k$. These edges are part of the BOSS representation and are needed by the navigation operations. Since we are just interested in the k-mers for the comparisons, we filtered out all the edges of the colored BOSS where KL$[j] < k$, for $1 \leq j \leq m$, during the BWSD computation.

From the colored BOSS shown in Fig. 3 filtering edges from k'-mers of size smaller than k we use the `colors` bitvector as the α bitvector of the BWSD, thus we have $\alpha = \{0, 0, 0, 1, 0, 0, 0, 1, 1, 0, 0, 1\}, r = 0^3 1^1 0^3 1^2 0^2 1^1, t_1 = 2, t_2 =$

i	$last$	$Node$	W	W^-	color	coverage	LCS	KL
1	1	$\$_1$	T	1	0	2	0	0
2	1	$\$_3$	G	1	1	1	0	0
3	1	ACA	C	1	0	1	0	3
4	1	TCA	$\$_3$	1	0	1	2	3
5	1	$\$_3$GA	C	1	1	1	1	2
6	1	$\$_1$TA	C	1	0	2	1	2
8	1	CAC	T	1	0	1	0	3
9	1	GAC	T	0	1	1	2	3
10	0	TAC	A	1	0	1	2	3
11	1	TAC	T	0	0	1	3	3
12	0	CTC	A	1	0	1	1	3
13	1	CTC	G	1	1	1	4	3
14	1	$\$_3$G	A	1	1	1	0	1
15	1	TCG	$\$_2$	1	1	1	1	3
16	1	$\$_1$T	A	1	0	1	0	1
17	0	ACT	$\$_1$	1	0	1	1	3
18	0	ACT	C	1	0	1	3	3
19	1	ACT	C	0	1	1	3	3

Fig. 3. $\mathcal{S}_1\mathcal{S}_2$ merged colored BOSS representation with $k = 3$. Lines where KL values are colored red represent edges from vertices not containing a symbol \$. In this example, we use $\$_i$ instead of k symbols \$. (Color figure online)

$2, t_3 = 2$ and $s = 6$. Hence, the BWSD$(\mathcal{S}_1, \mathcal{S}_2)$ is

$$P\{k_j = 1\} = \frac{2}{6}, P\{k_j = 2\} = \frac{2}{6}, P\{k_j = 3\} = \frac{2}{6}$$

Computing the distances we have $D_M(\mathcal{S}_1, \mathcal{S}_2) = 1$ and $D_E(\mathcal{S}_1, \mathcal{S}_2) = 1.584$.

Coverage Information. When handling a collection of reads from a real genome, each $(k + 1)$-mer may occur multiple times and we can use this information in the BWSD to weight the edges of the graph, aiming at improving the accuracy of the results. The same $(k + 1)$-mer from distinct genomes can be detected in the colored BOSS using the LCS array and the `colors` bitvector. These $(k+1)$-mers will appear in the α array with a 0 followed by a 1. Note that this happens only once independently of the number of times these $(k + 1)$-mers occurred in both genomes. Whenever these $(k + 1)$-mers occurred many times in both genomes, their similarity should be increased. To do that we also use the `coverage` array during the BWSD computation.

For example, we added the string ACTC in sets $\mathcal{S}_1$ and $\mathcal{S}_2$ from the previous example. Let $\mathcal{S}_1' = \{\text{TACTCA}, \text{TACACT}, \text{ACTC}, \text{ACTC}, \text{ACTC}\}$ and $\mathcal{S}_2' = \{\text{GACTCG}, \text{ACTC}, \text{ACTC}\}$. In both genomes we have to increment the coverage information of the k-mers ACT with the outgoing edge C. The updated lines of the BOSS representation are shown in Fig. 4.

Let $\alpha' = \{0, 0, 0, 1, 0, 0, 0, 1, 1, 0, 0, 1\}$ be a bitvector equal to α from the previous example. The last 0 and 1 values from α' represent the $(k + 1)$-mer

i	$last$	$Node$	W	W^-	color	coverage	LCS	KL
⋮	⋮	⋮	⋮	⋮	⋮	⋮	⋮	⋮
18	0	ACT	C	1	0	4	3	3
19	1	ACT	C	0	1	3	3	3

Fig. 4. Lines with `coverage` incremented for the $(k+1)$-mer `ACTC` in the colored BOSS for $\{S_1', S_2'\}$. (Color figure online)

`ACTC` from both genomes. We apply the coverage value to the positions of r' where these values occurred. That is, $r' = 0^3 1^1 0^3 1^2 0^{1+3} 1^{1+2} = 0^3 1^1 0^3 1^2 0^4 1^3$. Finally, we expand r' in the positions of the equal $(k+1)$-mers while merging them, that is $r' = 0^3 1^1 0^3 1^2 0^1 0^1 1^1 0^1 1^1 0^1 1^1 0^1 1^0$. Then, we have $t_1 = 8$, $t_2 = 1$, $t_3 = 2$ and $s = 11$. And the BWSD$(S_1' S_2')$ is

$$P\{k_j = 1\} = \frac{9}{12}, P\{k_j = 2\} = \frac{1}{12}, P\{k_j = 3\} = \frac{2}{12}$$

Computing the distances we have $D_M(S_1' S_2') = 0.41666$ and $D_E(S_1' S_2') = 1.04085$. The effect of coverage on the similarity was analysed in our experiments.

Time and Space Analysis. Let n_1 and n_2 be the sizes of two genomes.

Phase 1 takes $O((n_1 + n_2)\mathsf{maxlcp})$ time to construct and merge the BWT, LCP and CL in external memory with eGap, where maxlcp is the maximum in LCP.

Phase 2 takes $O(n_1 + n_2)$ time to construct the BOSS representation. Let m be the number of edges in the colored BOSS. The space required for the BOSS representation is $5m + 6\log m$ bits, as shown in Sect. 2.2. The `colors` bitvector and the `coverage` array require extra m bits and $4m$ bytes respectively. For reads with less than 65K symbols both LCS and KL can be stored in arrays of short integers, that is, $2m$ bytes for each one. Therefore, the overall space required is $6m$ bytes plus $6m + 6\log m$ bits.

Phase 3 takes $O(m)$ time to compute the BWSD from the `colors` bitvector, LCS and KL arrays. The arrays r and t require $O(m)$ bytes.

4   Experiments

We evaluated gcBB by reconstructing the phylogeny of the 12 Drosophila species in Table 1, obtained from FlyBase [21]. The reads were obtained with a NextSeq 500 sequencer[1] and have 302 bp on average, except that reads of *D. grimshawi* were obtained with a MinION sequencer[2] and have 6,520 bp on the average.

Our algorithm was implemented in C and compiled with `gcc` version 4.9.2. The source code can be accessed at github[3]. We used eGap [5] to construct

---

[1] https://www.illumina.com/systems/sequencing-platforms/nextseq.html.
[2] https://nanoporetech.com/products/minion.
[3] https://github.com/lucaspr98/gcBB.

and merge the data structures during Phase 1. The experiments were conducted on a system with Debian GNU/Linux 4.9.2 64 bits on Intel Xeon E5-2630 v3 20M Cache 2.40 GHz processors, 378 GB of RAM and 13 TB SAS storage. Our experiments were limited to 48 GB of RAM.

Table 1. Information on the genomes of Drosophilas, that can be accessed through their Run (SRR) or BioSample (SAMN) accessions at https://www.ncbi.nlm.nih.gov/genbank/. The Bases column has the number of sequenced bases in Gbp. The Reference column has the size of the complete genome in Mb. The sizes of data structures in Gb are shown in columns BWT, LCP and CL, and the average LCP is shown in column LCP avg.

Organism	Run	BioSample	Bases	Reference	BWT	LCP	CL	LCP avg
D. melanogaster	6702604	08511563	6.20	138.93	5.9	12	12	61.54
D. ananassae	6425991	08272423	7.13	215.47	6.7	14	14	55.56
D. simulans	6425999	08272428	9.22	131.66	8.7	18	18	58.96
D. virilis	6426000	08272429	11.16	189.44	11	21	21	55.06
D. willistoni	6426003	08272432	11.66	246.98	11	22	22	57.79
D. pseudoobscura	6426001	08272435	12.28	163.29	12	23	23	58.72
D. mojavensis	6425997	08272426	12.45	163.17	12	24	24	58.09
D. yakuba	6426004	08272438	12.78	147.90	12	24	24	59.98
D. persimilis	6425998	08272433	13.32	195.51	13	25	25	58.76
D. erecta	6425990	08272424	14.01	146.54	14	27	27	60.47
D. sechellia	6426002	08272427	14.44	154.19	14	27	27	61.25
D. grimshawi	13070661	16729613	14.50	191.38	14	27	27	43.07

During Phase 1, eGap was set to use up to 48 GB of RAM. The running time for each genome and the sizes of the arrays can be seen in Table 1. The longest running time was approximately 57 h, with the resulting arrays taking about 68 GB of space on disk.

The average time to merge the data structures of each pair in Phase 1 depends on their sizes. The fastest merge took approximately 27 h, between D. ananassae and D. melanogaster, while the longest merge took approximately 60 h, between D. grimshawi and D. sechellia. The size of the merged files was approximately the sum of the sizes of the input files. The document array file has the same size of the merged BWT file, since both store each value using one byte.

During Phase 2, the longest colored BOSS construction took approximately 1.5 h, while in Phase 3 the longest BWSD computation took less than 10 min.

The goal of our experiments was to analyse the ability of gcBB in reconstructing a phylogeny in agreement with the one by Hahn et al. [6], which we will refer to as *reference phylogeny* in the sequel. We ran gcBB for $k = 15$, 31 and 63, producing entropy and expectation BWSD distance matrices, with and without coverage information. From the output of our algorithm, we used

Table 2. Robinson-Foulds distances computed between phylogenies by gcBB and the reference phylogeny of Drosophila genomes in Table 1. The symbol c indicates the phylogenies constructed by gcBB using coverage information.

	15	15c	31	31c	63	63c
Entropy	7	2	2	1	2	1
Expectation	6	5	2	3	2	2

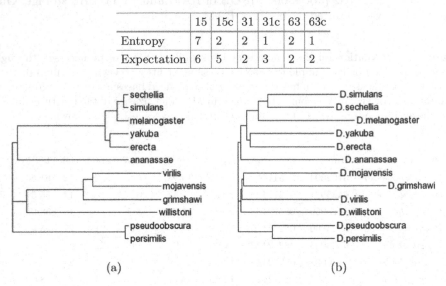

(a)   (b)

Fig. 5. (a) Reference phylogeny and (b) gcBB phylogeny for Drosophilas with $k = 31$, coverage information and entropy. These phylogenies were generated using T-Rex [1].

the Neighbor-Joining [19] to reconstruct the phylogenies. Finally, we used the Robinson-Foulds [18] distance to compare our phylogenies with the references.

Table 2 shows the Robinson-Foulds distance evaluated between the phylogenies by gcBB and the reference phylogeny. The phylogenies by gcBB which are closer to the reference were constructed using $k = 31$ and $k = 63$, with coverage information and the entropy measure. The reference phylogeny and the phylogeny produced by gcBB for $k = 31$ are shown in Fig. 5. There is one inconsistency involving *D. grimshawi* and *D. virilis*, which are swapped in our phylogeny, but in the same subtree. Nonetheless, the high level groups division agrees with the reference phylogeny.

In order to evaluate the effect of read sizes in the resulting phylogenies, we considered sequencing data from an Illumina HiSeq 2000 for *D. grimshawi*. The information on this genome, the running time taken by eGap to construct the data structures and their sizes are shown in Table 3.

We executed gcBB using the same parameters and values of k. The best phylogeny was obtained with $k = 15$ and using coverage information. By computing the Robinson-Foulds distance between these phylogenies and the reference phylogeny we obtained the values in Table 4.

We speculate that significantly different amounts of sequenced bases impairs gcBB in its current form. In this experiment *D. grimshawi* has 1.80G sequenced bases, while *D. sechellia* and *D. simulans* have more than 14G sequenced bases.

Table 3. Information on the genome of *D. grimshawi*, that can be accessed through its Run (SRR) or BioSample (SAMN) accessions at https://www.ncbi.nlm.nih.gov/genbank/. The Bases column has the number of sequenced bases in Gbp. The Reference column has the size of the complete genome in Mb. The sizes of data structures in Gb are shown in columns BWT, LCP and CL, and the average LCP is shown in column LCP avg.

Organism	Run	BioSample	Bases	Reference	BWT	LCP	CL	LCP avg
D. grimshawi	7642855	09764638	1.80	191.38	1.80	3.5	3.5	28.74

Table 4. Robinson-Foulds distance computed between phylogenies with *D. grimshawi* from another experiment and the reference phylogeny. The symbol c represents the phylogenies constructed using coverage information.

	15	15c	31	31c	63	63c
Entropy	7	5	5	5	5	5
Expectation	6	7	6	5	6	6

When constructing the BOSS representation for *D. grimshawi* and *D. sechellia* there will be much more edges from *D. sechellia* than from *D. grimshawi,* and the similarity between these genomes tends to be small. Moreover, when constructing the BOSS representation for *D. grimshawi* and *D. melanogaster* there will also be much more edges from *D. melanogaster*. The difference between the amount of bases from *D. melanogaster* and *D. grimshawi* is around 4 GB, while from *D. sechellia* to *D. grimshawi* is around 12 GB.

These results suggest that our method produces reasonable phylogenies when k is closer to the average LCP of the reads. Also, the usage of coverage information reduced the Robinson-Foulds distance to the reference in most cases. Finally, the fact that all reads in the dataset were obtained using similar sequencing protocols and, on average, have a similar number of sequenced bases may have helped obtaining favorable results.

5   Conclusions and Future Work

In this work we introduced a new method to compare genomes prior to assembly using space-efficient data structures implemented as gcBB, an algorithm to compare sets of reads of genomes using the BOSS representation and to compute the similarity measures based on the BWSD.

We evaluated our algorithm reconstructing the phylogeny of 12 *Drosophila* genomes. We used Neighbor-Joining over the matrices output by gcBB to reconstruct phylogenetic trees. Then we computed the Robinson-Foulds distance between the phylogenies by gcBB and a reference phylogeny. One issue when working with the de Bruijn graph is setting the value of k. We observed that values over the average LCP of the genomes lead to reasonable results. We observed better results using the entropy measure and coverage information in the BWSD computation.

Further experiments may help understanding the limits and the advantages of the approach introduced in this work. Future research may also investigate different strategies for dealing with coverage information, as the experiments indicate a positive contribution of coverage to the resulting phylogenies. The quality of sequenced bases may also be investigated in future work as a means to improve the method.

Acknowledgements. The authors thank Prof. Marinella Sciortino for helpful discussions and thank Prof. Nalvo Almeida for granting access to the computer used in the experiments.

Funding. L.P.R. acknowledges that this study was financed by Coordenação de Aperfeiçoamento de Pessoal de Nível Superior (CAPES), Brazil, Financing Code 001. F.A.L. acknowledges the financial support from CNPq (grant number 406418/2021-7) and FAPEMIG (grant number APQ-01217-22). G.P.T. acknowledges the financial support of Brazilian agencies CNPq and CAPES.

References

1. Boc, A., Diallo, A.B., Makarenkov, V.: T-REX: a web server for inferring, validating and visualizing phylogenetic trees and networks. Nucleic Acids Res. **40**(W1), W573–W579 (2012)
2. Bowe, A., Onodera, T., Sadakane, K., Shibuya, T.: Succinct de Bruijn graphs. In: Raphael, B., Tang, J. (eds.) WABI 2012. LNCS, vol. 7534, pp. 225–235. Springer, Heidelberg (2012). https://doi.org/10.1007/978-3-642-33122-0_18
3. Burrows, M., Wheeler, D.J.: A block-sorting lossless data compression algorithm. Technical report. 124, Systems Research Center (1994)
4. De Bruijn, N.G.: A combinatorial problem. In: Proceedings of the Koninklijke Nederlandse Academie van Wetenschappen, vol. 49, pp. 758–764 (1946)
5. Egidi, L., Louza, F.A., Manzini, G., Telles, G.P.: External memory BWT and LCP computation for sequence collections with applications. Algorithms Mol. Biol. **14**(1), 1–15 (2019)
6. Hahn, M.W., Han, M.V., Han, S.G.: Gene family evolution across 12 drosophila genomes. PLoS Genet. **3**(11), e197 (2007)
7. Kolmogorov, M., et al.: metaFlye: scalable long-read metagenome assembly using repeat graphs. Nat. Methods **17**(11), 1103–1110 (2020)
8. Langmead, B., Salzberg, S.L.: Fast gapped-read alignment with Bowtie 2. Nat. Methods **9**(4), 357 (2012)
9. Li, H., Durbin, R.: Fast and accurate short read alignment with Burrows-Wheeler transform. Bioinformatics **25**(14), 1754–1760 (2009)
10. Lyman, C.A., et al.: Whole genome phylogenetic tree reconstruction using colored de Bruijn graphs. In: 2017 IEEE 17th International Conference on Bioinformatics and Bioengineering (BIBE), pp. 260–265. IEEE (2017)
11. Manber, U., Myers, G.: Suffix arrays: a new method for on-line string searches. SIAM J. Comput. **22**(5), 935–948 (1993)

12. Mantaci, S., Restivo, A., Rosone, G., Sciortino, M.: An extension of the Burrows-Wheeler transform. Theor. Comput. Sci. **387**(3), 298–312 (2007)
13. Mantaci, S., Restivo, A., Sciortino, M.: Distance measures for biological sequences: some recent approaches. Int. J. Approximate Reasoning **47**(1), 109–124 (2008)
14. Navarro, G.: Compact Data Structures: A Practical Approach. Cambridge University Press, Cambridge (2016)
15. Polevikov, E., Kolmogorov, M.: Synteny paths for assembly graphs comparison. In: 19th International Workshop on Algorithms in Bioinformatics (WABI 2019). Schloss Dagstuhl-Leibniz-Zentrum fuer Informatik (2019)
16. Rice, E.S., Green, R.E.: New approaches for genome assembly and scaffolding. Ann. Rev. Animal Biosci. **7**(1), 17–40 (2019)
17. Rizzi, R., et al.: Overlap graphs and de Bruijn graphs: data structures for de novo genome assembly in the big data era. Quant. Biol. **7**(4), 278–292 (2019)
18. Robinson, D.F., Foulds, L.R.: Comparison of phylogenetic trees. Math. Biosci. **53**(1–2), 131–147 (1981)
19. Saitou, N., Nei, M.: The neighbor-joining method: a new method for reconstructing phylogenetic trees. Mol. Biol. Evol. **4**(4), 406–425 (1987)
20. Simpson, J.T., Durbin, R.: Efficient de novo assembly of large genomes using compressed data structures. Genome Res. **22**(3), 549–556 (2012)
21. Thurmond, J., et al.: FlyBase 2.0: the next generation. Nucleic Acids Res. **47**(D1), D759–D765 (2018)
22. Yang, L., Zhang, X., Wang, T.: The Burrows-Wheeler similarity distribution between biological sequences based on Burrows-Wheeler transform. J. Theor. Biol. **262**(4), 742–749 (2010)

Sorting Genomes by Prefix Double-Cut-and-Joins

Guillaume Fertin[1], Géraldine Jean[1], and Anthony Labarre[2]([✉])

[1] Nantes Université, CNRS, LS2N, UMR 6004, 44000 Nantes, France
{guillaume.fertin,geraldine.jean}@univ-nantes.fr
[2] LIGM, CNRS, Université Gustave Eiffel, 77454 Marne-la-Vallée, France
anthony.labarre@univ-eiffel.fr

Abstract. In this paper, we study the problem of sorting unichromosomal linear genomes by prefix double-cut-and-joins (or DCJs) in both
the signed and the unsigned settings. Prefix DCJs cut the leftmost segment of a genome and any other segment, and recombine the severed
endpoints in one of two possible ways: one of these options corresponds
to a prefix reversal, which reverses the order of elements between the two
cuts (as well as their signs in the signed case). Depending on whether
we consider both options or reversals only, our main results are: (1) new
structural lower bounds based on the breakpoint graph for sorting by
unsigned prefix reversals, unsigned prefix DCJs, or signed prefix DCJs;
(2) a polynomial-time algorithm for sorting by signed prefix DCJs, thus
answering an open question in [7]; (3) a 3/2-approximation for sorting
by unsigned prefix DCJs, which is, to the best of our knowledge, the
first sorting by *prefix* rearrangements problem that admits an approximation ratio strictly smaller than 2 (with the obvious exception of the
polynomial-time solvable problems); and finally, (4) an FPT algorithm
for sorting by unsigned prefix DCJs parameterised by the number of
breakpoints in the genome.

Keywords: Genome rearrangements · Prefix reversals · Prefix DCJs ·
Lower bounds · Algorithmics · FPT · Approximation algorithms

1 Introduction

Genome rearrangements is a classical paradigm to study evolution between
species. The rationale is to consider species by observing their genomes, which
are usually represented as ordered sets of elements (the genes) that can be signed
(according to gene strand when known). A genome can then evolve by changing
the order of its genes, through operations called *rearrangements*, which can be
generally described as cutting the genome at different locations, thus forming segments, and rearranging these segments in a different fashion. Given two genomes,
a *sorting scenario* is a sequence of rearrangements transforming the first genome
into the other. The length of a shortest such sequence of rearrangements is called
the rearrangement distance. Several specific rearrangements such as reversals,

© The Author(s), under exclusive license to Springer Nature Switzerland AG 2022
D. Arroyuelo and B. Poblete (Eds.): SPIRE 2022, LNCS 13617, pp. 178–190, 2022.
https://doi.org/10.1007/978-3-031-20643-6_13

translocations, fissions, fusions, transpositions, and block-interchanges have been defined, and the rearrangement distance together with its corresponding sorting problem have been widely studied either by considering one unique type of rearrangement or by allowing the combination of some of them [5]. The *double-cut-and-join* (or DCJ) operation introduced by Yancopoulos *et al.* [10] encompasses all the rearrangements mentioned above: it consists in cutting the genome in two different places and joining the four extremities in any possible way. A DCJ is a *prefix DCJ* whenever one cut is applied to the leftmost position of the genome. The prefix restriction can be applied to other rearrangements such as *prefix reversals*, which prefix DCJs generalise. Whereas the computational complexity of the sorting problems by unrestricted rearrangements has been thoroughly studied and pretty well characterised, there is still a lot of work to do to understand the corresponding prefix sorting problems (see Table 1 in [7] for a summary of existing results). Our interest in prefix rearrangements is therefore mostly theoretical: techniques that apply in the unrestricted setting do not directly apply under the prefix restriction, and new approaches are therefore needed to make progress on algorithmic issues and complexity aspects. Since DCJs generalise several other operations, we hope that the insight we gain through their study will shed light on other prefix rearrangement problems.

In this paper, we study the problem of SORTING BY PREFIX DCJs and, for the sake of simplicity, we consider the case where the source and the target genomes are unichromosomal and linear. This implies that genomes can be seen as (signed) permutations (depending on whether the gene orientation is known or not). Moreover, prefix DCJs applied to such genomes allow to exactly mimick three kinds of rearrangement: (i) a *prefix reversal* when the segment between the two cuts is reversed; (ii) a *cycle extraction* when the extremities of the segment between the two cuts are joined; (iii) a *cycle reincorporation* when the cut occurs in a cycle and the resulting linear segment is reincorporated at the beginning of the genome where the leftmost cut occurs.

Based on the study of the *breakpoint graph*, we first show new structural lower bounds for the problems SORTING BY UNSIGNED PREFIX DCJs and SORTING BY SIGNED PREFIX DCJs. Since prefix reversals are particular cases of prefix DCJs, we can extend this result to SORTING BY UNSIGNED PREFIX REVERSALS (it has been already shown for SORTING BY SIGNED PREFIX REVERSALS in [8]). Thanks to these preliminary results, we are able to answer an open question from [7] by proving that SORTING BY SIGNED PREFIX DCJs is in P just like the unrestricted case [10]. However, while sorting by unsigned DCJs is NP-hard [4], the computational complexity of the prefix-constrained version of this problem is still unknown. We provide two additional results: a 3/2-approximation algorithm, which is, to the best of our knowledge, the first sorting by *prefix* rearrangements problem that admits an approximation ratio strictly smaller than 2 (with the obvious exception of the polynomial-time solvable problems); and an FPT algorithm parameterised by the number of breakpoints in the genome. Due to space constraints, some proofs have been omitted.

1.1   Permutations, Genomes, and Rearrangements

We begin with the simplest models for representing organisms.

Definition 1. *A* (unsigned) permutation *of* $[n] = \{1, 2, \ldots, n\}$ *is a bijective application of* $[n]$ *onto itself. A* signed permutation *of* $\{\pm 1, \pm 2, \ldots, \pm n\}$ *is a bijective application of* $\{\pm 1, \pm 2, \ldots, \pm n\}$ *onto itself that satisfies* $\pi_{-i} = -\pi_i$. *The* identity permutation *is the permutation* $\iota = (1\ 2\ \cdots\ n)$.

We study transformations based on the following well-known operation.

Definition 2. *A* reversal $\rho(i, j)$ *with* $1 \le i < j \le n$ *is a permutation that reverses the order of elements between positions* i *and* j:

$$\rho(i, j) = \begin{pmatrix} 1 & \cdots & i-1 & i & i+1 & \cdots & j-1 & j & j+1 & \cdots & n \\ 1 & \cdots & i-1 & j & j-1 & \cdots & i+1 & i & j+1 & \cdots & n \end{pmatrix}.$$

A signed reversal $\overline{\rho}(i, j)$ *with* $1 \le i \le j \le n$ *is a signed permutation that reverses both the order and the signs of elements between positions* i *and* j:

$$\overline{\rho}(i, j) = \begin{pmatrix} 1 & \cdots & i-1 & i & i+1 & \cdots & j-1 & j & j+1 & \cdots & n \\ 1 & \cdots & i-1 & -j & -(j-1) & \cdots & -(i+1) & -i & j+1 & \cdots & n \end{pmatrix}.$$

If $i = 1$, *then* $\rho(i, j)$ *(resp.* $\overline{\rho}(i, j)$) *is called a* prefix (signed) reversal.

A reversal ρ applied to a permutation π transforms it into another permutation $\sigma = \pi\rho$. When the distinction matters, we mention whether objects or transformations are signed or unsigned; otherwise, we omit those qualifiers to lighten the presentation. The following model is a straightforward generalisation of unsigned permutations.

Definition 3. *A* genome G *is a collection of vertex-disjoint paths and cycles over* $\{0, 1, 2, \ldots, n+1\}$. *It is* linear *if it consists of a single path with endpoints* 0 *and* $n+1$. *The* identity genome *is the path induced by the sequence* $(0, 1, 2, \ldots, n+1)$.

Let us note that a genome may contain loops or parallel edges (see Fig. 1).

Definition 4. *Let* $e = \{u, v\}$ *be an edge of a genome* G. *Then* e *is a* breakpoint *if* $0 \notin e$ *and either* $|u - v| \ne 1$, *or* e *has multiplicity two. Otherwise,* e *is an* adjacency. *The number of breakpoints of* G *is denoted by* $b(G)$.

For instance, the genome with edge set $\{\{0, 4\}, \{4, 3\}, \underline{\{3, 6\}}, \{1, 2\}, \underline{\{2, 1\}}, \underline{\{5, 5\}}\}$ has three breakpoints (underlined). Note that permutations can be viewed as linear genomes using the following simple transformation: given a permutation π, extend it by adding two new elements $\pi_0 = 0$ and $\pi_{n+1} = n+1$, and build the linear genome G_π with edge set $\{\{\pi_i, \pi_{i+1}\} \mid 0 \le i \le n\}$. This allows us to use the notion of breakpoints on permutations as well, with the understanding that they apply to the extended permutation, and therefore $b(\pi) = b(G_\pi)$.

A reversal can be thought of as an operation that "cuts" (i.e., removes) two edges from a genome, then "joins" the severed endpoints (by adding two new edges) in such a way that the segment between the cuts is now reversed (see G_1 in Fig. 1). The following operation builds on that view to generalise reversals.

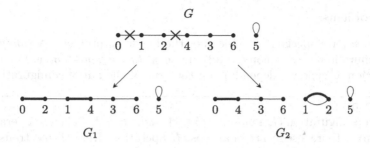

Fig. 1. Cutting edges $\{0,1\}$ and $\{2,4\}$ from the nonlinear genome G produces genome G_1 with a reversed segment, if we add edges $\{0,2\}$ and $\{1,4\}$, or genome G_2 with an extracted cycle if we add $\{0,4\}$ and $\{1,2\}$ instead.

Definition 5 [10]. *Let $e = \{u,v\} \neq f = \{w,x\}$ be two edges of a genome G. The* double-cut-and-join *(or DCJ for short) $\delta(e,f)$ applied to G transforms G into a genome G' by replacing edges e and f with either $\{\{u,w\},\{v,x\}\}$ or $\{\{u,x\},\{v,w\}\}$. δ is a* prefix DCJ *if either $0 \in e$ or $0 \in f$.*

DCJs that do not correspond to reversals extract paths from genomes and turn them into cycles (see G_2 in Fig. 1). Signed permutations can be generalised to signed genomes as well. The definition of a signed linear genome is more complicated than in the unsigned case, and is based on the following notion.

Definition 6. *Let π be a signed permutation. The* unsigned translation *of π is the unsigned permutation π' obtained by mapping π_i onto the sequence $(2\pi_i - 1, 2\pi_i)$ if $\pi_i > 0$, or $(2|\pi_i|, 2|\pi_i| - 1)$ if $\pi_i < 0$, for $1 \leq i \leq n$; and adding two new elements $\pi'_0 = 0$ and $\pi'_{2n+1} = 2n + 1$.*

Definition 7. *A* signed genome *G is a perfect matching over the set $\{0, 1, 2, \ldots, 2n + 1\}$. G is* linear *if there exists a signed permutation π such that $E(G) = \{\{\pi'_{2i}, \pi'_{2i+1}\} \mid 0 \leq i \leq n\}$. The* signed identity genome *is the perfect matching $\{\{2i, 2i + 1\} \mid 0 \leq i \leq n\}$.*

DCJs immediately generalise to signed genomes: they may cut any pair of edges of the perfect matching, and recombine their endpoints in one of two ways.

Finally, we will be using different kinds of graphs in this work with a common notation. The *length* of a cycle in a graph G is the number of elements[1] it contains, and a k-cycle is a cycle of length k: it is *trivial* if $k = 1$, and *nontrivial* otherwise. We let $c(G)$ (resp. $c_1(G)$) denote the number of cycles (resp. 1-cycles) in G.

---

[1] The definition of an element will depend on the graph structure and will be explicitly stressed.

1.2   Problems

We study several specialised versions of the following problem. A *configuration* is a permutation or a genome, and the *identity configuration* is the identity permutation or genome, depending on the type of the initial configuration.

SORTING BY Ω

Input: a configuration G, a number $K \in \mathbb{N}$, and a set Ω of allowed operations.
Question: is there a sequence of at most K operations from Ω that transforms G into the identity configuration?

Specific choices for Ω and the model chosen for G yield the following variants:

- SORTING BY UNSIGNED PREFIX DCJs, where G is a linear genome and Ω is the set of all prefix DCJs;
- SORTING BY SIGNED PREFIX DCJs, where G is a signed linear genome and Ω is the set of all prefix DCJs;
- SORTING BY UNSIGNED PREFIX REVERSALS, where G is an unsigned permutation and Ω is the set of all prefix reversals;
- SORTING BY SIGNED PREFIX REVERSALS, where G is a signed permutation and Ω is the set of all prefix signed reversals.

We refer to the smallest number of operations needed to transform G into the identity configuration as the Ω-*distance* of G. A specific distance is associated to each of the above problems; we use the following notation:

- $pdcj(G)$ for the prefix DCJ distance of an unsigned genome G, and $psdcj(G)$ for its signed version;
- $prd(\pi)$ for the prefix reversal distance of an unsigned permutation π, and $psrd(\pi)$ for its signed version.

2   A Generic Lower Bounding Technique

We present in this section a lower bounding technique which applies to both the signed and the unsigned models, and on which we will build in subsequent sections to obtain exact or approximation algorithms.

2.1   The Signed Case

We generalise a lower bounding technique introduced in the context of SORTING BY SIGNED PREFIX REVERSALS [8]. It is based on the following structure.

Definition 8 [2]. *Given a signed permutation π, let π' be its unsigned translation. The* breakpoint graph *of π is the undirected edge-bicoloured graph $BG(\pi)$ with ordered vertex set $(\pi'_0 = 0, \pi'_1, \pi'_2, \ldots, \pi'_{2n}, \pi'_{2n+1} = 2n + 1)$ and whose edge set consists of:*

– *black edges* $\{\pi'_{2i}, \pi'_{2i+1}\}$ *for* $0 \leq i \leq n$;
– *grey edges* $\{\pi'_{2i}, \pi'_{2i} + 1\}$ *for* $0 \leq i \leq n$.

See Fig. 2 for an example. Following Definition 7, the *breakpoint graph of a signed linear genome* is simply the union of that genome (which plays the role of black edges) and of the signed identity genome (which plays the role of grey edges). Breakpoint graphs are 2-regular and as such are the union of disjoint cycles whose edges alternate between both colours, thereby referred to as *alternating cycles*. Black edges play the role of elements in that graph, so the *length* of a cycle in a breakpoint graph is the number of black edges it contains.

$$0 \quad 2 \quad 1 \quad 7 \quad 8 \quad 6 \quad 5 \quad 4 \quad 3 \quad 9 \quad 10 \quad 11$$

Fig. 2. The breakpoint graph $BG(\pi)$ of $\pi = -1\ 4\ -3\ -2\ 5$.

To bound the prefix DCJ distance, we use a connection between the effect of a DCJ on the breakpoint graph and the effect of *algebraic transpositions*, or *exchanges*, on the classical cycles of a permutation.

Definition 9. *An* exchange $\varepsilon(i, j)$ *with* $1 \leq i < j \leq n$ *is a permutation that swaps elements in positions* i *and* j:

$$\varepsilon(i,j) = \begin{pmatrix} 1 & \cdots & i-1 & \boxed{i} & i+1 & \cdots & j-1 & \boxed{j} & j+1 & \cdots & n \\ 1 & \cdots & i-1 & \boxed{j} & i+1 & \cdots & j-1 & \boxed{i} & j+1 & \cdots & n \end{pmatrix}.$$

If $i = 1$, *then* $\varepsilon(i, j)$ *is called a* prefix exchange.

We let $\Gamma(\pi)$ denote the (directed) graph of a permutation π, with vertex set $[n]$ and which contains an arc (i, j) whenever $\pi_i = j$. Exchanges act on two elements that belong either to the same cycle in $\Gamma(\pi)$ or to two different cycles, and therefore $|c(\Gamma(\pi)) - c(\Gamma(\pi \varepsilon(i, j)))| \leq 1$. The following result allows the computation of the prefix exchange distance $ped(\pi)$ in polynomial time, and will be useful to our purposes.

Theorem 1 [1]. *For any unsigned permutation* π, *we have*

$$ped(\pi) = n + c(\Gamma(\pi)) - 2c_1(\Gamma(\pi)) - \begin{cases} 0 \text{ if } \pi_1 = 1, \\ 2 \text{ otherwise.} \end{cases}$$

Theorem 2. *For any signed linear genome* G, *we have*

$$psdcj(G) \geq n + 1 + c(BG(G)) - 2c_1(BG(G)) - \begin{cases} 0 \text{ if } \{0, 1\} \in G, \\ 2 \text{ otherwise.} \end{cases} \tag{1}$$

Proof. As observed in [10], a DCJ acts on at most two cycles of $BG(G)$ and can therefore change the number of cycles by at most one. This analogy with the effect of exchanges on the cycles of a permutation is preserved under the prefix constraint, and the lower bound then follows from Theorem 1.   □

Since (prefix) signed reversals are a subset of (prefix) signed DCJs, the result below from [8] is a simple corollary of Theorem 2.

Theorem 3 [8]. *For any signed permutation π, we have*

$$psrd(\pi) \geq n + 1 + c(BG(\pi)) - 2c_1(BG(\pi)) - \begin{cases} 0 \text{ if } \pi_1 = 1, \\ 2 \text{ otherwise.} \end{cases} \qquad (2)$$

2.2   The Unsigned Case

We now show that our lower bounds apply to the unsigned setting as well. The definition of the breakpoint graph in the unsigned case is slightly different, but the definition of the length of a cycle remains unchanged.

Definition 10 [2]. *The* unsigned breakpoint graph *of an unsigned permutation π is the undirected edge-bicoloured graph $UBG(\pi)$ with ordered vertex set ($\pi_0 = 0, \pi_1, \pi_2, \ldots, \pi_n, \pi_{n+1} = n + 1$) and whose edge set consists of:*

- *black edges $\{\pi_i, \pi_{i+1}\}$ for $0 \leq i \leq n$;*
- *grey edges $\{\pi_i, \pi_i + 1\}$ for $0 \leq i \leq n$.*

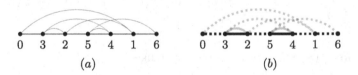

$$\begin{array}{ccccccc} 0 & 3 & 2 & 5 & 4 & 1 & 6 \\ & & & (a) & & & \end{array} \qquad \begin{array}{ccccccc} 0 & 3 & 2 & 5 & 4 & 1 & 6 \\ & & & (b) & & & \end{array}$$

Fig. 3. (a) The unsigned breakpoint graph $UBG(\pi)$ of $\pi = 3\ 2\ 5\ 4\ 1$; (b) an optimal decomposition of $UBG(\pi)$ into two trivial cycles (thick) and one 4-cycle (dotted).

Figure 3(a) shows an example of an unsigned breakpoint graph. Following Definition 3, the *breakpoint graph of an unsigned linear genome* is simply the union of that genome (which plays the role of black edges) and of the identity genome (which plays the role of grey edges). Vertices 0 and $n + 1$ in the unsigned breakpoint graph have degree 2, and all other vertices have degree 4. The unsigned breakpoint graph also decomposes into alternating cycles, but the decomposition is no longer unique.

For any genome G and an arbitrary decomposition $\mathscr{D}$ of $UBG(G)$, let $c^{\mathscr{D}}$ (resp. $c_1^{\mathscr{D}}$) denote the number of cycles (resp. trivial cycles) of $UBG(G)$ in $\mathscr{D}$. We call $\mathscr{D}$ *optimal* if it minimises $c^{\mathscr{D}} - 2c_1^{\mathscr{D}}$ (see Fig. 3(b)). The following result characterises optimal decompositions (proof omitted).

Lemma 1. *Let G be a genome and $\mathscr{D}$ be a decomposition of $UBG(G)$. Then $\mathscr{D}$ is optimal iff it maximises the number of trivial cycles and minimises the number of nontrivial cycles.*

As a result, we obtain the following lower bound on the prefix DCJ distance, where $c^*(UBG(G))$ and $c_1^*(UBG(G))$ denote, respectively, the number of cycles and the number of 1-cycles in an optimal decomposition of $UBG(G)$.

Theorem 4. *For any genome G, we have*

$$pdcj(G) \geq n + 1 + c^*(UBG(G)) - 2c_1^*(UBG(G)) - \begin{cases} 0 \text{ if } \{0,1\} \in G, \\ 2 \text{ otherwise.} \end{cases} \quad (3)$$

Proof. Follows from the fact that DCJs affect the number of cycles in a decomposition by at most one, Theorem 1, and Lemma 1.   □

As an immediate corollary, the above lower bound is also a lower bound on $prd(\pi)$, since (prefix) reversals are a subset of (prefix) DCJs.

Corollary 1. *For any unsigned permutation π, we have*

$$prd(\pi) \geq n + 1 + c^*(UBG(\pi)) - 2c_1^*(UBG(\pi)) - \begin{cases} 0 \text{ if } \pi_1 = 1, \\ 2 \text{ otherwise.} \end{cases} \quad (4)$$

We now show that an optimal decomposition can be found in polynomial time. This contrasts with the problem of finding an optimal decomposition in the case of sorting by unrestricted reversals, which was shown to be NP-complete [3] (note that in that context, an optimal decomposition *maximises* the number of cycles). Recall that an *alternating Eulerian cycle* in a bicoloured graph G is a cycle that traverses every edge of G exactly once and such that the colours of every pair of consecutive edges are distinct.

Corollary 2 [6,9]**.** *A bicoloured connected graph contains an alternating Eulerian cycle iff the number of incident edges of each colour is the same at every vertex.*

Proposition 1. *There exists a polynomial-time algorithm for computing an optimal decomposition for $UBG(G)$.*

Proof. Straightforward: extract all trivial cycles from $UBG(G)$. Each connected component in the resulting graph then corresponds to a cycle (Corollary 2).   □

Finally, we note that the lower bound of Theorem 4 is always at least as large as the number of breakpoints (proof omitted).

Proposition 2. *For any unsigned genome G, the lower bound from Eq. 3 is greater than or equal to $b(G)$, and the gap that separates both bounds can be arbitrarily large.*

3   Prefix DCJs

3.1   Signed Prefix DCJs

We give a polynomial-time algorithm for SORTING BY SIGNED PREFIX DCJs.

Theorem 5. *The* SORTING BY SIGNED PREFIX DCJs *problem is in* P.

Proof. We show that the lower bound of Theorem 2 is tight. For convenience, let $g(G)$ denote the right-hand side of Eq. 1, and let π' denote the unsigned translation of the underlying signed permutation π from which G is obtained (recall Definition 7 and the fact that G is linear):

– if $\pi'_1 \neq 1$: then the grey edge $\{\pi'_1, x\}$ connects by definition π'_1 to an element $x \in \{\pi'_1 - 1, \pi'_1 + 1\}$. Let $\{x, y\}$ be the black edge incident with x; then the prefix DCJ that replaces $\{0, \pi'_1\}$ and $\{x, y\}$ with $\{0, y\}$ and $\{\pi'_1, x\}$ creates one or two new 1-cycles, depending on the value of y. Let G' denote the resulting genome:

  1. if $y \neq 1$, then

$$g(G') - g(G) = n + 1 + c(BG(G)) + 1 - 2(c_1(BG(G)) + 1) - 2$$
$$- (n + 1 + c(BG(G)) - 2c_1(BG(G)) - 2)$$
$$= -1.$$

  2. if $y = 1$, then

$$g(G') - g(G) = n + 1 + c(BG(G)) + 1 - 2(c_1(BG(G) + 2))$$
$$- (n + 1 + c(BG(G)) - 2c_1(BG(G)) - 2)$$
$$= -1.$$

  Therefore, the value of the lower bound decreases by one in both cases.
– otherwise, let i be the smallest index such that $|\pi'_{2i-1} - \pi'_{2i}| \neq 1$. Then the prefix DCJ that replaces black edges $\{0, \pi'_1\}$ and $\{\pi'_{2i-1}, \pi'_{2i}\}$ with $\{0, \pi'_{2i-1}\}$ and $\{\pi'_1, \pi'_{2i}\}$ decreases the number of 1-cycles by 1. Let us again use G' to denote the resulting genome; then

$$g(G') - g(G) = n + 1 + c(BG(G)) - 1 - 2(c_1(BG(G)) - 1) - 2$$
$$- (n + 1 + c(BG(G)) - 2c_1(BG(G)))$$
$$= -1.$$

□

3.2   Unsigned Prefix DCJs

The complexity of the SORTING BY UNSIGNED PREFIX DCJs problem remains open, and we conjecture it to be NP-complete. Here, we prove two results, both based on the number of breakpoints. The first one is a 3/2-approximation algorithm for solving SORTING BY UNSIGNED PREFIX DCJs (Theorem 6), the second one is a FPT algorithm with respect to $b(G)$ (Theorem 7).

We start with our approximation algorithm. First, observe that prefix DCJs on linear genomes may produce nonlinear genomes, but the structure of these genomes is nonetheless not arbitrary. We characterise some of their properties in the following result, which will be useful later on.

Lemma 2. *Let G be a linear genome and S be an arbitrary sequence of prefix DCJs that transform G into a new genome G'. Then:*

1. *G' contains exactly one path, whose endpoints are 0 and $n + 1$;*
2. *if G' contains any other component, then that component is a cycle.*

Proof. By induction on $k = |S|$. If $k = 0$, then the claim clearly holds. Otherwise, let δ be a prefix DCJ that cuts edges $e = \{0, v\}$ and $f = \{w, x\}$ from a genome G'' obtained from G by $k - 1$ prefix DCJs; by hypothesis, 0 and $n + 1$ are the endpoints of the only path P of G''. If both e and f belong to P, then δ either extracts a subpath Q from P that will become a cycle, or reverses a subpath R of P; in both cases, neither Q nor R contains 0 nor $n + 1$, which become extremities of $P \setminus Q$ (or of the path obtained from P by reversing R). Otherwise, since $e = \{0, v\}$, by hypothesis f belongs to a cycle, and both ways of recombining the extremities of e and f yield a path starting with 0 and ending with $n + 1$, preserving any other cycle of G''. $\square$

We will need the following lower bound.

Lemma 3. *For any genome G, we have $pdcj(G) \geq b(G)$. Moreover, if G is unsorted and contains $\{0, 1\}$ and $\{1, 2\}$, then $pdcj(G) > b(G)$.*

Proof. The first claim follows directly from Theorem 4 and Proposition 2. For the second claim, if G is unsorted and contains $\{0, 1\}$ and $\{1, 2\}$, then any new edge $\{1, y\}$ that would replace $\{0, 1\}$ would yield a breakpoint — either because 1 and y cannot be consecutive in values or, in the event that $y = 2$, because edge $\{1, 2\}$ would get multiplicity 2 and thereby would also count as a breakpoint. $\square$

We are now ready to prove our upper bound on $pdcj(G)$.

Lemma 4. *For any linear genome G, we have $pdcj(G) \leq \frac{3b(G)}{2}$.*

Proof. Assume G is not the identity genome, in which case the claim trivially holds. We have two cases to consider:

1. if $\{0, v\} \in G$ with $v \neq 1$, then G contains an element $x \in \{v - 1, v + 1\}$ that is not adjacent to v. By Lemma 2, every vertex in G has degree 1 or 2, so x has a neighbour y such that $\{x, y\}$ is a breakpoint (either because $|x - y| \neq 1$ or because $\{x, y\}$ has multiplicity two). The prefix DCJ that replaces $\{0, v\}$ and the breakpoint $\{x, y\}$ with the adjacency $\{v, x\}$ and $\{0, y\}$ yields a genome G' with $b(G') = b(G) - 1$.

2. otherwise, $\{0,1\} \in G$. If $\{1,2\} \notin G$, then 2 has a neighbour y in G such that $\{2,y\}$ is a breakpoint, in which case the prefix DCJ that replaces $\{0,1\}$ and $\{2,y\}$ with $\{0,y\}$ and $\{1,2\}$ yields a genome G' with $b(G') = b(G) - 1$. If $\{1,2\} \in G$, then let k be the closest element to 0 in the only path of G such that the next vertex ℓ forms a breakpoint with k. Then the prefix DCJ δ_1 that replaces $\{0,1\}$ and $\{k,\ell\}$ with $\{0,\ell\}$ and $\{1,k\}$ yields a genome G' which contains the cycle $(1, 2, \ldots, k)$ and with $b(G') = b(G)$. Although δ_1 does not reduce the number of breakpoints, G' allows us to apply two subsequent operations that do:

(a) since $\{0,\ell\} \in G'$ with $\ell \neq 1$, the analysis of case 1 applies and guarantees the existence of a prefix DCJ δ_2 that produces a genome G'' with $b(G'') = b(G') - 1$.

(b) δ_2 replaces $\{0,\ell\}$ and breakpoint $\{a,b\}$ with $\{0,a\}$ and adjacency $\{b,\ell\}$. Since $\{k,\ell\}$ was a breakpoint in G, we have $k < \ell - 1$. Moreover, δ_1 extracted from G a cycle consisting of all elements in $\{1, 2, \ldots, k\}$. Therefore, the breakpoint $\{a,b\}$ cut by δ_2 belongs to a component of G'' different from that cycle, which means that $a > k > 1$ and in turn implies that case 1 applies again: there exists a third prefix DCJ δ_3 transforming G'' into a genome G''' such that $b(G''') = b(G'') - 1 = b(G) - 2$.

This implies that, in the worst case, i.e. when $\{0,1\} \in G$ and $\{1,2\} \in G$, there exists a sequence of three prefix DCJs that yields a genome G''' with $b(G''') = b(G) - 2$. Therefore, starting with $b(G)$ breakpoints, we can decrease this number by two using at most three prefix DCJs. Since the identity genome has no breakpoint, we conclude that $pdcj(G) \leq \frac{3b(G)}{2}$. □

Lemma 3 and Lemma 4 immediately imply the existence of a 3/2-approximation for sorting by prefix DCJs, as stated by the following theorem.

Theorem 6. *The* SORTING BY UNSIGNED PREFIX DCJs *problem is 3/2-approximable.*

Note that Lemma 4 also allows us to show that our approximation algorithm is tight for an unbounded number of genomes. Incidentally, this also shows that the lower bound of Eq. 3 is optimal for an unbounded number of genomes (proof omitted).

Observation 1. *There exists an unbounded number of genomes for which the algorithm described in proof of Lemma 4 is optimal.*

We now turn to proving that SORTING BY UNSIGNED PREFIX DCJs is FPT, as stated by the following theorem.

Theorem 7. *The* SORTING BY UNSIGNED PREFIX DCJs *problem is FPT parameterised by $b(G)$.*

Proof. The main idea is to use the search tree technique in a tree whose arity and depth are both bounded by a function of $b(G)$. For this, we will use the notion of *strip* in a genome G, which is defined as a maximal set of consecutive edges (in a path or a cycle of G) that contains no breakpoint. The *length* of a strip is the number of elements it contains, strips of length k are called k-strips; 1-strips are also called *singletons*, and strips of length > 2 are called *long strips*. We need the following result (proof omitted).

Observation 2. *For any instance of* SORTING BY UNSIGNED PREFIX DCJs, *there always exists a shortest sorting sequence of prefix DCJs that never cut a long strip.*

Now let us describe our search tree technique: at every iteration starting from G, guess in which location, among the available 2-strips and breakpoints, to operate the rightmost cut. Once this is done, guess among the two possibilities allowed by a DCJ to reconnect the genome. By definition, every strip is framed by breakpoints. Therefore, any genome G has at most $b(G)$ 2-strips (recall that $\{0, x\}$ is never a breakpoint). Altogether, this shows that, at each iteration, the rightmost cut has to be chosen among at most $2b(G)$ possibilities. Because there are two ways to reconnect the cuts in a DCJ, the associated search tree has arity at most $4b(G)$. Moreover, its depth is at most $\frac{3b(G)}{2}$ since $pdcj(G) \leq \frac{3b(G)}{2}$ (Theorem 6). Thus the above described algorithm uses a search tree whose size is a function of $b(G)$ only, which proves the result. More precisely, the overall complexity of the induced algorithm is in $O^*((4b(G))^{1.5b(G)})$. □

4   Conclusions and Future Work

In this paper, we focused on the problem of sorting genomes by prefix DCJs, a problem that had not yet been studied in its prefix-constrained version. We provided several algorithmic results for both signed and unsigned cases, including computational complexity, approximation and FPT algorithms. Nevertheless, several questions remain open: while we have shown that SORTING BY SIGNED PREFIX DCJs is a polynomial-time solvable problem, what about the computational complexity of SORTING BY UNSIGNED PREFIX DCJs? We were able to design a 3/2-approximation algorithm for the latter problem, which makes it to the best of our knowledge the first occurrence of a prefix rearrangement problem of unknown complexity where a ratio better than 2 has been obtained. Is it possible to improve it further, by making good use of the new lower bound introduced in Sect. 2? Whether or not this lower bound can help improve the 2-approximation ratios known for both SORTING BY UNSIGNED PREFIX REVERSALS and SORTING BY SIGNED PREFIX REVERSALS remains open. Finally, we have studied the case where both source and target genomes are unichromosomal and linear; it would be interesting to extend this study to a more general context where input genomes can be multichromosomal and not necessarily linear.

References

1. Akers, S.B., Krishnamurthy, B., Harel, D.: The star graph: an attractive alternative to the n-cube. In: Proceedings of the Fourth International Conference on Parallel Processing, pp. 393–400. Pennsylvania State University Press (1987)

2. Bafna, V., Pevzner, P.A.: Genome rearrangements and sorting by reversals. SIAM J. Comput. **25**(2), 272–289 (1996)
3. Caprara, A.: Sorting permutations by reversals and Eulerian cycle decompositions. SIAM J. Discret. Math. **12**(1), 91–110 (electronic) (1999)
4. Chen, X.: On sorting unsigned permutations by double-cut-and-joins. J. Comb. Optim. **25**(3), 339–351 (2013)
5. Fertin, G., Labarre, A., Rusu, I., Tannier, E., Vialette, S.: Combinatorics of Genome Rearrangements. Computational Molecular Biology. MIT Press (2009)
6. Kotzig, A.: Moves without forbidden transitions in a graph. Matematický časopis **18**(1), 76–80 (1968)
7. Labarre, A.: Sorting by prefix block-interchanges. In: Cao, Y., Cheng, S.-W., Li, M. (eds.) 31st International Symposium on Algorithms and Computation (ISAAC). Leibniz International Proceedings in Informatics (LIPIcs), Dagstuhl, Germany, vol. 181, pp. 55:1–55:15. Schloss Dagstuhl-Leibniz-Zentrum für Informatik (2020)
8. Labarre, A., Cibulka, J.: Polynomial-time sortable stacks of burnt pancakes. Theor. Comput. Sci. **412**(8–10), 695–702 (2011)
9. Pevzner, P.A.: DNA physical mapping and alternating Eulerian cycles in colored graphs. Algorithmica **13**(1/2), 77–105 (1995)
10. Yancopoulos, S., Attie, O., Friedberg, R.: Efficient sorting of genomic permutations by translocation, inversion and block interchange. Bioinformatics **21**(16), 3340–3346 (2005)

KATKA: A KRAKEN-Like Tool with k Given at Query Time

Travis Gagie[1(✉)], Sana Kashgouli[1], and Ben Langmead[2]

[1] Dalhousie University, Halifax, Canada
travis.gagie@dal.ca
[2] Johns Hopkins University, Baltimore, USA

Abstract. We describe a new tool, KATKA, that stores a phylogenetic tree T such that later, given a pattern $P[1..m]$ and an integer k, it can quickly return the root of the smallest subtree of T containing all the genomes in which the k-mer $P[i..i+k-1]$ occurs, for $1 \leq i \leq m-k+1$. This is similar to KRAKEN's functionality but with k given at query time instead of at construction time.

1 Introduction

KRAKEN [13,14] is a popular tool that addresses the basic problem of determining where a fragment of DNA occurs in the Tree of Life, which arises for every sequencing read in a metagenomic dataset. KRAKEN takes a phylogenetic tree T and an integer k and stores T such that later, given a pattern $P[1..m]$, it can quickly return the root of the smallest subtree of T containing all the genomes in which the k-mer $P[i..i+k-1]$ occurs, for $1 \leq i \leq m-k+1$. For example, if T is the small phylogenetic tree shown in Fig. 1, $k = 3$, and $P = \texttt{TAGACA}$, then KRAKEN returns

- 8 for `TAG` (which occurs in `GATTAGAT` and `GATTAGATA`),
- 6 for `AGA` (which occurs in `AGATACAT`, `GATTAGAT` and `GATTAGATA`),
- NULL for `GAC` (which does not occur in T),
- 2 for `ACA` (which occurs in `GATTACAT`, `AGATACAT` and `GATACAT`).

Notice that not all the genomes in the subtree returned for $P[i..i + k]$ need contain it: `AGA` does not occur in `GATTACAT` or `GATACAT`.

KRAKEN is widely used in metagenomic analyses, especially taxonomic classification, but there are some applications for which we would rather give k at query time instead of at construction time. For example, Nasko et al. [7] showed that "the [reference] database composition strongly influence[s] the performance", with larger k values generally working better as the database grows. When the representation of strains or species in the database is skewed, therefore, it may be hard to choose a single k that works well for all of them. In this paper we describe a new tool, KATKA, that allows k to be chosen at query time. We are still optimizing, testing and extending KATKA and will report experimental results in a future paper.

© The Author(s), under exclusive license to Springer Nature Switzerland AG 2022
D. Arroyuelo and B. Poblete (Eds.): SPIRE 2022, LNCS 13617, pp. 191–197, 2022.
https://doi.org/10.1007/978-3-031-20643-6_14

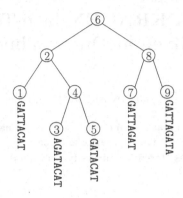

Fig. 1. A small phylogenetic tree.

2   Design

To simplify our presentation, in this paper we assume T is binary (although our approach generalizes to higher-degree trees). KATKA consists of three main components:

- a modified LZ77-index [6] for the concatenation of the genomes in T, in the order they appear from left to right in T and separated by copies of a special character $\$$;
- a modified LZ77-index for the reverse of that concatenation;
- a lowest common ancestor (LCA) data structure for T.

Given $P[1..m]$ and k, we use the first and second indexes to find the leftmost and rightmost genomes in T, respectively, that contain the k-mer $P[i..i+k-1]$, for $1 \le i \le m - k + 1$; we then use the LCA data structure to find the lowest common ancestor of those two genomes. Since the two indexes are symmetric and the LCA data structure takes only about 2 bits per vertex in T and has constant query time, we describe only the first index.

To build the index for the concatenation, we compute its LZ77 parse and consider the phrases and co-lexicographically sort the set of their maximal nonempty suffixes not containing $\$$, and consider the suffixes of the concatenation starting at phrase boundaries and lexicographically sort the set of their maximal prefixes not containing $\$$ (including the empty string ε after the last phrase boundary). We discard any of those maximal prefixes that do not occur starting at a phrase boundary immediately preceded by one of those maximal suffixes.

For our example, if the concatenation is

GATTACAT$AGATACAT$GATACAT$GATTAGAT$GATTAGATA,

then its LZ77 parse is

G A T TA C AT$ AG ATA CAT$G ATACAT$GATT AGAT$ GATTAGATA,

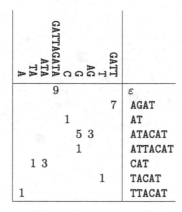

Fig. 2. The grid we build for the concatenation in our example.

the co-lexicographically sorted set of maximal suffixes is

$$A, TA, ATA, GATTAGATA, C, G, AG, T, GATT,$$

and the lexicographically sorted set of maximal prefixes is

$$\varepsilon, AGAT, AGATACAT, AT, ATACAT, ATTACAT, CAT,$$
$$GATTACAT, GATTAGATA, TACAT, TTACAT,$$

but we discard GATTACAT, AGATACAT and GATTAGATA because they do not occur starting at a phrase boundary immediately preceded by one of the maximal suffixes.

We build a grid with the number ℓ at position (x, y) if the genome at the ℓth vertex from the left in T is the first one in which there is a phrase boundary immediately preceded by the co-lexicographically xth of the maximal suffixes and immediately followed by the lexicographically yth of the maximal prefixes. Notice this grid will be of size at most $z \times z$ with at most z numbers on it, where z is the number of phrases in the LZ77 parse of the concatenation. Figure 2 shows the grid for our example.

We store data structures such that given strings α and β, we can quickly find the minimum number in the box $[x_1, x_2] \times [y_1, y_2]$ on the grid, where $[x_1, x_2]$ is the co-lexicographic range of the maximal suffixes ending with α and $[y_1, y_2]$ is the lexicographic range of the maximal prefixes starting with β. (For the index for the reversed concatenation, we find the maximum in the query box.) In our example, if $\alpha = G$ and $\beta = AT$, then we should find 1.

For example, we can store Patricia trees for the compact tries for the reversed maximal suffixes and the maximal prefixes, together with a data structure supporting fast sequential access to the concatenation starting at any phrase boundary. In the literature (see [8] and references therein), the latter is usually an augmented straight-line program (SLP) for the concatenation; if the genomes in T are similar enough, however, then in practice it could probably be simply a

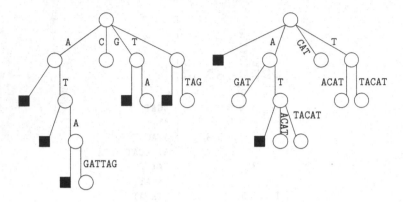

Fig. 3. The compact tries for the concatenation in our example.

VCF file. (We note that we can reuse the access data structure for the index for the reversed concatenation, augmented to support fast sequential access also at phrase boundaries in the reverse of the concatenation.) Figure 3 shows the compact tries for our example, with each black leaf indicating that one of the strings in the set ended at the parent of that leaf.

Nekrich [12] recently showed how to store the grid in $O(z)$ space and support 2-dimensional range-minimum queries in $O(\log^\epsilon z)$ time, for any constant $\epsilon > 0$. For simplicity, we consider his data structure in our analysis even though we are not aware of any implementation yet.

3   Queries

Given a pattern $P[1..m]$ and an integer k, for every substring $P[i..j]$ of P with length at most k, we find and verify the locus for the reverse of $P[i..j]$ in the compact trie for the reversed maximal suffixes, and the locus for $P[i..j]$ in the compact trie for the maximal prefixes. (Patricia trees can return false positives when the sought pattern does not occur, so we must verify the loci by, for example, extracting their path labels from the SLP).

By combining the searches for $P[i], P[i..i+1], \ldots, P[i..i+k-1]$, we make a total of $O(m)$ descents in the Patricia trees, each to a string-depth of at most k; we extract $O(m)$ substrings from the concatenation, each of length at most k and starting at a phrase boundary, to verify the loci. With care, this takes a total of $O(km)$ time in the worst case. When searching standard LZ77-indexes in practice, however, "queries often die in the Patricia trees" [10]—because of mismatches between characters in the pattern and the first characters in edge labels—which speeds up queries.

For each k-mer $P[i..i+k-1]$ in P and each way to split $P[i..i+k-1]$ into a non-empty prefix $P[i..j]$ and a suffix $P[j+1..i+k-1]$, we use a 2-dimensional range-minimum query to find the minimum number in the box for $\alpha = P[i..j]$ and $\beta = P[j+1..i+k-1]$ in $O(\log^\epsilon z)$ time.

By the definition of the LZ77 parse, the first occurrence of $P[i..i+k-1]$ in the concatenation crosses or ends at a phrase boundary. It follows that, by taking the minimum of the minima we find, in $O(k \log^{\epsilon} z)$ time we find the leftmost genome in T that contains $P[i..i + k - 1]$. Repeating this for every value of i takes $O(km \log^{\epsilon} z)$ time.

By storing symmetric data structures for the reverse of the concatenation and querying them, we can find the rightmost genome in T that contains $P[i..i+k-1]$, for $1 \leq i \leq m-k+1$. With the LCA data structure for T, we can find the lowest common ancestor of the two genomes, which is the root of the smallest subtree of T containing all the genomes in which the k-mer $P[i..i + k - 1]$ occurs.

For our example, if $P = $ TAGACA and $k = 3$, then we find and verify the loci for

$$T, A, AT, G, GA, GAT, A, AG, AGA, C, CA, CAG, A, AC, ACA$$

in the compact trie for the reversed maximal suffixes, and the loci for

$$A, AG, AGA, G, GA, GAC, A, AC, ACA, C, CA, A$$

in the compact trie for the maximal prefixes.

For $P[1..3] = $ TAG, we look up the minimum number 7 in the box for $\alpha = $ T and the locus $\beta = $ AGAT for AG; since G has no locus in the compact trie for the maximal prefixes and GAT has no locus in the compact trie for the maximal reversed suffixes, we correctly conclude that the leftmost genome in T containing TAG is at vertex 7. A symmetric process with the index for the reversed concatenation tells us the rightmost genome in T containing TAG is at vertex 9, and then an LCA query tells us that vertex 8 is the root of the smallest subtree containing all the genomes in which TAG occurs.

Theorem 1. *Given a phylogenetic tree T whose g genomes have total length n, we can store T in $O(z \log n + g/ \log n)$ space, where z is the number of phrases in the LZ77 parse of the concatenation of the genomes in T (separated by copies of a special character), such that when given a pattern $P[1..m]$ and an integer k, for $1 \leq i \leq m - k + 1$ we can find the root of the smallest subtree of T containing all genomes in which the k-mer $P[i..i + k - 1]$ of P occurs, in $O(km \log^{\epsilon} z)$ total time. If no two genomes in the tree are identical, our space bound simplifies to $O(z \log n)$.*

Proof. The LCA data structure takes $2g + o(g)$ bits, which is $O(g/ \log n)$ words (assuming $\Omega(\log n)$-bit words). If no two genomes are identical then each genome crosses or ends at a phrase boundary, so $g \leq z$ and our space bound simplifies to $O(z \log n)$ words. An SLP for the concatenation with bookmarks permitting sequential access with constant overhead from the phrase boundaries in the parses of the concatenation and its reverse, takes $O(z \log n)$ space. For the concatenation, the Patricia trees and the instance of Nekrich's 2-dimensional range-minimum data structure take $O(z)$ space; for the reverse of the concatenation, they take space proportional to the number of phrases in its LZ77 parse, which is $O(z \log n)$. In total, we use $O(z \log n + g/ \log n)$ space. As we have described, we make $O(m)$ descents in the Patricia trees, each to string-depth at most k, and extract only $O(m)$ substrings, each of length at most k, from the concatenation and its reverse. The time is dominated by the $O(km)$ range-minimum queries, which take $O(\log^\epsilon z)$ time each.

4 Future Work

In addition to optimizing and testing KATKA, we are also investigating adapting it to work with maximal exact matches (MEMs) instead of k-mers. For example, if we store $O(z)$-space z-fast tries [2] for the Patricia trees then, for each way to split P into a non-empty prefix $P[1..i]$ and a suffix $P[i+1..m]$, we can find the loci of $P[1..i]$ reversed and $P[i+1..m]$ in $O(\log m)$ time. We can verify those loci in $O(\log n)$ time by augmenting the SLP to return fingerprints, without changing its $O(z \log n)$ space bound [3].

With an $O(z)$-space data structure supporting heaviest induced ancestor queries in $O\left(\frac{\log^2 z}{\log \log z}\right)$ time [1,4,5], in that time we can find the longest substring $P[h..j]$ with $h \leq i \leq j$ that occurs in T with $P[h..i]$ immediately to the left of a phrase boundary and $P[h+1..j]$ immediately to its right. Note $P[h..j]$ must be a MEM. With 2-dimensional range-minimum and range-maximum queries, we can find the indexes of the leftmost and rightmost genomes in which $P[h..i]$ occurs immediately to the left of a phrase boundary and $P[h+1..j]$ immediately to its right. We still use a total of $O(z \log n + g/\log n)$ space and now we use a total of $O\left(m\left(\frac{\log^2 z}{\log \log n} + \log n\right)\right)$ time.

Unfortunately, we may not find every MEM this way: it may be that, for some MEM $P[h..j]$ and every i between h and j, either $P[h..j]$ is not split into $P[h..i]$ and $P[i+1..j]$ by any phrase boundary or some longer MEM is split into $P[h'..i]$ and $P[i+1..j']$ by a phrase boundary. For any MEM we do not find, however, we do find another MEM at least as long that overlaps it.

A more serious drawback to this scheme is that it is probably quite impractical (for example, we are not aware of any implementation of a data structure supporting fast heaviest induced ancestor queries, either). Fortunately, there is probably a simple and practical solution—although possibly without good worst-case bounds—since the problems we are considering are similar to those covered in Subsection 7.1 of Navarro's [11] survey on wavelet trees.

Finally, we are investigating adapting results [9] using LZ77-indexes for document-listing, in order to find the number of genomes in T in which each k-mer of P occurs. It is easy to store a small data structure that reports the number of genomes in the smallest subtree for a k-mer, so we may be able to determine what fraction contain that k-mer.

Acknowledgments. Many thanks to Nathaniel Brown, Younan Gao, Simon Gog, Meng He, Finlay Maguire and Gonzalo Navarro, for helpful discussions.

References

1. Abedin, P., Hooshmand, S., Ganguly, A., Thankachan, S.V.: The heaviest induced ancestors problem: better data structures and applications. Algorithmica 1–18 (2022). https://doi.org/10.1007/s00453-022-00955-7
2. Belazzougui, D., Boldi, P., Pagh, R., Vigna, S.: Fast prefix search in little space, with applications. In: de Berg, M., Meyer, U. (eds.) ESA 2010. LNCS, vol. 6346, pp. 427–438. Springer, Heidelberg (2010). https://doi.org/10.1007/978-3-642-15775-2_37
3. Bille, P., Gørtz, I.L., Cording, P.H., Sach, B., Vildhøj, H.W., Vind, S.: Fingerprints in compressed strings. J. Comput. Syst. Sci. **86**, 171–180 (2017)
4. Gagie, T., Gawrychowski, P., Nekrich, Y.: Heaviest induced ancestors and longest common substrings. In: Proceedings of the CCCG (2013)
5. Gao, Y.: Computing matching statistics on repetitive texts. In: Proceedings of the DCC (2022)
6. Kreft, S., Navarro, G.: On compressing and indexing repetitive sequences. Theor. Comput. Sci. **483**, 115–133 (2013)
7. Nasko, D.J., Koren, S., Phillippy, A.M., Treangen, T.J.: RefSeq database growth influences the accuracy of k-mer-based lowest common ancestor species identification. Genome Biol. **19**(1), 1–10 (2018)
8. Navarro, G.: Compact Data Structures: A Practical Approach. Cambridge University Press, Cambridge (2016)
9. Navarro, G.: Document listing on repetitive collections with guaranteed performance. Theor. Comput. Sci. **772**, 58–72 (2019)
10. Navarro, G.: Personal communication (2013)
11. Navarro, G.: Wavelet trees for all. J. Discret. Algorithms **25**, 2–20 (2014)
12. Nekrich, Y.: New data structures for orthogonal range reporting and range minima queries. In: Proceedings of the SODA (2021)
13. Wood, D.E., Lu, J., Langmead, B.: Improved metagenomic analysis with KRAKEN 2. Genome Biol. **20**(1), 1–13 (2019)
14. Wood, D.E., Salzberg, S.L.: KRAKEN: ultrafast metagenomic sequence classification using exact alignments. Genome Biol. **15**(3), 1–12 (2014)

Computing All-vs-All MEMs
in Run-Length-Encoded Collections
of HiFi Reads

Diego Díaz-Domínguez[✉], Simon J. Puglisi, and Leena Salmela

Department of Computer Science, University of Helsinki, Helsinki, Finland
{diego.diaz,simon.puglisi,leena.salmela}@helsinki.fi

Abstract. We describe an algorithm to find maximal exact matches
(MEMs) among HiFi reads with homopolymer errors. The main novelty
in our work is that we resort to run-length compression to help deal with
errors. Our method receives as input a run-length-encoded string collection containing the HiFi reads along with their reverse complements.
Subsequently, it splits the encoding into two arrays, one storing the
sequence of symbols for equal-symbol runs and another storing the run
lengths. The purpose of the split is to get the BWT of the run symbols
and reorder their lengths accordingly. We show that this special BWT,
as it encodes the HiFi reads and their reverse complements, supports
bi-directional queries for the HiFi reads. Then, we propose a variation of
the MEM algorithm of Belazzougui et al. (2013) that exploits the run-
length encoding and the implicit bi-directional property of our BWT to
compute approximate MEMs. Concretely, if the algorithm finds that two
substrings, $a_1 \ldots a_p$ and $b_1 \ldots b_p$, have a MEM, then it reports the MEM
only if their corresponding length sequences, $\ell_1^a \ldots \ell_p^a$ and $\ell_1^b \ldots \ell_p^b$, do not
differ beyond an input threshold. We use a simple metric to calculate the
similarity of the length sequences that we call the *run-length excess*. Our
technique facilitates the detection of MEMs with homopolymer errors as
it does not require dynamic programming to find approximate matches
where the only edits are the lengths of the equal-symbol runs. Finally,
we present a method that relies on a geometric data structure to report
the text occurrences of the MEMs detected by our algorithm.

Keywords: Genomics · Text indexing · Compact data structures

1 Introduction

HiFi reads are a new type of DNA sequencing data developed by PacBio [30].
They are long overlapping strings with error rates (mismatches) comparable
to those of Illumina data. They have become popular in recent years as their
features improve the accuracy of biological analyses [21]. Still, mapping a collection of HiFi reads against a reference genome or computing suffix-prefix overlaps among them for *de novo* assembly remain important challenges as these

Supported by Academy of Finland Grants 323233 and 339070.

D. Arroyuelo and B. Poblete (Eds.): SPIRE 2022, LNCS 13617, pp. 198–213, 2022.
https://doi.org/10.1007/978-3-031-20643-6_15

tasks require approximate alignments of millions of long strings. Popular tools that address these problems use seed-and-extend algorithms with minimizers as seeds [19] for the alignments. This technique is a cheap solution that makes the processing of HiFi reads feasible.

An alternative approach is to use *maximal exact matches* (MEMs) as seeds. A MEM is a match $S[a, b] = S'[a', b']$ between two strings S and S' that cannot be extended to the left or to the right without introducing mismatches or reaching an end in one of the sequences. MEMs are preferable over minimizers because they can capture long exact matches between the reads, thus reducing the computational costs of extending the alignments with dynamic programming. However, they are expensive to find in big collections.

A classical solution to detect MEMs among strings of a large collection is to concatenate the strings in one sequence S (separated by sentinel symbols), construct the suffix tree of S, and then traverse its topology to find MEMs in linear time [13]. Still, producing the suffix tree of a massive collection, although linear in time and space, is expensive for practical purposes. Common approaches to deal with the space overhead are sparse suffix trees [16,28], hash tables with k-mers [11,17], and Bloom filters [20].

Another way to deal with the space overhead is to find MEMs on top of a compact suffix tree [23,27]. For instance, Ohlebusch et al. [24] described a method that computes MEMs between two strings via matching statistics [6]. Their technique requires only one of the strings to be indexed using a compact suffix tree while the other is kept in plain format. Other more recent methods [3,25,26] follow an approach similar to that of Ohlebusch et al., but they use the r-index [9] instead of the compact suffix tree.

The problem with the algorithms that rely on matching statistics is that they consider input collections with two strings (one indexed and the other in plain format). It is not clear how to generalize these techniques to compute all-vs-all MEMs in a collection with an arbitrary number of sequences. A simple solution would be to implement classical MEM algorithms on top of the compact suffix tree. Still, producing a full compact suffix tree is expensive for genomic applications as it requires producing a sampled version of the suffix array [22], the Burrows–Wheeler transform [4], and the longest common prefix array [15]. Although it is possible to obtain these composite data structures in linear time and space, in practice, they might require an amount of working memory that is several times the input size. In this regard, Belazzougui et al. [2] proposed a MEM algorithm that only uses the bi-directional BWT of the text, although their idea reports the sequences for the MEM, not their occurrences in the text.

Besides the input size, there is another relevant issue when computing MEMs in HiFi data: homopolymer errors. Concretely, if a segment of the DNA being sequenced has an equal-symbol run of length ℓ, then the PacBio sequencer might spell imprecise copies of that run in the reads that overlap the segment. These copies have a correct[1] DNA symbol, but the value ℓ might be incorrect. In gen-

---

[1] The symbol correctly represents the nucleotide that was read from the DNA molecule.

eral, homopolymer errors shorten the alignment seeds, which means that the pattern matching algorithm will spend more time performing dynamic programming operations to extend the alignments. In this work, we study the problem of finding MEMs in HiFi reads efficiently. Our strategy is to use run-length encoding to remove the homopolymer errors, and then try to filter out the matches between different sequences that, by chance, were compressed to the same run-length encoded string.

Our Contribution. We present a set of techniques to compute all-vs-all MEMs in a collection of HiFi reads of n symbols. We build on the MEM algorithm of Belazzougui et al. [2] that uses the bi-directional BWT, a versatile succinct text representation that uses $2n \log \sigma + o(2n \log \sigma)$ bits of space. Strings in a DNA collection have two complementary sequences that we need to consider for the matches, meaning that we need to create a copy of the input with the complementary strings and merge all in one collection $\mathcal{R}$. We describe a framework that exploits the properties of these DNA complementary sequences to produce an implicit bi-directional BWT for $\mathcal{R}$ without increasing the input size by a factor of 4x. In addition, we define parameters to detect MEMs in a run-length-encoded representation of $\mathcal{R}$. Concretely, we propose the concept of run-length excess, which we use to differentiate homopolymer errors from sporadic matches generated by the run-length compression. Finally, we describe our variation of the algorithm of Belazzougui et al. [2] for computing MEMs using our implicit bi-directional BWT constructed on a run-length-encoded version of $\mathcal{R}$, denoted $\mathcal{R}^h$. Let S be a sequence of length $d = |S|$ that has x occurrences in $\mathcal{R}^h$, with $l \leq x$ of them having MEMs with other positions of $\mathcal{R}^h$. Once our algorithm detects S, it can report its MEMs in $O(\sigma^2 \log \sigma + x^2 d)$ time, where σ is the alphabet of $\mathcal{R}^h$. We also propose an alternative solution that uses a geometric data structure, and report the MEMs of S in $O((x + \sigma)(1 + \log n_h / \log \log n_h) + l^2 d)$ time, where n_h is the number of symbols in $\mathcal{R}^h$.

2 Preliminaries

Rank and Select Data Structures. Given a sequence $B[1, n]$ of symbols over the alphabet $\Sigma = [1, \sigma]$, the operation $\mathsf{rank}_a(B, i)$, with $i \in [1, n]$ and $a \in \Sigma$, returns the number of times the symbol a occurs in the prefix $B[1, i]$. On the other hand, the operation $\mathsf{select}_a(B, r)$ returns the position of the rth occurrence of a in B. For binary alphabets, B can be represented in $n + o(n)$ bits so that it is possible to solve rank_a and select_a, with $a \in \{0, 1\}$, in constant time [7,14].

Wavelet Trees. Let $S[1, n]$ be a string of length n over the alphabet $\Sigma = [1, \sigma]$. A wavelet tree [12] is a tree data structure W that encodes S in $n \log \sigma + o(n \log \sigma)$ bits of space and supports several queries in $O(\log \sigma)$ time. Among them, the following are of interest for this work:

- $\mathsf{access}(W, i)$: retrieves the symbol at position $T[i]$
- $\mathsf{rank}_a(W, i)$: number of symbols a in the prefix $T[1, i]$

- $select_a(W, r)$: position j where the rth symbol a lies in S

The wavelet tree can also answer more elaborated queries efficiently [10]. From them, the following are relevant:

- rangeList(W, i, j) : the list of all triplets (c, r_i^c, r_j^c) such that c is one of the distinct symbols within $S[i, j]$, r_i^c is the rank of c in $S[1, i - 1]$, and r_j^c is the rank of c in $S[1, j]$.
- rangeCount(W, i, j, l, r) : number of symbols $y \in S[i, j]$ such that $l \leq y \leq r$.

It is possible to answer rangeList in $O(u \log \frac{\sigma}{u})$ time, where u is the number of distinct symbols in $S[i, j]$, and rangeCount in $O(\log \sigma)$ time.

Suffix Arrays and Suffix Trees. Consider a string $S[1, n - 1]$ over alphabet $\Sigma[2, \sigma]$, and the sentinel symbol $\Sigma[1] = \$$, which we insert at $S[n]$. The *suffix array* [22] of S is a permutation $SA[1, n]$ that enumerates the suffixes $S[i, n]$ of S in increasing lexicographic order, $S[SA[i], n] < S[SA[i + 1], n]$, for $i \in [1, n - 1]$.

The suffix trie [8] is the trie T induced by the suffixes of S. For every $S[i, n]$, there is a path $U = v_1, v_2, \ldots, v_p$ of length $p = n - i + 2$ in T, where v_1 is the root and v_p is a leaf. Each edge (v_j, v_{j+1}) in U is labeled with a symbol in Σ, and concatenating the edge labels from v_1 to v_p produces $S[i, n]$. The child nodes of each internal node v are sorted from left to right according to the ranks of the labels in the edges that connect them to v. Further, when two or more suffixes of S have the same j-prefix, their paths in T share the first $j + 1$ nodes.

It is possible to compact T by discarding each unary path $U = v_i, \ldots, v_j$ where every node $v_i, v_{i-1}, \ldots, v_{j-1}$ has exactly one child. The procedure consists of removing the subpath $U' = v_{i+1}, \ldots, v_{j-1}$ and connect v_i with v_j by an edge labeled with the concatenation of the labels in U'. The result is a compact trie of n leaves and less than n internal nodes called the *suffix tree* [29].

The suffix tree can contain other special edges that connect nodes from different parts of the tree, not necessarily a parent with its children. These edges are called suffix and Weiner links. Let us denote label(v) the string that labels the path starting at the root and ending at v. Two nodes u and v are connected by a suffix link (u, v) if label$(u) = aW$ and label$(v) = W$. Similarly, an explicit Weiner link (u, v) labeled a occurs if label$(u) = W$ and label$(v) = aW$. A Weiner link is implicit when, for label$(u) = W$, the sequence aW matches a proper prefix of a node label (i.e., there is no node labeled aW). The suffix and Weiner links along with the suffix tree nodes yield another tree called the suffix link tree.

The Burrows–Wheeler Transform. The *Burrows–Wheeler transform* (BWT) [4] is a reversible string transformation that stores in $BWT[i]$ the symbol that precedes the ith suffix of S in lexicographical order, i.e., $BWT[i] = S[SA[i] - 1]$ (assuming $S[0] = S[n] = \$$).

The mechanism to revert the transformation is the so-called LF mapping. Given an input position $BWT[j]$ that maps a symbol $S[i]$, $\mathsf{LF}(j) = j'$ returns the index j' such that $BWT[j'] = S[i - 1]$ maps the preceding symbol of $S[i]$. Thus, spelling S reduces to continuously applying LF from $BWT[1]$, the symbol to the left of $T[n] = \$$, until reaching $BWT[j] = \$$.

Implementing LF requires to encode BWT with a data structure that supports rank_a. A standard solution is to use the wavelet tree of Sect. 2, which enables to answer LF in $O(\log \sigma)$ time. It is also necessary to have an array $C[1, \sigma]$ storing in $C[c]$ the number of symbols in S that are lexicographically smaller than c. This enables the implementation of the inverse function for LF (denoted as LF^{-1}). That is, given the position $BWT[j]$ that maps $S[i]$, $\text{LF}^{-1}(j) = j'$ returns the index j' such that $BWT[j']$ maps $S[i+1]$.

The BWT also allows to count the number of occurrences of a pattern $P[1, m]$ in S in $\mathcal{O}(m \log \sigma)$ time. The method, called backwardsearch, builds on the observation that if the range $SA[s_j, e_j]$ encoding the suffixes of S prefixed by $P[j, m]$ is known, then it is possible to compute the next range $SA[s_{j-1}, e_{j-1}]$ with the suffixes of S prefixed by $P[j-1, m]$. This computation, or *step*, requires two operations: $s_{j-1} = C[P[j-1]] + \text{rank}_{P[j-1]}(BWT, s_j - 1) + 1$ and $e_{j-1} = C[P[j-1]] + \text{rank}_{P[j-1]}(BWT, e_j)$. Thus, after m steps of $O(\log \sigma)$ time each, backwardsearch will find the range $SA[s_1, e_1]$ encoding the suffixes of S prefixed by $P[1, m]$ (provided P exists as substring in S).

Bi-directional BWT. The bi-directional BWT [18] of a string $S[1, n]$ is a data structure that maintains the BWT of S and the BWT of the reverse of S (denoted here as $\bar{S}$). Belazzougui et al. [2] demonstrated that it is possible to use this representation to visit the internal nodes in the suffix tree T of S in $O(n \log \sigma)$ time.

The work of Belazzougui et al. exploits the fact that the suffixes of S prefixed by the label of an internal node v in T are stored in a consecutive range $SA[s_v, e_v]$, and that $BWT[s_v, e_v]$ encodes the labels for the Weiner links of v.

Let SA_S and BWT_S be the suffix array and BWT for S (respectively). Equivalently, let $SA_{\bar{S}}$ and $BWT_{\bar{S}}$ be the suffix array and BWT for $\bar{S}$. For any sequence X, Belazzougui et al. maintain two pairs: (s_X, e_X) and $(s_{\bar{X}}, e_{\bar{X}})$, where $SA_S[s_X, e_X]$ stores the suffixes of S prefixed by X and $SA_{\bar{S}}[s_{\bar{X}}, e_{\bar{X}}]$ stores the suffixes of $\bar{S}$ prefixed by $\bar{X}$. They also define a set of primitives for the encoding $(s_X, e_X), (s_{\bar{X}}, e_{\bar{X}})$ of X:

- isLeftMaximal(X): 1 if $BWT_S[s_X, e_X]$ contains more than one distinct symbol, and 0 otherwise.
- isRightMaximal(X): 1 if $BWT_{\bar{S}}[s_{\bar{X}}, s_{\bar{X}}]$ contains more than one distinct symbol, and 0 otherwise.
- enumerateLeft(X): list of distinct symbols in $BWT_S[s_X, e_X]$.
- enumerateRight(X): list of distinct symbols in $BWT_{\bar{S}}[s_{\bar{X}}, e_{\bar{X}}]$
- extendLeft(X, c): list $(i, j), (i', j')$ where $SA_S[i, j]$ is the range for cX and $SA_{\bar{S}}[i', j']$ is the range for $\bar{X}c$
- extendRight(X, c): list $(i, j), (i', j')$ where $SA_S[i, j]$ is the range for Xc and $SA_{\bar{S}}[i', j']$ is the range for $c\bar{X}$

The key aspect of the bi-directional BWT is that, every time it performs a left or a right extension in (s_X, e_X) (respectively, $(s_{\bar{X}}, e_{\bar{X}})$), it also synchronizes $(s_{\bar{X}}, e_{\bar{X}})$ (respectively, (s_X, e_X)). By encoding BWT_S and $BWT_{\bar{S}}$ as wavelet trees (Sect. 2), it is possible to perform extendLeft and extendRight in $O(\log \sigma)$

time using a backward search step (Sect. 2), and then synchronizing the other range with rangeCount. The functions enumerateLeft and enumerateRight take $O(u \log \frac{\sigma}{u})$ time as they are equivalent to rangeList. Finally, both isLeftMaximal and isRightMaximal run in $O(\log \sigma)$ time.

Belazzougui et al. use the primitives above to traverse the suffix link tree and thus visiting the internal nodes of T in $O(n \log \sigma)$ time.

3   Our Contribution

3.1   Definitions

We consider the set $\{A, C, G, T\}$ to be the *DNA alphabet*. For practical reasons, we compact it to the set $\Sigma = [2, 5]$, and regard $\Sigma[1] = \$$ as a *sentinel* that is lexicographically smaller than any other symbol. Given a string R in Σ^*, we define an *homopolymer* as an equal-symbol run $R[i, j] = (c, \ell)$ of maximal length storing $\ell = j - i + 1 > 1$ consecutive copies of a symbol c. Maximal length means that $i = 1$ or $R[i - 1] \neq c$, and $j = |R|$ or $R[j + 1] \neq c$.

We regard the DNA *complement* as a permutation $\pi[1, \sigma]$ that reorders the symbols in Σ, exchanging 2 (A) with 5 (T) and 3 (C) with 4 (G). The permutation does not modify 1 (\$) as it does not represent a nucleotide (i.e., $\pi(1) = 1$). The *reverse complement* of R, denoted $\hat{R}$, is the string formed by reversing R and replacing every symbol $R[i]$ by its complement $\pi(R[i])$. Given a DNA symbol $a \in \Sigma$, let us define the operator $\underline{a} = \pi(a)$ to denote the DNA complement of a.

The input for our algorithm is a collection $\mathcal{X} = \{R_1, R_2, \ldots, R_k\}$ of k HiFi reads over the alphabet Σ. However, we operate over the expanded collection $\mathcal{R} = \{R_1\$, \hat{R}_1\$, \ldots, R_k\$, \hat{R}_k\$\}$ storing the reads of $\mathcal{X}$ along with their reverse complements, where all the strings have a sentinel appended at the end. $\mathcal{R}$ has $2k$ strings, with a total of $n = \Sigma_{i=1}^{k} 2(|R_i| + 1)$ symbols. We refer to every possible sequence over the DNA alphabet that label a MEM in $\mathcal{R}$ as a *MEM sequence*.

3.2   Description of the Problem

Before developing our ideas, we formalize our problem as follows.

Definition 1. *Let $\mathcal{S} = \{S_1, S_2, \ldots, S_k\}$ be a string collection of k strings and n total symbols. The problem of all-vs-all MEMs consists in reporting every possible pair $(S_x[a, b], S_y[a', b'])$ such that $S_x, S_y \in \mathcal{S}$, $S_x \neq S_y$, and $S_x[a, b] = S_y[a', b']$ is a MEM of length $b - a + 1 = b' - a' + 1 \geq \tau$, where τ is a parameter.*

HiFi data is usually strand unspecific, meaning that, for any two reads $R_a, R_b \in \mathcal{X}$, there are four possible combinations in which we can have MEMs: (R_a, R_b), $(\hat{R}_a, R_b)$, $(R_a, \hat{R}_b)$, $(\hat{R}_a, \hat{R}_b)$. We can access all such combinations in $\mathcal{R}$, but not in $\mathcal{X}$. Hence, our algorithmic framework solves the problem of Definition 1 using $\mathcal{R}$ as input.

4   Bi-directional BWT and DNA Reverse Complements

In this section, we explain how to exploit the properties of the DNA reverse complements to implement an *implicit* bi-directional BWT for $\mathcal{R}$ that does not require the BWT of the reverse sequences of $\mathcal{R}$ (see Sect. 2). We assume the BWT of $\mathcal{R}$ is the BCR BWT [1], a variation for string collections. This decision is for technical convenience, and does not affect the output of our framework. We begin by describing the key property in our implicit bi-directional representation:

Lemma 1. *Let X be a string over alphabet Σ that appears as a substring in $\mathcal{R}$. Additionally, let the pairs (s_X, e_X) and $(s_{\hat{X}}, e_{\hat{X}})$ be the ranges in SA of $\mathcal{R}$ storing all suffixes prefixed by X and $\hat{X}$, respectively. It holds that the sorted sequence of DNA complement symbols in $BWT[s_X, e_X]$ matches the right-context symbols of the occurrences of $\hat{X}$ when sorted in lexicographical order. This relationship applies symmetrically to $BWT[s_{\hat{X}}, e_{\hat{X}}]$ and the sorted occurrences of X.*

Proof. Assume the symbol $a \in \Sigma$ appears to the left of p occurrences of Xb in $\mathcal{R}$. We know that for each occurrence of aXb in $\mathcal{R}$ there will be also an occurrence of $\underline{b}\hat{X}\underline{a}$ as we enforce that property by including the DNA reverse complements of the original reads (collection $\mathcal{X}$ of Sect. 3.1). As a result, $BWT[s_{Xb}, e_{Xb}]$ will contain p copies of a and $BWT[s_{\hat{X}\underline{a}}, e_{\hat{X}\underline{a}}]$ will contain p copies of $\underline{b}$.

We will use Lemma 1 to implement the functions enumerateRight, extendRight and isRightMaximal (Sect. 2) on top of the BWT of the text. Unlike the technique of Belazzougui et al., we synchronize the pairs $(s_X, e_X), (s_{\hat{X}}, e_{\hat{X}})$. Another difference is that both pairs $(s_X, e_X), (s_{\hat{X}}, e_{\hat{X}})$ map to the suffix array of the text. In the original version, the second pair maps to the suffix array of the reverse text. To implement the functions above, we need to update both pairs every time we perform extendLeft and extendRight.

Belazzougui et al. implement extendLeft(X, c) by performing a backward search step over $BWT[s_X, e_X]$ using the symbol c. The result of this operation is the suffix array range for cX. To modify $(s_{\hat{X}}, e_{\hat{X}})$ so it maps to the suffix array range for $\hat{X}\underline{c}$, we sum the frequencies of the distinct symbols within $BWT[s_X, e_X]$ whose DNA reverse complements are lexicographically smaller than $\underline{c}$. This operation comes directly from Lemma 1. Assume the sum is y and that the frequency of c in $BWT[s_X, e_X]$ is z, then we compute $s_{\hat{X}\underline{c}} = s_{\hat{X}} + y$ and $e_{\hat{X}\underline{c}} = s_{\hat{X}} + y + z$. We use a special form of rangeCount to get the value for y. If $c < \underline{c}$, then we will use $y = $ rangeCount$(BWT, s_X, e_X, c+1, \sigma)$. In the other case, $c > \underline{c}$, we use rangeCount$(BWT, s_X, e_X, 1, c-1)$. The rationale for computing rangeCount comes from the relationship between complementary nucleotides in the permutation π of Sect. 3.1. The operation extendRight(X, c) is analogous; we perform the backwardsearch step over $BWT[s_{\hat{X}}, e_{\hat{X}}]$ using $\underline{c}$ as input, and then we count the number of symbols that are lexicographically smaller than c.

The functions enumerateRight(X) and isRightMaximal(X) are implemented with minor changes. The only caveat is that, when we use enumerateRight, we need to spell the DNA reverse complements of the symbols returned by rangeList.

Corollary 1. *Given a collection $\mathcal{X}$ of DNA sequences and its expanded version $\mathcal{R}$ that contains the strings of $\mathcal{X}$ along their reverse complement sequences, we can construct an implicit bi-directional BWT index that does not require the BWT of the reverse of $\mathcal{R}$ and that answers the queries* enumerateRight, extendRight *and* isRightMaximal *in $O(u \log \frac{\sigma}{u})$ and $O(\log \sigma)$ time, respectively, where u is the number of distinct symbols within the input range for* extendRight.

Observe the BWT for $\mathcal{R}$ is implicitly bi-directional as the DNA reverse complements are just the reverse strings with their symbols permuted according to π (see Definitions). However, in the case of $\mathcal{R}$, both BWTs are merged in a single representation. Producing a standard bi-directional BWT would increase the size of $\mathcal{X}$ by a factor of 4. In real applications where the data is a multiset of DNA sequencing reads, we have to transform $\mathcal{X}$ into $\mathcal{R}$ regardless if we construct a bi-directional BWT as the reads are strand-unspecific (see Sect. 3.2).

Contraction Operations in the Implicit Bi-directional BWT. Given a range $SA[i, j]$ of suffixes prefixed by a string X, and a parameter $w \leq |X|$, a *contraction* operation returns the range $i' \geq i, j \leq j'$ in SA storing the suffixes of the text prefixed by $X[1, w]$. It is possible to solve this query efficiently with either the wavelet tree of the LCP or with a compact data structure that encodes the suffix tree's topology. The problem with those solutions is that we have to deal with the overhead of constructing and storing those representations. We describe how to use our implicit bi-directional BWT to visit the ancestors of a node v in the suffix tree in $O(|\mathsf{label}(v)| \log \sigma)$ time to solve contraction operations. This idea is slower than using the LCP or the suffix tree's topology, but it does not require extra space, and it is faster than the quadratic cost of using a regular BWT. Our technique is a byproduct of our framework, and it is of independent interest. The inputs for the ancestors' traversal are the range $SA[s_v, e_v]$ for v, and its string depth $d = |\mathsf{label}(v)|$. The procedure is as follows: starting from $BWT[s_v]$, we perform d LF^{-1} operations to spell $\mathsf{label}(v)$. Simultaneously as we spell the sequence, we also perform backward search steps using the DNA complement of the symbols we obtain with LF^{-1}. We use Lemma 1 to keep the ranges of the backward search steps synchronised with the ranges for the distinct prefixes of $\mathsf{label}(v)$. Recall that backwardsearch consumes the input from right to left. In our case, this input is a sequence W that matches the DNA reverse complement of $\mathsf{label}(v)$. Thus, by Lemma 1, we know that visiting the SA ranges for the suffixes of W is equivalent to visit the SA ranges for the prefixes of $\mathsf{label}(v)$. Finally, each time we obtain a new range $SA[i', j']$ with the backward search step, we use isLeftmaximal to check if $BWT[i', j']$ is unary. If that is the case, then we report the synchronized range of $SA[i', j']$ as an ancestor of v. The rationale is that if W is left-maximal, then $\hat{W} = \mathsf{label}(v)[1, |W|]$ is right-maximal too, and hence, its sequence is the label of an ancestor of v in the suffix tree.

4.1   Homopolymer Errors and MEM Sequences

A MEM algorithm that runs on top of the suffix tree of $\mathcal{R}$ is unlikely to report all the real[2] matches if the input collection is HiFi data. The difficulty is that some of the MEMs are "masked" in the suffix tree. More specifically, suppose we have two nodes v and u, with $\mathsf{label}(v) \neq \mathsf{label}(u)$. It might happen that, by removing or adding copies of symbols in the equal-symbol runs of $\mathsf{label}(u)$, we can produce $\mathsf{label}(v)$. If those edits are small enough for the PacBio machine to produce them during the sequencing process, then it is plausible to assume that $\mathsf{label}(u)$ is an homopolymer error of $\mathsf{label}(v)$. This situation becomes even more likely if $\mathsf{label}(u)$ is long and its frequency is low in the collection.

Looking for all the possible suffix tree nodes that only have small differences in the length of homopolymer runs similar to v and u could be expensive. A simple workaround is to run-length compress $\mathcal{R}$ and execute the suffix-tree-based MEM algorithm with that as input. Now the problem is that we can report false positive MEMs between different sequences that have the same run-length representation but that are not homopolymer errors. Fortunately, filtering those false positive is not so difficult. Before explaining our idea, we formally define the notion of equivalence between sequences.

Definition 2. *Let A be a string whose run-length encoding is the sequence of pairs $A = (a_1, \ell_1), (a_2, \ell_2), \ldots, (a_p, \ell_p)$, where a_i is the symbol of the ith equal-symbol run, and $\ell_i \geq 1$ is its length. Additionally, let the operator $\mathsf{rlc}(A) = a_1, a_2, \ldots, a_p$ denote the sequence of run heads for A. We say that two strings A and B are equivalent iff $\mathsf{rlc}(A) = \mathsf{rlc}(B)$.*

We use equivalent sequences (Definition 2) to define a filtering parameter to discard false positive MEMs. We call this parameter the *run-length excess*:

Definition 3. *Let A and B be two distinct strings with $\mathsf{rlc}(A) = \mathsf{rlc}(B)$. Additionally, let the pair sequences $A = (x_1, \ell_1^a), (x_2, \ell_2^a), \ldots, (x_p, \ell_p^a)$ and $B = (x_1, \ell_1^b), (x_2, \ell_2^b), \ldots, (x_p, \ell_p^b)$ be the run-length encoding for A and B, respectively. Each $x_i \in \Sigma$ is the ith run head, and $\ell_i^a, \ell_i^b \geq 1$ are the lengths for x_i in A and B, respectively. Now consider the string $E = |\ell_1^a - \ell_1^b|, \ldots, |\ell_n^a - \ell_n^b|$ storing the absolute differences between the run lengths of A and B. We define the run-length excess as $\mathsf{rlexcess}(A, B) = \max(E[1], E[2], \ldots, E[n])$.*

Intuitively, equivalent sequences that have a high run-length excess are unlikely to have a masked MEM. The reason is because, although the PacBio sequencing process makes mistakes estimating the lengths of the equal-symbol runs, the error in the estimation is unlikely to be high.

Now that we have a framework to detect MEMs in run-length-compressed space, we construct a new collection $\mathcal{R}^h$ of $n_h \leq n$ symbols encoding the same strings of $\mathcal{R}$ but with their homopolymers compacted. Namely, every equal-symbol run $R_u[i, j] = (c, \ell)$ of maximal length $\ell > 1$ in $\mathcal{R}$ is represented with

---

[2] Those we would obtain in a collection with no homopolymer errors.

a special metasymbol $c^* \notin \Sigma$ in $\mathcal{R}^h$. We store the ℓ values in another list H, sorted as their respective homopolymers occur in $\mathcal{R}$. Each element of Σ has its own metasymbol, including the sentinel. We reorder the alphabet $\Sigma \cup \Sigma^h$ of $\mathcal{R}^h$ to the set $\{\$, \mathtt{A}, \mathtt{A}^*, \mathtt{C}, \mathtt{C}^*, \mathtt{G}^*, \mathtt{G}, \mathtt{T}^*, \mathtt{T}, \$^*\}$, which we map to its compact version $\Sigma^{hp} = [1, 10]$. This reordering will facilitate the synchronization of ranges when we perform extendLeft or extendRight in our implicit bi-directional BWT.

Recall from Sect. 4 that, when we call the operation extendLeft(X, c) (respectively, extendRight(X, c)), we need to perform rangeCount(BWT, s_X, e_X) to get the number of symbols within $BWT[s_X, e_X]$ whose DNA complements are smaller than c. For this counting operation to serve to synchronize $BWT[s_{\hat{X}}, e_{\hat{X}}]$ in constant time, we need the BWT alphabet to be symmetric. Concretely, the permutation π for the DNA complements has to exchange $\Sigma^{hp}[1]$ with $\Sigma^{hp}[\sigma]$, $\Sigma^{hp}[2]$ with $\Sigma^{hp}[\sigma-1]$, and so on. This is the reason why the sentinel has a metasymbol too, even though there are no sentinel homopolymers in $\mathcal{R}$. Additionally, we define a function $g : \Sigma^{hp} \to \Sigma$ to map metasymbols back to their nucleotides in Σ. When the input for g is not a metasymbol, g returns the nucleotide itself.

The next step is to run the suffix-tree-based algorithm to solve the all-vs-all MEM problem of Definition 1 (see Sect. 2) using $\mathcal{R}^h$ as input. However, we add one extra step. For each candidate MEM $(R_a[i, j], R_b[i', j'])$, with $R_a, R_b \in \mathcal{R}^h$, reported by the algorithm, we check if the run-length excess between $R_a[i, j]$ and $R_b[i, j]$ is below some minimum threshold e. If that is not the case, then we discard that pair as a MEM. We can easily compute the run-length excess value using the suffix array of $\mathcal{R}^h$ and the vector H. If the MEM algorithm detects that an internal node v of the suffix tree encodes a list of MEMs, then we use the suffix array of $\mathcal{R}^h$ to access the text positions label(v). Subsequently, we map those positions to H to get the lengths of the distinct variations of label(v) on the text, and thus compute excess among them.

4.2 Computing MEMs in Compressed Space

We now have all the elements to solve Problem 1 in run-length-compressed space using our implicit bi-directional BWT. Our input is the BWT of $\mathcal{R}^h$ (encoded as a wavelet tree BWT), the array H storing the lengths of the homopolymers in the HiFi reads, and the parameters τ and e for, respectively, the minimum MEM length and the maximum run-length excess (see Sect. 4.1).

We resort to the algorithm of Belazzougui et al. [2] to visit the internal nodes in the suffix tree T of $\mathcal{R}^h$ in $O(n_h \log |\Sigma^{hp}|)$ time, with $n_h = \Sigma_1^{|\mathcal{R}_h|}|R_i|$ (see Sect. 2). The advantage of their method is that we can use backward search operations over BWT to navigate T without visiting its edge labels (i.e., unary paths in the suffix trie of $\mathcal{R}^h$). Algorithm 1 describes the procedure.

Each internal node v of T with more than one Weiner link (i.e., $BWT[s_v, e_v]$ is not unary) encodes a group of MEMs. This property holds because label(v) has more than one left-context symbol and more than one right-context symbol in the text. Thus, any possible combination of strings $a \cdot$label$(v) \cdot b$ and $y \cdot$label$(v) \cdot z$ we can decode from v, with $a, b, y, z \in \Sigma^{hp}$, $a \neq y$, and $b \neq z$, corresponds to a MEM sequence (see Definitions). The sequences we obtain from v can have multiple occurrences in $\mathcal{R}^h$, and we need to report all of them. However, some of them might be false positives. For instance, the pair of text positions conforming a MEM are in the same string, or in strings that are DNA reverse complements of each other. We filter those cases as they are artefacts in our model.

When we visit a node v with more than one Weiner link during the traversal of T, we access its MEM sequences as follows: we use enumerateRight and extendRight to compute every range $SA[s_u, e_u]$, with $s_v \leq s_u \leq e_u \leq e_v$, encoding a child u of v. Then, over each $SA[s_u, e_u]$, we perform enumerateLeft and extendLeft to compute every range $SA[s_u^c, e_u^c]$ encoding a Weiner link c of u. This procedure yields a set $\mathcal{M} = \{\mathcal{I}_1, \mathcal{I}_2, \ldots, \mathcal{I}_p\}$, where p is the number of children of v, and $\mathcal{I}_q$, with $q \in [1, p]$, is the set of ranges in SA for the Weiner links of the qth child of v (from left to right).

The next step is to report the text positions of the MEM sequences encoded by $\mathcal{M}$. For this purpose, we consider the list of pairs $\{(\mathcal{I}_e, \mathcal{I}_g) \mid \mathcal{I}_e, \mathcal{I}_g \in \mathcal{M}$ and $\mathcal{I}_e \neq \mathcal{I}_g\}$. Every element $(SA[i, j], SA[i', j']) \in \mathcal{I}_e \times \mathcal{I}_g$ is a pair of ranges such that $SA[i, j]$ stores the suffixes of $\mathcal{R}^h$ prefixed by a label $a \cdot \mathsf{label}(v) \cdot b$ and $SA[i', j']$ stores the suffixes of $\mathcal{R}^h$ prefixed by another label $y \cdot \mathsf{label}(v) \cdot z$. We know that b and z are different as they come from different children of v. However, the symbols a and y might be equal, which means $\mathsf{label}(v)$ is not a MEM sequence when we match $a \cdot \mathsf{label}(v) \cdot b$ and $y \cdot \mathsf{label}(v) \cdot z$. We can find out this information easily: if $SA[i, j]$ and $SA[i', j']$ come from different buckets[3], then $a \neq y$. If that is the case, we have to report the MEMs associated to $(SA[i, j], SA[i', j'])$. For doing so, we first get the string depth $d = |\mathsf{label}(v)|$ of v. Then, we regard $X = \{i, \ldots, j\}$ and $O = \{i', \ldots, j'\}$ as two different sequences of consecutive indexes in SA, and iterate over their Cartesian product $X \times O$. When we access a pair $(SA[x], SA[o])$, with $(x, o) \in X \times O$, we compute the run-length excess e' between $\mathcal{R}^h[SA[x] + 1, SA[x] + d]$ and $\mathcal{R}^h[SA[o] + 1, SA[o] + d]$ as described in Sect. 4.1, and discard the MEM in $(SA[x], SA[o])$ if $e' \geq e$. We also discard it if $SA[x]$ and $SA[o]$ map the same string or map different strings that are reverse complements between each other. This procedure is described in Algorithm 2.

Theorem 1. *Let $\mathcal{R}^h$ be the run-length encoded collection of HiFi reads, with an alphabet of $\sigma_h = |\Sigma^{hp}|$ symbols. Additionally, let v be an internal node in the suffix tree of $\mathcal{R}^h$ that has more than one Weiner link. The string depth of v is $d = |\mathsf{label}(v)|$ and its associated range $SA[i, j]$ has length $x = j - i + 1$. We can compute all the MEMs encoded by v in $O(\sigma_h^2 \log \sigma_h + x^2 d)$ time.*

Proof. We first compute the ranges for the children of v with the operations enumerateRight and extendRight. These two operations take $O(\sigma_h \log \sigma_h)$ time. Then, for every child, we compute its Weiner links. The node v has up to σ_h children, each child has up to σ_h Weiner links, and to compute each of these takes $\log \sigma_h$ time via extendLeft, making $O(\sigma_h^2 \log \sigma_h)$ time in total. The number of suffixes of $\mathcal{R}^h$ in $\mathcal{M}$ is x, and the total number of suffix pairs we visit during the scans of the Cartesian products between sets of $\mathcal{M}$ is bound by x^2. Each time

---

[3] The bth bucket of SA is the range containing all suffixes prefixed by symbol $b \in \Sigma$.

we visit a pair of suffixes, computing the run-length excess between them takes us $O(d)$ time. Thus, the time for reporting the MEMs from v is $O(\sigma_h^2 \log \sigma_h + x^2 d)$.

$\square$

4.3 Improving the Time Complexity for Reporting MEMs

We can think of the problem of reporting MEMs from v as two-dimensional sorting. We need the occurrences of label(v) to be sorted by their left and right contexts at the same time (the dimensions) to report the MEMs from v efficiently. We can implement this idea using a grid $\mathcal{G}$ with dimensions $n_h \times n_h$. We (logically) label the rows of $\mathcal{G}$ with the suffixes of $\mathcal{R}^h$ sorted in lexicographical order, and do the same with the columns. We then store the values of SA in the grid cells, with the (row,column) coordinate for each $SA[j]$ being $(j, \mathsf{LF}(j))$. We encode $\mathcal{G}$ with the data structure of Chan et al. [5] that increases the space by $O(n_h \log n_h) + o(n_h \log n_h)$ bits and allows reporting of the occ points in the area $[x_1, x_2], [y_1, y_2]$ of $\mathcal{G}$ in $O((occ+1)(1+\log n_h / \log \log n_h))$ time. In exchange, we no longer require SA.

The procedure to report MEMs is now as follows. When we reach v during the suffix tree traversal, we perform extendLeft with each of v's Weiner links. This produces a list $\mathcal{L}$ of up to σ_h non-overlapping ranges in SA. We then create another list $\mathcal{Q}$ with the ranges obtained by following v's children. Notice that the ranges of $\mathcal{Q}$ are a partition of the range $[i, j]$ in SA for label(v). For every $[l_1, l_2] \in \mathcal{L}$, we extract the points in $\mathcal{G}$ in the area $[l_1, l_2], [i, j]$, and partition the result into subsets according to $\mathcal{Q}$. The partition is simple as the points can be reported in increasing order of the y coordinates (range $[i, j]$). The idea is to generate a list $\mathcal{I} = \{I_1, I_2, \ldots, I_x\}$ of at most σ_h^2 elements, where each element is a point set for an area $[l_1, l_2], [q_1, q_2] \in \mathcal{L} \times \mathcal{Q}$. Finally, we scan all possible distinct pairs of $\mathcal{I}$ that yield MEMs, processing suffixes as in lines 18–23 of Algorithm 2. Let $I_i, I_j \in \mathcal{I}$ be two point sets, extracted from the areas $[l_1, l_2], [q_1, q_2]$ and $[l_1', l_2'], [q_1', q_2']$ of $\mathcal{G}$, respectively. The points of I_i will have MEMs with the points of I_j if $[l_1, l_2] \neq [l_1', l_2']$ and $[q_1, q_2] \neq [q_1', q_2']$. See Fig. 1.

Corollary 2. *By replacing SA with the grid of Chan et al. [5], reporting the MEMs associated with internal node v of the suffix tree of $\mathcal{R}^h$ takes $O((x + \sigma)(1 + \log n_h / \log \log n_h) + l^2 d)$ time, where x is the number of occurrences of* label(v) *in $\mathcal{R}^h$, $l \leq x$ is the number of those occurrences that have MEMs, and $d =$* label(v).

5   Concluding Remarks

We presented a framework to compute all-vs-all MEMs in a collection of run-length encoded HiFi reads. Our techniques can be adapted to other types of collections with properties similar to that of HiFi data (e.g., Nanopore sequencing data, proteins, Phred scores, among others). The larger alphabet of proteins and Phred scores make our MEM reporting algorithm that uses the geometric data structure more relevant (as it avoids the σ^2 complexity of our first method). We are also applying these techniques to *de novo* assembly of HiFi reads.

Appendix

Algorithm 1. Computing MEMs in one traversal of the suffix tree T of $\mathcal{R}^h$. Arrays BWT, SA, and H are implicit in the pseudo-code. Each node $v \in T$ is encoded by the pair (v, d), where $v = (i, j), (i', j')$ are the ranges in SA for label(v) and its DNA reverse complement label$(\hat{v})$, and d is the string depth.

Input: Suffix tree T for $\mathcal{R}^h$ encoded by the implicit bi-directional BWT.
Output: MEMs as described in Definition 1.
```
 1: S ← ∅                                              ▷ Empty stack
 2: r ← (1, n + 1), (1, n + 1)                         ▷ The root of T
 3: push(S, (r, 0))
 4: while S ≠ ∅ do
 5:     (v, d) ← top(S)        ▷ Extract suffix tree node v from the top of the stack
 6:     pop(S)
 7:     if d ≥ τ and isLeftMaximal(v) and isRightMaximal(v) then
 8:         repMEM(v, e, d)
 9:     end if
10:     for c ∈ enumerateLeft(v) do    ▷ Continue visiting other suffix tree nodes
11:         u ← extendLeft(v, c)
12:         if isLeftMaximal(u) then
13:             insert(S, (u, d + 1))
14:         end if
15:     end for
16: end while
```

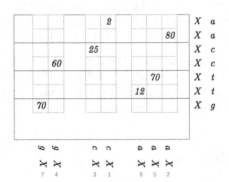

Fig. 1. Reporting MEMs from an internal node v labeled label$(v) = X$ using the grid $\mathcal{G}$. The rows are labeled with the suffixes prefixed by X, while the column are labeled with the suffixes prefixed with the labels of v's Weiner links. The horizontal red lines represents the partition of the SA range for X induced by the children of v. The grey numbers below the column labels are the LF^{-1} values. For each column j', its associated SA value is in the row $\mathsf{LF}^{-1}(j') = j$.

Algorithm 2. Report all-vs-all MEMs from a suffix tree node v. Arrays BWT and H for $\mathcal{R}^h$ are implicit in the pseudo-code. Node v is encoded as described in Algorithm 1

Input: An internal node $v \in T$ with more than one Weiner link, run-length excess threshold e, and $d = |\mathsf{label}(v)|$.

Output: List of MEMs among strings of $\mathcal{R}^h$ that can be computed from v.

```
 1: procedure repMEM(v, d, e)
 2:     C ← ∅
 3:     for c ∈ enumerateRight(v) do        ▷ Partition SA[v.i, v.j] according v's children
 4:         C ← C ∪ {extendRight(v, c)}
 5:     end for
 6:     M ← ∅
 7:     for x ← 1 to |C| do                  ▷ Get Weiner links for every child of v
 8:         I_x ← ∅
 9:         for d ← enumerateLeft(C[x]) do
10:             I_x ← I_x ∪ {extendLeft(C[x], d)}
11:         end for
12:         M ← M ∪ I_x
13:     end for
14:     for I_a, I_b ∈ M with I_a ≠ I_b do ▷ I_a and I_b come from different children of v
15:         for (X, Y) ∈ I_a × I_b do
16:             if X and Y belong to distinct SA bucket then
17:                 for (q, e) ∈ X × Y do
18:                     R_q ← string in R^h for SA[q] + 1
19:                     R_e ← string in R^h for SA[e] + 1
20:                     e' ← rlExcess(SA[q] + 1, SA[e] + 1, d)
21:                     if e' ≤ e then
22:                         Report MEM in (q, e)
23:                     end if
24:                 end for
25:             end if
26:         end for
27:     end for
28: end procedure
```

References

1. Bauer, M.J., Cox, A.J., Rosone, G.: Lightweight algorithms for constructing and inverting the BWT of string collections. Theor. Comput. Sci. **483**, 134–148 (2013)
2. Belazzougui, D., Cunial, F., Kärkkäinen, J., Mäkinen, V.: Versatile succinct representations of the bidirectional Burrows-Wheeler transform. In: Bodlaender, H.L., Italiano, G.F. (eds.) ESA 2013. LNCS, vol. 8125, pp. 133–144. Springer, Heidelberg (2013). https://doi.org/10.1007/978-3-642-40450-4_12
3. Boucher, C., et al.: PHONI: streamed matching statistics with multi-genome references. In: Proceedings of the 21st Data Compression Conference (DCC), pp. 193–202 (2021)
4. Burrows, M., Wheeler, D.: A block sorting lossless data compression algorithm. Technical report 124, Digital Equipment Corporation (1994)

5. Chan, T., Larsen, K.G., Pătraşcu, M.: Orthogonal range searching on the RAM, revisited. In: Proceedings of the 27th Annual Symposium on Computational Geometry (SoCG), pp. 1–10 (2011)

6. Chang, W.I., Lawler, E.L.: Sublinear approximate string matching and biological applications. Algorithmica **12**(4), 327–344 (1994)

7. Clark, D.: Compact PAT trees. Ph.D. thesis, University of Waterloo, Canada (1996)

8. Fredkin, E.: Trie memory. Commun. ACM **3**(9), 490–499 (1960)

9. Gagie, T., Navarro, G., Prezza, N.: Fully functional suffix trees and optimal text searching in BWT-runs bounded space. J. ACM (JACM) **67**(1), 1–54 (2020)

10. Gagie, T., Navarro, G., Puglisi, S.J.: New algorithms on wavelet trees and applications to information retrieval. Theor. Comput. Sci. **426**, 25–41 (2012)

11. Grabowski, S., Bieniecki, W.: copMEM: finding maximal exact matches via sampling both genomes. Bioinformatics **35**(4), 677–678 (2019)

12. Grossi, R., Gupta, A., Vitter, J.S.: High-order entropy-compressed text indexes. In: Proceedings of the 14th Annual ACM-SIAM Symposium on Discrete Algorithms (SODA), pp. 841–850 (2003)

13. Gusfield, D.: Algorithms on Strings, Trees, and Sequences: Computer Science and Computational Biology. Cambridge University Press, Cambridge (1997)

14. Jacobson, G.: Space-efficient static trees and graphs. In: Proceedings of the 30th Annual Symposium on Foundations of Computer Science (FOCS), pp. 549–554 (1989)

15. Kasai, T., Lee, G., Arimura, H., Arikawa, S., Park, K.: Linear-time longest-common-prefix computation in suffix arrays and its applications. In: Amir, A. (ed.) CPM 2001. LNCS, vol. 2089, pp. 181–192. Springer, Heidelberg (2001). https://doi.org/10.1007/3-540-48194-X_17

16. Khan, Z., Bloom, J.S., Kruglyak, L., Singh, M.: A practical algorithm for finding maximal exact matches in large sequence datasets using sparse suffix arrays. Bioinformatics **25**(13), 1609–1616 (2009)

17. Khiste, N., Ilie, L.: E-MEM: efficient computation of maximal exact matches for very large genomes. Bioinformatics **31**(4), 509–514 (2015)

18. Lam, T.W., Li, R., Tam, A., Wong, S., Wu, E., Yiu, S.-M.: High throughput short read alignment via bi-directional BWT. In: Proceedings of the 3rd International Conference on Bioinformatics and Biomedicine (BIBM), pp. 31–36 (2009)

19. Li, H.: Minimap2: pairwise alignment for nucleotide sequences. Bioinformatics **34**(18), 3094–3100 (2018)

20. Liu, Y., Zhang, L.Y., Li, J.: Fast detection of maximal exact matches via fixed sampling of query k-mers and bloom filtering of index k-mers. Bioinformatics **35**(22), 4560–4567 (2019)

21. Logsdon, G.A., Vollger, M.R., Eichler, E.E.: Long-read human genome sequencing and its applications. Nat. Rev. Genet. **21**(10), 597–614 (2020)

22. Manber, U., Myers, G.: Suffix arrays: a new method for on-line string searches. SIAM J. Comput. **22**(5), 935–948 (1993)

23. Ohlebusch, E., Fischer, J., Gog, S.: CST++. In: Proceedings of the 17th International Symposium on String Processing and Information Retrieval (SPIRE), pp. 322–333 (2010)

24. Ohlebusch, E., Gog, S., Kügel, A.: Computing matching statistics and maximal exact matches on compressed full-text indexes. In: Chavez, E., Lonardi, S. (eds.) SPIRE 2010. LNCS, vol. 6393, pp. 347–358. Springer, Heidelberg (2010). https://doi.org/10.1007/978-3-642-16321-0_36

25. Rossi, M., Oliva, M., Bonizzoni, P., Langmead, B., Gagie, T., Boucher, C.: Finding maximal exact matches using the r-index. J. Comput. Biol. **29**(2), 188–194 (2022)

26. Rossi, M., Oliva, M., Langmead, B., Gagie, T., Boucher, C.: MONI: a pangenomic index for finding maximal exact matches. J. Comput. Biol. **29**(2), 169–187 (2022)
27. Sadakane, K.: Compressed suffix trees with full functionality. Theory Comput. Syst. **41**(4), 589–607 (2007)
28. Vyverman, M., De Baets, B., Fack, V., Dawyndt, P.: essaMEM: finding maximal exact matches using enhanced sparse suffix arrays. Bioinformatics **29**(6), 802–804 (2013)
29. Weiner, P.: Linear pattern matching algorithms. In: Proceedings of the 14th Annual Symposium on Switching and Automata Theory (SWAT), pp. 1–11 (1973)
30. Wenger, A.M., et al.: Accurate circular consensus long-read sequencing improves variant detection and assembly of a human genome. Nat. Biotechnol. **37**(10), 1155–1162 (2019)

Space-Efficient Data Structures

Internal Masked Prefix Sums and Its Connection to Fully Internal Measurement Queries

Rathish Das[1], Meng He[2], Eitan Kondratovsky[1(✉)], J. Ian Munro[1], and Kaiyu Wu[1]

[1] Cheriton School of Computer Science, University of Waterloo, Waterloo, Canada
{rathish.das,eitan.kondratovsky,imunro,k29wu}@uwaterloo.ca
[2] Faculty of Computer Science, Dalhousie University, Halifax, Canada
mhe@cs.dal.ca
https://cs.uwaterloo.ca/~imunro/, https://web.cs.dal.ca/~mhe/

Abstract. We define a generalization of the prefix sum problem in which the vector can be masked by segments of a second (Boolean) vector. This problem is shown to be related to several other prefix sum, set intersection and approximate string match problems, via specific algorithms, reductions and conditional lower bounds. To our knowledge, we are the first to consider the fully internal measurement queries and prove lower bounds for them. We also discuss the hardness of the sparse variation in both static and dynamic settings. Finally, we provide a parallel algorithm to compute the answers to all possible queries when both vectors are fixed.

1 Introduction

The prefix sums problem (also known as *scans*) is one in which one seeks to preprocess an array of n numbers to answer prefix sum queries. This problem is widely known and has been motivated by many fields such as parallel algorithm, graph theory, and more [5,10]. Pătrașcu et al. [23] proved tight lower bounds for the query time when using linear space. Later, Bille et al. [4] provided tight lower bounds in the dynamic model where insertion, deletion, and modifications of the array's numbers are allowed. On the implementation side, Pibiri and Venturini [24] give an overview of current techniques and how well they perform in practice.

The internal model has received much attention in recent years. In this setting, the input is given as a sequential string and the queries are asked on different substrings of the input. One of the first problems in this research field was the *internal pattern matching problem*. In this problem, a string S is preprocessed to answer queries of the form: "report all occurrences where a substring of S is located in another substring of S" [15,17]. Since then, many internal query problems were introduced. Examples include the longest common prefix of two substrings of S, computing the periods of a substring of S, etc. We refer the interested reader to [16], which contains an overview of the literature.

© The Author(s), under exclusive license to Springer Nature Switzerland AG 2022
D. Arroyuelo and B. Poblete (Eds.): SPIRE 2022, LNCS 13617, pp. 217–232, 2022.
https://doi.org/10.1007/978-3-031-20643-6_16

The *internal masked prefix sum problem* takes as input an m-bit mask B and an array A of n numbers, where $m \geq n$, and supports the query: "report the prefix sum of the numbers against some substring of the mask". It is easy to see that the subtraction of two such prefix sums supports any masked sum of a substring of A. That is, subtracting the prefix sum until position i from the prefix sum until position $j > i$, where i, j are the substring positions. A simplified case of the problem is when A consist of binary values, i.e. $A \in \{0, 1\}^*$. In this case, the masked prefix sums are the inner products of the substrings.

Clifford et al. [7] introduced the problem of dynamic data structure that supports the inner product (and other measurements) between an m-length pattern and any m-length substring of the text, where m is fixed. Here dynamic means that substitutions of letters in the pattern and in the text are allowed. The additional measurements that were considered are the Hamming distance and exact matching with wildcards. It was shown that in the dynamic setting, both query and update cannot be done in $O(n^{\frac{1}{2}-\varepsilon})$ time, for some $\varepsilon > 0$, unless the OMv (Online Boolean Matrix-Vector Multiplication) conjecture is false.

Our contributions are the following.

1. We give a preprocess-query time trade-off that matches the lower bound of the batched problem up to logarithmic factors. The trade-off algorithm works in $O(\frac{nm}{f(n)} \log f(n))^1$ preprocessing time and $O(\frac{nm}{f(n)})$ space and answers queries in $O(f(n))$ time, for any $1 \leq f(n) \leq n$.
2. We prove a lower bound for the batched problem in which the algorithm preprocesses the data to answer a batch of n queries, where n is the length of the input. We show a lower bound of $p(n) + nq(n) = \Omega(n^{\frac{3}{2}-\varepsilon})$, for any $\varepsilon > 0$, where $p(n)$ is the preprocessing time and $q(n)$ is the query time. This lower bound illustrates that queries that consist of substrings from both A and B even in a static setting, have similar hardness as the dynamic setting but on substrings with fixed length.
3. We show a $(1 + \varepsilon)$-approximation algorithm for the *internal masked prefix sums* that works in near constant time for any $\varepsilon > 0$.
4. We give a parallel algorithm that computes all the internal prefix sums in $O(\log n + \log m)$ span and $O(nm)$ work to answer all the possible queries that are stored explicitly, or $O(\log n + \log m)$ span and $O(\frac{nm}{\log n})$ work for (implicit) constant-time queries.
5. We consider the *sparse internal inner product* and we show conditional lower bounds from SETDISJOINTNESS and 3SUM in the static and dynamic settings, respectively.

The paper is organized as follows: In Sect. 2 we give the formal definition of our problems and other data structures that we will be using as building blocks. In Sect. 3 we study the *internal masked prefix sum*. We give the preprocess-query trade-off data structure for the problem along with the conditional lower bound. Furthermore, we study an approximate form of the problem and finally give a

---

[1] We will use log to denote $\log_2$, though as our log all eventually end up in asymptotic notation, the constant bases are irrelevant.

parallel algorithm that can be used in the preprocessing of the data structures. In Sect. 4 we study the *sparse internal inner product* and give conditional lower bounds for the problem. Finally, in Sect. 5 we discuss a related problem, of calculating Hamming distances and how it can be reduced to the *internal masked prefix sum* problem.

2   Preliminaries

Let Σ be an alphabet. A *string* S over Σ is a finite sequence of letters from Σ. By $S[i]$, for $1 \leq i \leq |S|$, we denote the ith letter of S. The *empty string* is denoted by ε. By $S[i..j]$ we denote the string $S[i] \cdots S[j]$, called a *substring* of S (if $i > j$, then the substring is the empty string). A substring is called a *prefix* if $i = 1$ and a *suffix* if $j = |S|$.

In the paper, we assume that Σ consists of nonnegative numbers. The non-negativity ensures that binary search on the partial prefix sums is well defined.

Definition 1 (`MaskedPrefixSumProblem`).
Input: *A bit vector (mask) B and an array of numbers A of lengths m and n, respectively. The goal is to preprocess a data structure that answers the masked prefix sum queries of the following form.*
Query: *Given i and k, where $1 \leq k \leq |A|, 1 \leq i \leq |B| - k + 1$. We wish to report the sum of the first k numbers of A that corresponds to 1s in the k-length submask of B located at position i. Formally,*

$$\sum_{j=1}^{k} A[j] \cdot B[i+j-1] = \langle A[1 \ldots k], B[i \ldots i+k-1] \rangle.$$

Example 1. Let $A = [1, 0, 2, 0, 0, 0, 4, 8]$ and $B = 10100111001$. For query $k = 7$ and $i = 5$ the answer is 6. To see this, we look at the first 7 numbers of A against the 7-length submask starting at position 5. After masking out the elements of A, which are against 0s, we are left with 2 and 4 which sum to 6.

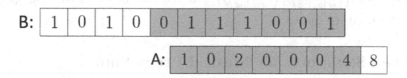

Fig. 1. Highlighted cells are those used in the query.

Definition 2 (`SparseMaskedPrefixInnerProduct`).
Input: *A bit vector (mask) B and a bit vector (numbers) A such that the sum of the number of 1s in A and the number of 1s in B is n, i.e. the problem can be represented by writing down the n positions of the ones. The goal is to*

preprocess a data structure that answers the sparse internal inner product queries of the following form.

Query: *Given i and k, where $1 \leq k \leq |A|, 1 \leq i \leq |B| - k + 1$. We wish to report the inner product of the first k bits of A against the k-length submask of B located at position i.*

In the paper, we consider the RAM model with word size $w = \log n$. We assume that each integer fits into one $\log n$-length RAM word. Thus the sum of n integers is at most n^2, and each prefix sum fits into constant number of RAM words (more specifically 2). Notice that in some problems $w = \log U$, where U is the universe size of the problem. When the context is clear, we write a capital U to indicate that this is the case.

In our solutions to the above `MaskedPrefixSumProblem` and the `SparseMaskedPrefixInnerProduct` problems, we will extensively use data structures for the predecessor/successor problem.

The predecessor problem is to preprocess a set S of N integers from a universe of size U, to answer queries of the form: "Given an input i, return the largest integer in S smaller than i". The successor problem is analogous. We only state the results for the predecessor query but all of the following data structures can handle the successor query as well with the same complexities.

Willard's y-fast tries [25] gives a $O(N)$ word space solution to the problem with $O(\log \log U)$ query time. Another approach is Fredman and Willard's fusion trees [11], which also uses $O(N)$ words of space and supports the predecessor query in $O(\sqrt{\log N})$ amortized time. Andersson [1] showed how to make the query in worst case time $O(\sqrt{\log N})$.

Combining these results, we obtain the following lemma. The decision on which data structure to use follows from the values of U and N.

Lemma 1 (Predecessor Query). *There is a data structure for the predecessor problem that uses $O(N)$ words of space and query time $O(\min\{\log \log U, \sqrt{\log N}\})$.*

We note that the data structure of Beame and Fich [3] gives a better running time of $O(\min\{\frac{\log \log U}{\log \log \log U}, \sqrt{\frac{\log N}{\log \log N}}\})$, at the cost of increasing the space to $O(N^{O(1)})$ words of space.

3   Data Structures for Masked Prefix Sum

In this section, we design data structures for masked prefix sum and some of its variants. More specifically, we present a time-space trade-off for this problem (Sect. 3.1) and prove a conditional lower bound (Sect. 3.2). We also design data structures for the dynamic version (Sect. 3.3) and the approximate version (Sect. 3.4) of the masked prefix sum problem. Finally, we consider a related problem of designing a parallel algorithms to compute the answers to all possible queries over a given instance of this problem (Sect. 3.5).

3.1   Time-Space Trade-off

We now present a data structure for the masked prefix sum problem. With it, we achieve a trade-off between space/preprocessing cost and query time. We note that the data structure can be immediately generalized to solve other internal measurements - such as those studied by Clifford et al. [7] by replacing the inner product by any linear function, i.e., any function g over two strings that can be written as

$$g(S, T) = \sum_i g(S[i], T[i]).$$

The main result can be stated as:

Theorem 1. *Given a bit vector B of length m and an array A of length n, there is a data structure that uses $\frac{mn}{f(n)}$ words of space that can answer masked prefix sum queries in $O(f(n))$ time, for any function $f(n)$ with $0 < f(n) < n$. The preprocessing time is $O(\frac{mn}{f(n)} \log f(n))$.*

Furthermore, for any $c > 0$, there is a data structure that uses $\frac{mn}{f(n)} + \frac{n^{1+c}}{c \log n}$ words of space and can answer masked prefix sum queries in $O(\frac{f(n)}{c \log n})$ time. The preprocessing time is $O(\frac{mn}{f(n)} \log f(n) + \frac{n^{1+c}}{c \log n})$.

Proof. We do this by writing down the answer to the queries for each ending position in A that is a multiple of $f(n)$ and each offset in B, using $\frac{mn}{f(n)}$ words of space. That is, we create a table D such that

$$D[i, k] := \sum_{j=1}^{k f(n)} A[j] \cdot B[i + j - 1]$$

For every query length k not a multiple of $f(n)$, we find the largest multiple of $f(n)$ and write $k = k' f(n) + \ell$, where $k' = \lfloor k/f(n) \rfloor$. We then break the sum into

$$\sum_{j=1}^{k} A[j] \cdot B[i + j - 1] = \sum_{j=1}^{k' f(n)} A[j] \cdot B[i + j - 1] + \sum_{j=k' f(n)+1}^{k} A[j] \cdot B[i + j - 1]$$

The first term is $D[i, k']$, and we compute the second term by computing $A[j] \cdot B[i + j - 1]$ one by one, which requires $O(f(n))$ time. Thus, the query time is $O(f(n))$.

To slightly improve the time, we consider all bit masks of length $c \log n$, of which there are n^c such bit masks. For each index that is a multiple of $c \log n$ in A and each possible bit mask M of length $c \log n$, we write down

$$D'[M, j] := \sum_{k=0}^{c \log n - 1} A[j + k] \cdot M[k]$$

and we store these values in the table D', using $\frac{n^{1+c}}{c \log n}$ space.

This allows us to compute the sum of $c \log n$ elements at a time. To see this, consider the second term in the previous sum $\sum_{j=k'f(n)+1}^{k} A[j] \cdot B[i+j-1]$, which we summed one term at a time. Let $h_1 = \lfloor \frac{k'f(n)+1}{c \log n} \rfloor + 1$ and $h_2 = \lfloor \frac{k}{c \log n} \rfloor$. Then we have $k'f(n) + 1 < h_1 c \log n < (h_1 + 1)c \log n < \cdots < h_2 c \log n \leq k$. Create the masks $M_{h_1-1} = ..000B[k'f(n)+1, h_1 c \log n - 1]$, $M_{h_1} = B[h_1 c \log n, (h_1 + 1)c \log n - 1]$, ..., $M_{h_2} = B[h_2 c \log n, k]000..$, where M_{h_1-1} and M_{h_2} are padded with 0s so that their lengths are $c \log n$. We can then write the sum as

$$\sum_{j=k'f(n)+1}^{k} A[j] \cdot B[i+j-1] = \sum_{i=h_1-1}^{h_2} D'[M_i, ic \log n]$$

Thus the time to sum the remainder would be $O(\frac{f(n)}{c \log n})$.

We now show how to build these data structures. Computing all $D[i,k] = \sum_{j=1}^{kf(n)+1} A[j] \cdot B[i+j-1]$, where $i \in [m - kf(n)]$, $1 \leq k \leq \frac{n}{f(n)}$ can be done in $O(\frac{mn}{f(n)} \log f(n))$ time. The computation is done in two phases. First, for every $1 \leq k \leq \frac{n}{f(n)}$, we consider the subarray $A[(k-1)f(n)+1 \ldots kf(n)]$, we call it A_k for short. In the first stage of the computation, we wish to compute the following sum, for every $1 \leq k \leq \frac{n}{f(n)}$, and for every $1 \leq i \leq m - f(n) + 1$.

$$\sum_{j=1}^{f(n)} A_k[j] \cdot B[i+j-1]$$

Such sum is called a convolution of A_k and B as one can think of A_k slides over B. It is well known that performing the *generalized fast Fourier transform* (see for example, [8,14]) between each A_k and B computes such convolution. For short, we refer to such an algorithm as FFT. The FFT computes the masked sum of each A_k against every $f(n)$-length submask of B. Each of these FFTs is done in $O(m \log f(n))$ time. Thus, $O(\frac{nm}{f(n)} \log f(n))$ overall time is required. The second stage aggregates the sums of the smaller intervals for each offset in B. Let i be an offset in B. We know $\langle B[i \ldots i+f(n)-1], A_1\rangle, \langle B[i+f(n) \ldots i+2f(n)-1], A_2\rangle, \langle B[i+2f(n) \ldots i+3f(n)-1], A_3\rangle \ldots \langle B[i+n-f(n) \ldots i+n-1], A_{\frac{n}{f(n)}-1}\rangle$. Thus we partially sum the entries, i.e. for array of values P, we compute $\Sigma_{q=1}^{r} P[q]$, for each $1 \leq r \leq |P|$ in linear time. The first stage requires $O(\frac{nm}{f(n)} \log f(n))$ time and the second stage requires $O(\frac{nm}{f(n)})$ time. The preprocessing times given in the first half of this theorem thus follows.

Finally, as D' has $\frac{n^{1+c}}{c \log n}$ entries and each entry can be computed in $O(1)$ time, D' can be constructed in $O(\frac{n^{1+c}}{c \log n})$ time, and we obtain the preprocessing time in the second half of this theorem. To see this, fix a particular value of j and compute values for different masks by ascending number of 1s in the mask. Mask pattern M would then only introduce 1 new term over a previously computed mask pattern. □

Finally we note that for $m = O(n)$, we may set $f(n) = \sqrt{n} \log n$ and $c = 1/2$ to obtain an $O(n^{3/2}/\log n)$-word data structure with $O(\sqrt{n})$ query time and $O(n^{3/2})$ preprocessing time.

3.2 A Conditional Lower Bound

We now give a conditional lower bound to show the hardness of this problem.

Theorem 2. *Let $p(n)$ and $q(n)$ respectively denote the preprocessing time and query time of a masked prefix sum data structure constructed over a bit vector mask of length n and an array of n bits. Then Boolean matrix multiplication over two $\sqrt{n/2} \times \sqrt{n/2}$ matrices can be solved in $p(n) + nq(n) + O(n)$ time.*

Proof. Let X and Y be two $\sqrt{n/2} \times \sqrt{n/2}$ Boolean matrices and $Z = X \times Y$. Let $x_{i,j}$, $y_{i,j}$ and $z_{i,j}$ denote the elements in row i and column j of matrices X, Y and Z, respectively. We then construct an array A of n bits and a bit mask B of length n as follows. A is obtained by storing the bits in X in row-major order in its first half, i.e., $A[(i - 1)\sqrt{n/2} + j] = x_{i,j}$ for any $1 \le i, j \le \sqrt{n/2}$, and storing all 0s in its second half. The first $n/2$ bits of the mask B are also all 0s. Then we store the content of Y in $B[n/2 + 1..n]$ in column-major order, i.e., $B[n/2 + (j - 1)\sqrt{n/2} + i] = y_{i,j}$ for any $1 \le i, j \le \sqrt{n/2}$.

To compute $z_{i,j}$ for any $1 \le i, j \le \sqrt{n/2}$, observe that the ith row of X form the content of $A[(i - 1)\sqrt{n/2} + 1..i\sqrt{n/2}]$ while the jth column of Y is in $B[n/2 + (j - 1)\sqrt{n/2} + 1..n/2 + j\sqrt{n/2}]$. It is then sufficient to answer the following two masked prefix sum queries: The first query uses $(i - 1)\sqrt{n/2}$ and $n/2 + (j-1)\sqrt{n/2} - (i-1)\sqrt{n/2}$ as the mask length and offset of B, respectively, while the second uses $i\sqrt{n/2}$ and $n/2 + (j - 1)\sqrt{n/2} - (i - 1)\sqrt{n/2}$. Note that by setting the offset of B to be $n/2 + (j-1)\sqrt{n/2} - (i-1)\sqrt{n/2}$ in both queries, we ensure that $A[(i - 1)\sqrt{n/2} + 1..i\sqrt{n/2}]$ (storing the ith row of X) is always masked by $B[n/2 + (j - 1)\sqrt{n/2} + 1..n/2 + j\sqrt{n/2}]$ (storing the jth column of Y). Hence the answer to the second query subtracted by that to the first will give us the dot product of the ith row of X and the jth column of Y. Then, $z_{i,j} = 0$ if this product is 0, and $z_{i,j} = 1$ otherwise. Hence, we can compute Z by answering n masked prefix sum queries, and the theorem follows. $\square$

As the current best algebraic method of multiplying two $n \times n$ Boolean matrices has complexity $O(n^\omega)$ with $\omega < 2.3727$ [26], two $\sqrt{n/2} \times \sqrt{n/2}$ matrices can be multiplied in $O(n^{\omega/2})$ time. This implies that, with current knowledge, either the preprocessing time $p(n)$ must be $\Omega(n^{\omega/2}) = \Omega(n^{1.18635})$, or the query time $q(n)$ must be $\Omega(n^{\omega/2-1}) = \Omega(n^{0.18635})$. Furthermore, the running time of the best known combinatorial approach for multiplying two $n \times n$ Boolean matrices is only polylogarithmically faster than cubic [2,6,27]. Hence, by purely combinatorial methods with the current best knowledge, either the preprocessing time $p(n)$ must be $\Omega(n^{3/2})$ or the query time $q(n)$ must be $\Omega(\sqrt{n})$, save for polylogarithmic speed-ups. On the other hand, the specific trade-off given in Theorem 1 gives a data structure with $O(\sqrt{n})$ query time and $O(n^{3/2})$ preprocessing time, for any $m = O(n)$. Hence, it can be used to multiply two $\sqrt{n} \times \sqrt{n}$ Boolean matrices in $O(n^{3/2})$ time, matching the time required for the best known combinatorial algorithm for Boolean matrix multiplication within polylogarithmic factors.

3.3   Dynamic Masked Prefix Sum

In dynamic settings, we support the update to any entry of A or B by assigning a new value to it. The following theorem presents our result.

Theorem 3. *Given a bit vector B of length m and an array A of length n, there is a data structure that uses $O(\frac{mn}{f(n)} + m + n)$ words of space that can answer masked prefix sum queries in $O(f(n) + g(n))$ time and support updates in $O(\frac{mn \log f(n)}{g(n)f(n)} + g(n))$ time, for any functions $f(n)$ and $g(n)$ with $0 < f(n) < n$ and $0 < g(n) < m + n$.*

Alternatively, for any $c > 0$, there is a data structure that uses $O(\frac{mn}{f(n)} + \frac{n^{1+c}}{c \log n} + m + n)$ words of space and can answer masked prefix sum queries in $O(\frac{f(n)}{c \log n} + g(n))$ time and support updates in $O(\frac{mn \log f(n)}{g(n)f(n)} + \frac{n^{1+c}}{g(n)} + g(n))$ time. If $m = O(n)$, setting $f(n) = n^{2/3} \log n$, $g(n) = n^{2/3}$ and $c = 1/3$ yields an $O(n^{4/3}/\log n)$-word data structure with $O(n^{2/3})$ query and update times.

For the full proof, see Appendix A.

3.4   Approximate Masked Prefix Sum

To achieve faster query time and decrease the space cost, we consider the problem of building a data structure to answer the masked prefix sum problem approximately.

Theorem 4. *Given a bit vector B of length m and an array A of length n, there is a data structure that uses $O((m \log n)/\varepsilon)$ words of space and can answer $(1+\varepsilon)$-approximate masked prefix sum queries in $O(\min\{\log \log n, \sqrt{\log(\log n/\varepsilon)}\})$ time for any $\varepsilon \in (0, 1]$.*

Proof. We first consider the approximate prefix sum solution on just the integer vector A. Consider the n prefix sums $S[j] = \sum_{k=1}^{j} A[k]$. We build a mapping P such that $j \in P$ if $(1+\varepsilon)^i \le S[j] < (1+\varepsilon)^{i+1}$ for some i and $S[j-1] < (1+\varepsilon)^i$. That is, whenever the prefix sum reaches a power of $(1 + \varepsilon)$, we write down the index where the prefix sum reaches it. Furthermore, for these indices, we write down the actual prefix sum, so that $P[j] = S[j]$. We may store this mapping in linear space with a hash table. To answer the approximate query for an index ℓ, we find the predecessor of ℓ in P and report the prefix sum at the predecessor.

By construction, if the predecessor of ℓ is j and $(1 + \varepsilon)^i \le S[j] < (1 + \varepsilon)^{i+1}$, then $S[\ell] < (1 + \varepsilon)^{i+1}$, as otherwise there would be a predecessor of ℓ where the prefix sum reaches $(1 + \varepsilon)^{i+1}$. Similarly, $S[\ell] \ge (1 + \varepsilon)^i$. Thus our output gives a $(1 + \varepsilon)$-approximation to the actual result.

We note that the universe of the integers for the predecessor problem is $U = n$, and the number of elements in the set is at most $N = \log_{(1+\varepsilon)} n$. Since $0 < \epsilon \le 1$, by Taylor series expansion, $\log(1 + \varepsilon) = \Theta(\varepsilon)$. Hence, $N = O((\log n)/\varepsilon)$. Thus by Lemma 1 the predecessor data structure takes $O((\log n)/\varepsilon))$ words of space. The query time is $O(\min\{\log \log n, \sqrt{\log(\log n/\varepsilon)})$.

We now apply this solution to solve the masked prefix sum problem approximately, by building the above data structure for each index i of B on the integer vector A_i defined as $A_i[j] = A[j] \cdot B[i+j-1]$ (we mask the integer vector by the length n bit vector obtained from B starting at index i). This gives a solution of $O((m \log n)/\varepsilon)$ words of space.   □

3.5 Parallel Algorithms

We now consider a related problem which is to design a parallel algorithm to answer all queries of an instance of the masked prefix sum problem in the PRAM model.

Lemma 2. *Let A be the array of numbers of length n and let B be the bit vector of length m in the masked prefix sum problem. Then there is an optimal span (parallel running time) and work parallel algorithm that stores explicitly the answers of all mn queries in $O(\log n + \log m)$ span and performs $\Theta(mn)$ work in the PRAM model. In the implicit model, the work can be improved to be $O(\frac{mn}{\log n})$.*

The proof will appear in the journal version.

4 Data Structures for Sparse Internal Inner Product

In this section, we study the `SparseMaskedPrefixInnerProduct` problem, in both static (Sect. 4.1) and dynamic (Sect. 4.2) settings.

4.1 Static Sparse Internal Inner Product

We first present conditional lower bounds for the sparse internal inner product problem, SPARSEIIP for short, by giving a reduction from the SETDISJOINTNESS problem, which is defined as follows.

Definition 3 (SETDISJOINTNESS Problem). *Preprocess a family F of m sets, all from universe U, with total size $n = \bigcup_{S \in F} |S|$ so that given two query sets $S, S' \in F$ one can determine if $S \cap S' = \emptyset$.*

The following conjecture addresses the hardness of this problem.

Conjecture 1 (SETDISJOINTNESS Conjecture [19]). Any data structure for the SETDISJOINTNESS problem with constant query time must use $\tilde{\Omega}(n^{2-\varepsilon})$ space, while any data structure for this same problem that uses $O(n)$ space must have $\tilde{\Omega}(n^{1/2-\varepsilon})$ query time, where ε is an arbitrary small positive constant. This conjecture is true unless the 3SUM conjecture is false. Where $\tilde{\Omega}(f(n))$ means $\Omega(\frac{f(n)}{polylog(n)})$.

Recently, a stronger conjecture was proposed. A matching upper bound exists for Conjecture 2 by generalizing the ideas from [9, 18].

Conjecture 2 (Strong SETDISJOINTNESS *Conjecture* [12]). Any data structure for the SETDISJOINTNESS problem that answers the query in x time must use $S = \tilde{\Omega}(\frac{n^2}{x^2})$ space for any $x \in (0, n]$.

We now show our reduction from SETDISJOINTNESS to SPARSEIIP to prove the following conditional lower bound:

Theorem 5. *Unless the* SETDISJOINTNESS *Conjecture is false, any data structure for the* SparseMaskedPrefixInnerProduct *problem with constant query time must use* $\tilde{\Omega}(n^{2-\varepsilon})$ *space, while any data structure for this same problem that uses* $O(n)$ *space must have* $\tilde{\Omega}(n^{1/2-\varepsilon})$ *query time, where* ε *is an arbitrary small positive constant. Furthermore, unless the* Strong SETDISJOINTNESS *Conjecture is false, any data structure for the* SparseMaskedPrefixInnerProduct *problem with query time* x *must use* $\tilde{\Omega}(\frac{n^2}{x^2})$ *space for any* $x \in (0, n]$.

Proof. Let $U = \{1, \ldots, u\} \subseteq \mathbb{N}$, where $u = |U|$. Let $F = \{S_1, \ldots, S_m\}$, such that $n = \bigcup_{S_i \in F} |S_i|$. Each $S_i \in F$ is a set represented by a sparse bit vector B_i of size u, where $B_i[j] = 1$ if and only if $j \in S_i$. Let $A = B = B_1 \cdot B_2 \cdots B_m$, i.e. the concatenation of all B_i one after another. It is easy to see that A and B have n ones. We treat them as sparse bit vectors. A and B are the inputs to the SparseMaskedPrefixInnerProduct.

The query $S_i \cap S_j$ for any $i, j \in [m]$ with $i < j$ is done by performing two prefix sum queries. The first has offset $(j - i) \cdot u + 1$ and length $(i - 1) \cdot u$, while the second has offset $(j - i) \cdot u + 1$ and length $(i) \cdot u$. We subtract the first query result from the second query result and check if the result is zero or not. The answer is zero if and only if $S_i \cap S_j = \emptyset$.

A and B are represented using the predecessor data structure. Thus, the reduction takes time linear in the number of ones. $\square$

We now design a quadratic space data structure with polylogarithmic query time. Thus it matches the conditional lower bounds proved under the Strong SETDISJOINTNESS Conjecture within a polylogarithmic factor in query time.

Theorem 6. *Let A and B be two sparse U-bit vectors, and let n represent the sum of the number of 1s in A and the number of 1s in B. There is a data structure that uses $O(n^2)$ words of space that can answer a* SparseMaskedPrefixInnerProduct *query in $O(\min\{\log \log U, \sqrt{\log n}\})$ time for any $p \in [1, n]$.*

Proof. For each position k of B, create the vector A_k as we defined in Sect. 3.4, with $A_k[j] = A[j] \cdot B[j + k]$. Since there are at most n 1s in A and at most n 1s in B, the positions of the 1 bits in A can only form at most n^2 pairs with the positions of the 1 bits in B. Therefore, all these bit vectors, $A_1, A_2, \cdots, A_U$ have $O(n^2)$ 1 bits in total, where U is the length of A and B. If a bit vector A_k does not have any 1s, we do not store it at all. Otherwise, we represent A_k using Lemma 1 to answer predecessor queries, by viewing the position of each 1 bit as an element of a subset of $\{1, 2, \cdots, U\}$. We also augment this data structure by storing the rank of each element present in the subset, so that, given an index i, we can compute the number of 1s in $A_k[1..i]$ in $O(\min\{\log \log U, \sqrt{\log n}\})$ time. Since there are at most n^2 1s in all A_k's, these data structures use $O(n^2)$ space in total. We further build a perfect hash table T of $O(n^2)$ space to record which of these bit vectors have at least a 1, and for each such bit vector, a pointer to its predecessor data structure.

With these data structures, we can answer a query as follows: Suppose we need to compute the inner product of the first j bits of A against the j-length submask of B starting at position k. Then we check whether A_k has at least a 1 bit using T. If it does not, we return 0. Otherwise, we find the predecessor of j in A_k and return its rank as the answer, which requires $O(\min\{\log \log U, \sqrt{\log n}\})$ time. Thus, we have an $O(n^2)$-space data structure with $O(\min\{\log \log U, \sqrt{\log n}\})$ query time.   □

4.2   Dynamic Sparse Internal Inner Product

In this section, we assume that the sparse bit vectors support updates. That is, we support the operation update(V, i, x) which sets the vector V (which is either A or B) at position $1 \le i \le |V|$ to value $x \in \{0, 1\}$. We prove conditional lower bounds and show tight upper bounds up to polylogarithmic factors.

Definition 4 (3SUM [22]). *Let A, B, and C three sets of numbers in $[-n^3, n^3]$, where $|A| + |B| + |C| = n$. The goal is to determine whether there is a triple $a \in A, b \in B, c \in C$ such that $a + b = c$.*

The 3SUM conjecture claims that it is not possible to solve the 3SUM problem in $O(n^{2-\varepsilon})$ time, for any $\varepsilon > 0$. It is believed that even when relaxing the range to be $[-n^2, n^2]$ the problem has the same lower bound. It was shown that even if one can preprocess A or B (but not both [13,20]) the lower bound holds [21]. It follows that the lower bound that we will prove holds for the case in which updates are allowed in only one of the bit vectors.

Lemma 3. *Unless the 3SUM conjecture is false, the* SPARSEIIP *problem in the dynamic setting must have at least query or update in $\Omega(n^{1-\varepsilon})$ time, for $\varepsilon > 0$.*

Proof. We initialize two empty bit vectors A' and C' of length $N = 2n^3 + 1$ corresponding to the range $[-n^3, n^3]$. In order to handle negative number, we begin by setting bits in A' to 1s at positions $a + n^3 + 1$, for any $a \in A$. Similarly, we set $C'[c + n^3 + 1]$ to 1 for any $c \in C$. For each $b \in B$, we perform an internal query in the following way. We ask for b as the offset and $N - b$ as the length of the inner product. If the inner product is not zero then we have a 3SUM triple. Finding such a triple $a + b = c$ is done by a binary search on the staring and ending positions of the interval until one triple is left. Note that b is known, thus, we only need to find the corresponding a and c. Overall the reduction uses $|A| + |C| = O(n)$ updates with $|B| = O(n \log n)$ queries. From the 3SUM conjecture we have that either query or update uses $O(n^{1-\varepsilon})$ time, for any $\varepsilon > 0$.   □

Lemma 4. *The lower bounds of the dynamic* SPARSEIIP *are tight up to poly-logarithmic factors.*

The proof will appear in the journal version.

5 The Connections Between the Problems and the Internal Measurements

It follows directly from the definition, solving the internal prefix sums also solves the internal inner product problem. Thus, all the lower bound on the SPARSEIIP apply on the sparse internal prefix sum. Moreover, all upper bound algorithms for the internal prefix sums problem apply on the internal inner product problem. In this section, we emphasize the connection of these two problems to the internal measurements. The considered measurements are Hamming distance and Exact Matching with wildcards.

Definition 5 (INTERNALHAMMINGDISTANCE and INTERNALEMWW).
Let S, T be two strings of lengths n and m, respectively. The problem of INTER-NALHAMMINGDISTANCE *is to preprocess S and T to answer Hamming distance queries between any equal-length substrings of S and T, where Hamming distance counts the number of mismatches between the two substrings.*

Similarly, the problem of INTERNALEMWW *is to preprocess S and T to answer exact matching with wildcards queries between any equal-length substrings of S and T, where the query counts the number of mismatches between the two substrings. However, mismatches with wildcards are not counted.*

Lemma 5. *Assume a constant-size alphabet Σ. Then, there is a linear-time reductions from the* INTERNALHAMMINGDISTANCE *to the internal inner product problem, and vice versa. Moreover, there is a linear-time reductions from the* INTERNALEMWW *problem to the internal inner product problem, and vice versa.*

For the full proof, see Appendix B.

A   Details Omitted from Sect. 3

Proof of Theorem 3. Given a bit vector B of length m and an array A of length n, there is a data structure that uses $O(\frac{mn}{f(n)} + m + n)$ words of space that can answer masked prefix sum queries in $O(f(n) + g(n))$ time and support updates in $O(\frac{mn \log f(n)}{g(n) f(n)} + g(n))$ time, for any functions $f(n)$ and $g(n)$ with $0 < f(n) < n$ and $0 < g(n) < m + n$.

Alternatively, for any $c > 0$, there is a data structure that uses $O(\frac{mn}{f(n)} + \frac{n^{1+c}}{c \log n} + m + n)$ words of space and can answer masked prefix sum queries in $O(\frac{f(n)}{c \log n} + g(n))$ time and support updates in $O(\frac{mn \log f(n)}{g(n) f(n)} + \frac{n^{1+c}}{g(n)} + g(n))$ time. If $m = O(n)$, setting $f(n) = n^{2/3} \log n$, $g(n) = n^{2/3}$ and $c = 1/3$ yields an $O(n^{4/3}/\log n)$-word data structure with $O(n^{2/3})$ query and update times.

Proof. We first present a data structure with amortized bounds on update operations. The main idea is to rebuild the data structures from Theorem 1 every $g(n)$ updates. Since Theorem 1 presents multiple trade-offs, in the rest of the proof, we use $s(m, n)$, $p(m, n)$ and $q(n)$ to represent the space cost, preprocessing time and query time of the data structures in that theorem. Before a rebuilding is triggered, we maintain two copies of the array and the bit mask: A and B store the current content of this array and the bit mask, respectively, while A' and B' store their content when the previous rebuilding happened. Thus, the data structure, D, constructed in the previous rebuilding, can be used to answer masked prefix sum queries over A' and B'. For the updates arrived after the previous rebuilding, we maintain two lists: a list L_A that stores a sorted list of the indexes of the entries of A that have been updated since the previous rebuilding, and a list L_B that stores a sorted list of the indexes of the entries of B that have been updated since the previous rebuilding. Since the length of either list is at most $g(n) < m + n$, all the data structures occupy $O(s(m, n) + m + n)$ words.

We then answer a masked prefix sum query as follows. Let k and i be the parameters of the query, i.e., we aim at computing $\sum_{j=1}^{k} A[j] \cdot B[i + j - 1]$. We first perform such a query using D in $q(n)$ time and get what the answer would be if there had been no updates since the last rebuilding. Since both L_A and L_B are sorted, we can walk through them to compute the indexes of the elements of A that have either been updated since the last rebuilding, or it is mapped by the query to a bit in B that has been updated since the last rebuilding. This uses $O(g(n))$ time. Then, for each such index d, we consult A, A', B and B' to compute how much the update, to either $A[d]$ or $B[d + i - 1]$, affects the answer to the query compared to the answer given by D. This again requires $O(g(n))$ time over all these indexes. This entire process then answers a query in $O(q(n) + g(n))$ time.

For each update, it requires $O(1)$ time to keep A and B up-to-date. It also requires an update to the sorted list L_A or L_B, which can be done in $O(g(n))$ time. Finally, since the rebuilding requires $O(p(m, n))$ time and it is done every $g(n)$ updates, the amortized cost of each update is then $O(p(m, n)/g(n) + g(n))$.

The bounds in this theorem thus follows from the specific bounds on $s(m, n)$, $p(m, n)$ and $q(n)$ in Theorem 1.

Finally, to deamortize using the global rebuilding approach, instead of rebuilding this data structure entirely during the update operation that triggers the rebuilding, we rebuild it over the next $g(n)$ updates. This requires us to create two additional lists L'_A and L'_B: Each time a rebuilding starts, we rename L_A and L_B to L'_A and L'_B, and create new empty lists L_A and L_B to maintain indexes of the updates that arrive after the rebuilding starts. To answer a query, we cannot use the data structure that is currently being rebuilt since it is not complete, but we use the previous version of it and consult L_A, L_B, L'_A and L'_B to compute the answer using ideas similar to those described in previous paragraphs. □

B   Details Omitted from Sect. 5

Proof of Lemma 5. Assume a constant-size alphabet Σ. Then, there is a linear-time reductions from the INTERNALHAMMINGDISTANCE to the internal inner product problem, and vice versa. Moreover, there is a linear-time reductions from the INTERNALEMWW problem to the internal inner product problem, and vice versa.

Proof. **The reduction from the INTERNALHAMMINGDISTANCE to the internal inner product.** For each letter $\sigma \in \Sigma$, we change S and T to be bit vectors: σ in T become 1 and $\Sigma \setminus \{\sigma\}$ become 0, while in S, σ become 0 and $\Sigma \setminus \{\sigma\}$ become 1. That is, the Hamming distance query sums a constant number of internal inner products in order to answer the query.

The reduction from the internal inner product problem to the INTERNALHAMMINGDISTANCE. Assume we have two bit vectors A and B. Every 1 in A is transferred to 001, and 0 to 010, while in B, each 1 is transferred to 001, and 0 to 100. Let S and T be the transformed strings from A and B, respectively. It is easy to see that only 1 against 1 in A against B causes 0 mismatches between the corresponding substrings of S and T and any of the other 3 combinations results in 2 mismatches. Where corresponding substrings means that the starting and ending positions of the substrings are chosen to fit the original query, i.e. by multiplying the query indices by 3. Note that this reduction transfers the internal inner product to the INTERNALEMWW, as well.

The reduction from the INTERNALEMWW problem to the internal inner product. In a similar way, the inner product solves the exact matching with wildcards problem. We repeat the same process as described previously for Hamming distance but this time, wildcards are always transferred to 0 in both S and T. It is easy to see that when the sum over all the inner products is 0, there is an exact match with wildcards.

$\square$

References

1. Andersson, A.: Faster deterministic sorting and searching in linear space. In 37th Annual Symposium on Foundations of Computer Science, FOCS 1996, Burlington, Vermont, USA, 14–16 October 1996, pp. 135–141. IEEE Computer Society (1996)
2. Bansal, N., Williams, R.: Regularity lemmas and combinatorial algorithms. Theory Comput. **8**(1), 69–94 (2012)
3. Beame, P., Fich, F.E.: Optimal bounds for the predecessor problem and related problems. J. Comput. Syst. Sci. **65**(1), 38–72 (2002)
4. Bille, P., et al.: Dynamic relative compression, dynamic partial sums, and substring concatenation. Algorithmica **80**(11), 3207–3224 (2017). https://doi.org/10.1007/s00453-017-0380-7
5. Blelloch Guy, E.: Prefix sums and their applications. In: Synthesis of Parallel Algorithms, vol. 1, pp. 35–60. M. Kaufmann (1993)
6. Chan, T.M.: Speeding up the four Russians algorithm by about one more logarithmic factor. In: SODA, pp. 212–217 (2015)

7. Clifford, R., Grønlund, A., Larsen, K.G., Starikovskaya, T.: Upper and lower bounds for dynamic data structures on strings. In: Niedermeier, R., Vallée, B. (eds.) 35th Symposium on Theoretical Aspects of Computer Science, STACS 2018, 28 February–3 March 2018, Caen, France, vol. 96, pp. 22:1–22:14. LIPIcs, Schloss Dagstuhl - Leibniz-Zentrum für Informatik (2018)
8. Clifford, R., Iliopoulos, C.S.: Approximate string matching for music analysis. Soft. Comput. 8(9), 597–603 (2004). https://doi.org/10.1007/s00500-004-0384-5
9. Cohen, H., Porat, E.: Fast set intersection and two-patterns matching. Theor. Comput. Sci. 411(40–42), 3795–3800 (2010)
10. Dhulipala, L., Blelloch, G.E., Shun, J.: Theoretically efficient parallel graph algorithms can be fast and scalable. ACM Trans. Parallel Comput. 8(1), 1–70 (2021)
11. Fredman, M.L., Willard, D.E.: Surpassing the information theoretic bound with fusion trees. J. Comput. Syst. Sci. 47(3), 424–436 (1993)
12. Goldstein, I., Lewenstein, M., Porat, E.: On the hardness of set disjointness and set intersection with bounded universe. In: Lu, P., Zhang, G. (eds.) 30th International Symposium on Algorithms and Computation (ISAAC 2019), 8–11 December 2019, Shanghai University of Finance and Economics, Shanghai, China, vol. 149, pp. 7:1–7:22. LIPIcs, Schloss Dagstuhl - Leibniz-Zentrum für Informatik (2019)
13. Golovnev, A., Guo, S., Horel, T., Park, S., Vaikuntanathan, V.: Data structures meet cryptography: 3SUM with preprocessing. In: Makarychev, K., Makarychev, Y., Tulsiani, M., Kamath, G., Chuzhoy, J. (eds.) Proceedings of the 52nd Annual ACM SIGACT Symposium on Theory of Computing, STOC 2020, Chicago, IL, USA, 22–26 June 2020, pp. 294–307. ACM (2020)
14. Kalai, A.: Efficient pattern-matching with don't cares. In: Eppstein, D. (ed.) Proceedings of the Thirteenth Annual ACM-SIAM Symposium on Discrete Algorithms, 6–8 January 2002, San Francisco, CA, USA, pp. 655–656. ACM/SIAM (2002)
15. Keller, O., Kopelowitz, T., Feibish, S.L., Lewenstein, M.: Generalized substring compression. Theor. Comput. Sci. 525, 42–54 (2014)
16. Kociumaka, T.: Efficient data structures for internal queries in texts. PhD Thesis. University of Warsaw (2019)
17. Kociumaka, T., Radoszewski, J., Rytter, W., Walen, T.: Internal pattern matching queries in a text and applications. In: Indyk, P. (ed.) Proceedings of the Twenty-Sixth Annual ACM-SIAM Symposium on Discrete Algorithms, SODA 2015, San Diego, CA, USA, 4–6 January 2015, pp. 532–551. SIAM (2015)
18. Kopelowitz, T., Pettie, S., Porat, E.: Dynamic set intersection. In: Dehne, F., Sack, J.-R., Stege, U. (eds.) WADS 2015. LNCS, vol. 9214, pp. 470–481. Springer, Cham (2015). https://doi.org/10.1007/978-3-319-21840-3_39
19. Kopelowitz, T., Pettie, S., Porat, E.: Higher lower bounds from the 3SUM conjecture. In: Krauthgamer, R. (ed.) Proceedings of the Twenty-Seventh Annual ACM-SIAM Symposium on Discrete Algorithms, SODA 2016, Arlington, VA, USA, 10–12 January 2016, pp. 1272–1287. SIAM (2016)
20. Kopelowitz, T., Porat, E.: The strong 3SUM-INDEXING conjecture is false. arXiv preprint arXiv:1907.11206 (2019)
21. Green Larsen, K.: Personal communication
22. Patrascu, M.: Towards polynomial lower bounds for dynamic problems. In: Schulman, L.J. (ed.) Proceedings of the 42nd ACM Symposium on Theory of Computing, STOC 2010, Cambridge, Massachusetts, USA, 5–8 June 2010, pp. 603–610. ACM (2010)

23. Patrascu, M., Demaine, E.D.: Tight bounds for the partial-sums problem. In: Ian Munro, J. (ed.) Proceedings of the Fifteenth Annual ACM-SIAM Symposium on Discrete Algorithms, SODA 2004, New Orleans, Louisiana, USA, 11–14 January 2004, pp. 20–29. SIAM (2004)
24. Pibiri, G.E., Venturini, R.: Practical trade-offs for the prefix-sum problem. Softw. Pract. Exp. **51**(5), 921–949 (2021)
25. Willard, D.E.: Log-logarithmic worst-case range queries are possible in space $\theta(N)$. Inf. Process. Lett. **17**(2), 81–84 (1983)
26. Williams, V.V.: Multiplying matrices faster than Coppersmith-Winograd. In: STOC, pp. 887–898 (2012)
27. Huacheng, Yu.: An improved combinatorial algorithm for Boolean matrix multiplication. Inf. Comput. **261**, 240–247 (2018)

Compressed String Dictionaries via Data-Aware Subtrie Compaction

Antonio Boffa$^{(\boxtimes)}$ [iD], Paolo Ferragina [iD], Francesco Tosoni [iD],
and Giorgio Vinciguerra [iD]

Department of Computer Science, University of Pisa, Pisa, Italy
{antonio.boffa,francesco.tosoni}@phd.unipi.it,
paolo.ferragina@unipi.it, giorgio.vinciguerra@di.unipi.it

Abstract. String dictionaries are a core component of a plethora of applications, so it is not surprising that they have been widely and deeply investigated in the literature since the introduction of tries in the '60s.

We introduce a new approach to trie compression, called COmpressed COllapsed Trie (CoCo-trie), that hinges upon a data-aware optimisation scheme that selects the best subtries to collapse based on a pool of succinct encoding schemes in order to minimise the overall space occupancy. CoCo-trie supports not only the classic lookup query but also the more sophisticated rank operation, formulated over a sorted set of strings.

We corroborate our theoretical achievements with a large set of experiments over datasets originating from a variety of sources, e.g., URLs, DNA sequences, and databases. We show that our CoCo-trie provides improved space-time trade-offs on all those datasets when compared against well-established and highly-engineered trie-based string dictionaries.

Keywords: String dictionaries · Tries · Compressed data structures

1 Introduction

Let S be a *sorted set* of n variable-length strings $s_1, s_2, \ldots, s_n$ drawn from an alphabet $\Sigma = \{1, 2 \ldots, \sigma\}$. The String Dictionary problem consists of storing S in a *compressed format* while supporting the rank operation that returns the number of strings in S lexicographically smaller than or equal to a pattern $P[1, p]$.

Some other classic operations such as lookup(P) (returning a unique stringID for P if $P \in S$, and -1 otherwise), access(i) (returning the string in S having stringID i), predecessor(P) (returning the lexicographically largest string in S smaller than P), prefix_range(P) (returning all strings in S that are prefixed by P), longest_prefix_match(P) (returning the longest prefix of P which is shared with one of the strings in S) can be implemented through the rank operation, possibly using compact auxiliary data structures [19].

String dictionaries constitute a core component of a plethora of applications such as query auto-completion engines [23,26,29], RDF and key-value stores [36, 50], computational biology tools [5,33], and n-gram language models [27,42], just

D. Arroyuelo and B. Poblete (Eds.): SPIRE 2022, LNCS 13617, pp. 233–249, 2022.
https://doi.org/10.1007/978-3-031-20643-6_17

to mention a few. They are typically approached via the *trie* data structure, which dates back to the '60s [31, §6.3]. Since then, researchers have put a lot of effort to improve the time and space efficiency of the naïve pointer-based implementation. Some solutions compact paths [19, 24, 28, 37] or subtrees [4, 7, 10, 25, 40, 45, 46], succinctly encode node fan-outs [9, 15, 32, 34, 43, 48], apply sophisticated string transformations [20, 21] or proper disk-based layouts [8, 18, 22]. Many recent results aim at reducing further the space occupancy of tries without impairing their efficient query time via sophisticated compression techniques (see e.g. [4, 6, 10, 11, 14, 28, 35, 43, 45, 46, 50]). As a result, this plethora of proposals offers different space-time trade-offs over various datasets, but without a clear winner. Choosing the appropriate storage solution is indeed quite a daunting task, requiring specific expertise and accurate analysis of the input datasets.

In this paper, we tackle this long-standing problem by introducing a fully-new approach that exploits a principled and data-aware optimisation strategy to collapse and compress subtries. More precisely, we make the following contributions:

- We revisit the subtrie compaction technique by introducing a novel representation that encodes a collapsed subtrie via standard integer compressors. Then, by means of a concrete motivating example, we observe that the effectiveness of this compressed representation depends upon the "shape" of the collapsed subtrie and its possibly long "edge labels" (Sect. 2).
- In light of this, we propose a new data structure, called CoCo-trie, which stands for COmpressed and COllapsed trie.[1] It orchestrates three main tools: the above novel representation for collapsed subtries, a pool of succinct encoding schemes to compress the edge labels, and an optimisation procedure that selects the *best* subtries to collapse into macro-nodes to minimise the overall occupied space while guaranteeing efficient queries due to the shorter trie traversal and the efficiently-searchable encoding schemes (Sect. 3).
- We corroborate our theoretical results with an experimental evaluation on several datasets with different characteristics originating from a variety of sources (e.g. URLs, XML, DNA sequences, and databases) and against five well-established and highly-engineered competitors (namely, ART [32], CART [49], ctrie++ [46], FST [50], and PDT [24]). To the best of our knowledge, this is the very first work experimenting with all these implementations together over a wide variety of datasets. Our results show that CoCo-trie is a robust and flexible data structure since: in two cases, it significantly improves the space-time performance of all competitors; in two other cases, it is on the Pareto frontier of the best approaches (thus offering new competitive space-time trade-offs); and, in the last case, it is very close to the Pareto frontier (Sect. 4).

2   A Motivating Example

"Subtrie compaction" is a common technique in the design of compressed string dictionaries. However, it has been mainly investigated in the restricted context

---

[1] The source code is publicly available at https://github.com/aboffa/CoCo-trie.

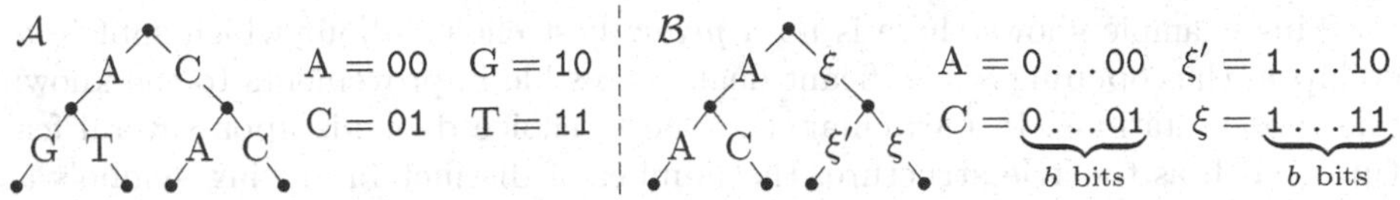

Fig. 1. Two tries $\mathcal{A}$ and $\mathcal{B}$ built on two sets of four strings each: $\{AG, AT, CA, CC\}$ on the left, and $\{AA, AC, \xi\xi', \xi\xi\}$ on the right. $\mathcal{A}$ uses just four alphabet symbols, and $\mathcal{B}$ uses a much larger alphabet in which ξ' and ξ are the last two symbols.

of either bounding the subtrie height, to fit the branching substring into one machine word [10,45,46]; or when bounding the macro-node fan-out so that more space-time efficient data structures can be used for it [7,11,25].

In what follows, we firstly introduce a novel macro-node representation, and then we provide a concrete example of the impact this technique can have on the space-time efficiency of the resulting trie representation. Our technique consists of properly choosing (i) the heights of the subtries to collapse into macro-nodes, and (ii) the coding mechanisms to represent the corresponding branching substrings (associated with the collapsed edge labels).

Consider the tries $\mathcal{A}$ and $\mathcal{B}$ of Fig. 1 built respectively on the string sets $S_1 = \{AG, AT, CA, CC\}$ and $S_2 = \{AA, AC, \xi\xi', \xi\xi\}$, where ξ denotes the last symbol in a (potentially large) alphabet Σ, and ξ' denotes the symbol preceding ξ in Σ. In $\mathcal{A}$, the alphabet $\{A, C, G, T\}$ consists of just 4 symbols, so we need 2 bits to represent them. In $\mathcal{B}$, the alphabet is assumed to be $\Sigma = \{A, C, \dots, \xi', \xi\}$ and its symbols can be represented with $b = \lceil \log_2 |\Sigma| \rceil$ bits.

Let us now consider two scenarios for both of the tries above: one in which the trie $\mathcal{T} \in \{\mathcal{A}, \mathcal{B}\}$ succinctly encodes the individual branching symbols; the other one in which the two levels of $\mathcal{T}$ are *collapsed* at the root node, thereby creating a macro-root $\mathcal{T}^c$ with branching macro-symbols of length 2 symbols. For evaluating the space cost of encoding $\mathcal{T}$ and $\mathcal{T}^c$ we consider the following succinct scheme: for every node in level order, we store the first branching symbol explicitly and then encode the *gap* between successive symbols using some coding tool, say γ-code (see Appendix A for the definition of γ-code and the full calculations). If we refer, say, to the root of $\mathcal{A}$, its two branching symbols, namely A and C, are encoded in 3 bits as $enc(A) = 00$, followed by $\gamma(C - A) = \gamma(01 - 00) = 1$.

The encoding of $\mathcal{A}$ takes 9 bits, while if we collapse the two levels of $\mathcal{A}$ in the root of $\mathcal{A}^c$, this root gets four children whose edge labels are $\{AG, AT, CA, CC\}$, and their succinct representation takes 7 bits. Hence, the representation of $\mathcal{A}$ takes more space than the one of $\mathcal{A}^c$. Surprisingly, one comes to the opposite conclusion with $\mathcal{B}$, despite having the same topology of $\mathcal{A}$. Here, the larger alphabet together with the different distribution of the edge labels changes the optimal choice. Indeed, the encoding of $\mathcal{B}$ takes at most $5b+1$ bits, while the one of $\mathcal{B}^c$ takes $6b - 1$ bits. Hence, it is better not to collapse $\mathcal{B}$ because its succinct encoding takes b bits less than the one of $\mathcal{B}^c$, and b can make this gap arbitrarily large.

This example shows there is no *a priori* best choice about which subtrie to collapse, thus opening a significant deal of possible improvements to the known trie representations. In particular, the "best" choice depends upon several features, such as the trie structure, the number of distinct branching symbols at each node and their distribution among the trie edges. Consequently, designing a principled approach to finding that "best" choice for each trie node is quite a complex task, that we rigorously investigate throughout the rest of the paper.

3   CoCo-trie: Compressed Collapsed Trie

The simplest and most used approach to collapsing tries is to obtain the trie T_ℓ by collapsing ℓ levels of the subtries rooted at the nodes whose distances from the root of T are multiple of ℓ. In this way, one can seek for a pattern $P[1, p]$ over T_ℓ by traversing at most p/ℓ (macro-)nodes and edges, that is, p/ℓ branches over (macro-)characters (e.g., in [10, 45, 46], ℓ is the number of characters that fit into a RAM word). Obviously, increasing ℓ reduces the number of branching steps, but it may increase (i) the computational cost of each individual step, given that the number and the length of the branching characters increase; and, (ii) the space occupancy of the overall trie, given that shared paths within the collapsed subtries are turned into distinct substrings by macro-characters (see e.g. the paths "e$" and "es" descending from v in Fig. 2, which share "e").

Our proposal addresses three main questions:

Q1: Can we tackle in a principled algorithmic way the issues (i) and (ii) above as ℓ increases?

Q2: How does the choice about the number ℓ of levels to collapse depend on the dictionary of strings?

Q3: Should the choice of ℓ be *global*, and thus unique to the entire trie, or should it be *local*, and thus vary among trie nodes?

These questions admit surprising answers in theory, which have equally-surprising impacts in practice. In particular, we will:

- answer Q1 affirmatively, by resorting to a pool of succinct encoding schemes to compress the possibly long edge labels (i.e., branching macro-characters);
- show for Q2 that the choice for ℓ has to account for the topology and edge labelling of the trie T, and thus the characteristics of its indexed strings;
- show for Q3 that one has to find locally, i.e., node by node, the best value of ℓ, via a suitably-designed optimisation procedure aimed at minimising the overall space occupancy.

Our algorithmic answer consists of six main steps. Firstly, we introduce a novel compressed encoding for the collapsed subtries (Sect. 3.1). Secondly, we provide an optimisation procedure to choose the subtries to collapse (Sect. 3.2). Thirdly, we show how to select the best compression scheme for each collapsed subtrie in a data-aware manner (Sect. 3.3). Fourthly, we present a further compression step that exploits the local alphabet of the edge labels in the collapsed

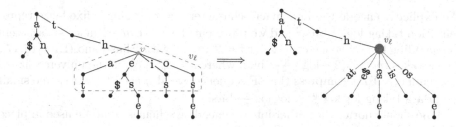

Fig. 2. Collapsing $\ell = 2$ levels of the subtrie rooted at v.

subtrie (Sect. 3.4). Fifthly, we show how to trade space occupancy with query time (Sect. 3.5). Sixthly, we describe how to implement the `rank` operation over the resulting compressed trie structure (Sect. 3.6).

3.1 Compressed Encoding of Collapsed Subtries

Let us be given a trie $\mathcal{T}$ whose edges are drawn from an integer alphabet $\Sigma = \{0, \ldots, \sigma - 1\}$ and sorted increasingly at each node. The special character 0 (indicated with $) is the string terminator. We formalise the notion of collapsed subtries as follows.

Definition 1. *Given an internal node v of a trie $\mathcal{T}$ and an integer $\ell \geq 1$, the collapsing of ℓ levels of the subtrie of $\mathcal{T}$ rooted at v consists in replacing this subtrie with a macro-node v_ℓ such that (i) the substrings branching out of v_ℓ are the ones corresponding to the paths of length ℓ descending from v in $\mathcal{T}$, and (ii) the children of v_ℓ are the nodes at distance ℓ from v. If branching substrings are shorter than ℓ, we pad them with the character $.*

This is depicted in Fig. 2, where five paths of length $\ell = 2$ are collapsed to form the five branching edges $\{at, e\$, es, is, os\}$ of v_ℓ.

To encode a string s branching out of v_ℓ, we initially right-pad it with $\ell - |s|$ characters $ if $|s| < \ell$; then, we assign it the integer

$$enc_\ell(s) = \sum_{i=1}^{\ell} s[i] \cdot \sigma^{\ell-i}. \tag{1}$$

Intuitively, we interpret $enc_\ell(s)$ as a branching *macro-character* of v_ℓ drawn from the integer alphabet $\Sigma^\ell = \{0, \ldots, \sigma^\ell - 1\}$.

Furthermore, we observe that enc_ℓ is monotonic, i.e., given two strings s' and s'' such that s' is lexicographically smaller than s'', then $enc_\ell(s') < enc_\ell(s'')$.

After computing each branching macro-character, we need to define a compression scheme that guarantees efficient access to edge labels, so to support fast pattern searches over the resulting collapsed trie. Let us assume that a macro-node v_ℓ has m branching macro-characters, indicated with $c_{v_\ell}^i$ for $i = 1, 2, \ldots, m$.

We explicitly encode the first macro-character $x = c_{v_\ell}^1$ using a fixed-size representation taking $\log \sigma^\ell$ bits,[2] and we represent the other $m-1$ macro-characters by encoding the sequence $c_{v_\ell}^i - x$ for $i = 2, \ldots, m$ with Elias-Fano (EF) [16, 17], which takes $(m-1)(2 + \log \frac{u}{m-1})$ bits, where $u = c_{v_\ell}^m - c_{v_\ell}^1$ is the universe size of the sequence. To decompress the EF sequence, we also need to store some small metadata taking $\log \log \frac{u}{m} \leq \log \log \frac{\sigma^\ell}{m}$ bits.

One should notice that other integer encoding schemes could be used in place of EF, and indeed we do so in Sect. 3.3.

Summing up, the space occupancy of the collapsed and compressed macro-node v_ℓ is (excluding EF's metadata)

$$C(v_\ell) = \log \sigma^\ell + (m-1)\left(2 + \log \frac{u}{m-1}\right) + 2 \text{ bits}, \qquad (2)$$

where the first term corresponds to the space for the first macro-character, the second term accounts for the space to store the $(m-1)$ EF-coded integers, and the last 2 bits account for the contribution of the node v_ℓ to the space required by a succinct trie representation (we use LOUDS [39, §8.1]).

We underline that the subtraction of x has a subtle (yet paramount) impact on the space occupancy of our trie representation. It indeed removes any possible redundancy given by the longest common prefix (shortly, lcp) among the branching macro-characters. For instance, if we have $\ell = 2$ and the four branching macro-characters {ha, he, hi, ho}, then our encoding scheme stores $x = enc_\ell(\text{ha})$ explicitly as the integer $\text{h} \cdot \sigma^1 + \text{a} \cdot \sigma^0$, and it encodes the following three branching macro-characters {he, hi, ho} as the difference with x, i.e., it encodes "he" as $enc_\ell(\text{he}) = (\text{h} \cdot \sigma^1 + \text{e} \cdot \sigma^0) - x = (\text{h} \cdot \sigma^1 + \text{e} \cdot \sigma^0) - (\text{h} \cdot \sigma^1 + \text{a} \cdot \sigma^0) = \text{e} - \text{a}$. So our encoding scheme stores the lcp "h" only once in x, thereby getting rid of much redundancy in the edge labels, and saving a big deal of space, especially when ℓ gets longer. As a matter of fact, we are reducing the value of the integers $enc_\ell(c_{v_\ell}^i)$, which are upper-bounded by σ^ℓ, to the values $enc_\ell(c_{v_\ell}^i) - enc_\ell(c_{v_\ell}^1)$, which are upper-bounded by $\sigma^{\ell - lcp}$.

3.2 On the Choice of the Subtries to Collapse

We now get down to the details of our algorithm that, given an input trie $\mathcal{T}$, identifies which subtries of $\mathcal{T}$ to collapse (and for which height ℓ each one), in order to minimise the space occupancy of the resulting representation.

Our algorithm performs a post-order traversal of $\mathcal{T}$, starting from the root. Let $h(v)$ denote the height of the subtrie rooted at v (and reaching its descending leaves in $\mathcal{T}$). For each node v, the algorithm evaluates the cost of encoding the entire subtrie descending from v by taking into account the space cost $C(v_\ell)$ of Eq. (2) referring to the subtrie of v limited to height ℓ, plus the optimal space cost $C^*(d)$ of encoding recursively the entire subtries hanging from the nodes d descending from v at distance ℓ. We vary $\ell = 1, \ldots, h(v)$, thereby determining the

---

[2] We omit ceilings for the sake of simplicity.

minimum space occupancy $C^*(v)$. Formally, if $desc(v, \ell)$ is the set of descendants of v at distance ℓ (recall that $\ell \leq h(v)$), we have

$$C^*(v) = \min_{\ell=1,\ldots,h(v)} \left\{ C(v_\ell) + \sum_{d \in desc(v,\ell)} C^*(d) \right\}. \tag{3}$$

Note that if v is a leaf, we simply set $C^*(v) = C(v_1) = 2$, since a leaf cannot be collapsed and its cost in the LOUDS representation is 2 bits. Clearly, because of the post-order visit, the values $C^*(d)$ are available whenever we compute $C^*(v)$.

When the root of $\mathcal{T}$ is eventually visited, the topology and the encoding of all (macro-)nodes of our CoCo-trie have already been fully determined. Thus, we know which subtries to collapse and for which height ℓ, which may vary from one subtrie to another. By Eq. (3), the resulting data structure is the space-optimal one using the selected encoding scheme. The following result, proved in Appendix B, bounds the space-time efficiency of this approach.

Theorem 1. *The CoCo-trie of a given input trie $\mathcal{T}$ of height h and N nodes can be computed in $\mathcal{O}(Nh)$ time and $\mathcal{O}(N)$ space.*

We finally remark that in the above optimisation process we can upper bound the maximum number ℓ of collapsed levels so that the above time cost becomes $\mathcal{O}(N)$. This is actually the approach we take in our experimental section, where we bound ℓ for each node v by setting $h(v) = w/\log \sigma$ in Eq. (3), where w is the RAM word size in bits (see also Sect. 3.4). This feature may remind similar mechanisms adopted in ctrie [10, 45] and ctrie++ [46], where a subtrie is packed into a machine word. However, our approach is more powerful because the height of the subtrie to collapse is not chosen in advance and equal over the whole trie, but it is adaptively chosen on a single-node basis and in a data-aware manner according to the subtrie topology and the distribution of its edge labels.

3.3 A Pool of Succinct Encoding Schemes

Thus far, we represented the $m - 1$ branching macro-characters $c_{v_\ell}^i$ of a macro-node v_ℓ via the EF-encoding of the increasing integers $c_{v_\ell}^i - x$, for $i = 2, \ldots, m$, where $x = c_{v_\ell}^1$ is the first branching character we stored explicitly. This sequence of $m - 1$ macro-characters is drawn from a universe of size $u = c_{v_\ell}^m - c_{v_\ell}^1 + 1$. Depending on m, u and the values of the branching macro-characters, it may be beneficial in time, in space, or both, to resort to other kinds of encodings.

On the grounds of this observation and inspired by the hybrid integer-encoding literature [12, 30, 41, 44], we now equip the CoCo-trie optimisation algorithm of the previous section with an assortment of encoding mechanisms so that the compressed representation of every single node can be chosen in a *data-aware* manner. This amounts to redefining the bit cost $C(v_\ell)$ of storing the macro-node v_ℓ so as to consider the cost in bits of other compression schemes besides EF. Specifically, when evaluating $C(v_\ell)$ during the traversal, whichever compression scheme gives the minimum bit-representation size for all collapsed subtries descending from v is selected and returned as the result of $C(v_\ell)$ (see Eq. (3)).

For our experimental study of Sect. 4, we follow [41] and, alongside EF, we adopt packed encoding (PA), characteristic bitvectors (BV), and dense encoding (DE). PA uses a fixed amount $\log u$ of bits for each $c_{v_\ell}^i$ for a total of $(m-1)\log u$ bits. BV uses u bits initially set to 0, and then sets to 1 the $m-1$ bits corresponding to each $c_{v_\ell}^i$. DE comes into use whenever $u = m - 1$, i.e. for representing a complete sequence of consecutive macro-characters; in this case, no additional bits are required. These encoding schemes allow to implement the predecessor search easily, as needed by the rank of CoCo-trie (see Sect. 3.6).

3.4   Squeezing the Universe of Branching Labels

We now describe an optimisation to further decrease the space requirements for the macro-characters by means of an alphabet-aware encoding. The idea lies in replacing the encoding function enc_ℓ defined in Eq. (1) with a new one that depends on the alphabet of the branching macro-characters $c_{v_\ell}^i$ local to each macro-node v_ℓ rather than on the global alphabet Σ of the whole trie.

Let $\Sigma_{v_\ell} \subseteq \Sigma$ be the alphabet of symbols occurring in the edge labels of the collapsed macro-node v_ℓ. By changing $\sigma = |\Sigma|$ in Eq. (1) with $\sigma_{v_\ell} = |\Sigma_{v_\ell}|$, we can squeeze the size of the universe of the branching macro-characters of v_ℓ from σ^ℓ to $\sigma_{v_\ell}^\ell$. This, in turn, reduces the magnitude and the distance between consecutive integers associated with the branching macro-characters and thus allows a more effective compression. Also, we reduce the first space term of Eq. (2) to $\log \sigma_{v_\ell}^\ell$.

Clearly, each macro-node v_ℓ adopting this optimisation must store a mapping between Σ and the local alphabet Σ_{v_ℓ}, e.g., via a bitvector $B[0, \sigma - 1]$ where $B[i] = 1$ if symbol i appears in Σ_{v_ℓ}. We observe this optimisation requires modifying $C(v_\ell)$ to account for both the more efficient macro-characters representation due to the squeezed universe and the size of the alphabet mapping (i.e. σ bits).

Overall, the time complexity for building the CoCo-trie becomes $\mathcal{O}(Nh^2)$ since we cannot compute u_ℓ incrementally as described in the proof of Theorem 1 (in Appendix B); the space complexity instead does not change, as we do not store the bitmaps $B_\ell[1, \sigma]$, but we compute them incrementally while visiting the subtries as in the proof of Theorem 1.

3.5   On the Space-Time Trade-Off

Under some scenarios, it might be of interest to slightly readjust the optimisation procedure to take into account query performance too, while possibly giving up the space optimality. To accomplish this space-time trade-off, we rely on the intuition that collapsing more levels generally improves the query time. As a matter of fact, the more levels are collapsed, the faster each trie traversal will be. But, on the other hand, as we collapse more levels, the fan-out of each macro-node increases and so the time to traverse each individual macro-node increases as well. However, we experimentally observed that this is not a major concern, since our compressed encoding of collapsed subtries and our succinct encoding schemes are in practice extremely efficient to be navigated; thus the time reduction given by increasing the number of collapsed levels dominates the increased access time due to the bigger node fan-out.

With this in mind, we modify the algorithm of Sect. 3.2 as follows. At each visited internal node v, we compute $C^*(v)$ as usual and denote by ℓ^* the value of ℓ minimising the right-hand side of Eq. (3). Then, we find the largest value $\ell \in \{\ell^*, \ell^* + 1, \ldots, h(v)\}$ that allows to represent the collapsed node adding just a constant factor $\alpha \geq 0$ overhead over the optimal space $C^*(v)$. We observe this new approach has no impact on the construction complexity. We experiment with it in Sect. 4, where α is expressed as a percentage.

3.6 Trie Operations

The lookup(P) in the CoCo-trie begins from the root macro-node r_ℓ, by computing the integer $x = enc_\ell(P[1, \ell]) - c_{r_\ell}^1$. Then, we seek for x into the increasing sequence $c_{r_\ell}^i$, for $i = 2, \ldots, m$: if the search fails, we return -1; otherwise, we obtain an index j of x, and proceed with the recursion in the j-th child of the macro-node. The compressed and indexed macro-node encoding guarantees the search for x is very efficient. We iteratively consume multiple characters at once from the pattern P as we descend the CoCo-trie downwards via LOUDS. When P is exhausted, we return the unique LOUDS index of the node we reach.

As for rank(P), we switch to the DFUDS encoding for the trie topology as it allows us to compute the rank of a leaf efficiently, takes the same space of LOUDS, and is still efficient in navigating the trie downwards [39, §8.3]. At each macro-node v_ℓ, we seek for the largest index j such that $c_{v_\ell}^j \leq x$, and we keep searching recursively into the j-th child of r_ℓ. Again, the EF and hybrid schemes for the $c_{v_\ell}^j$s result in fast branching operations. If P is exhausted at an internal node, the navigation proceeds downwards until the leftmost descendant of that node. In any case, we eventually reach a leaf node and return its rank.

4 Experiments

We run our experiments on a machine equipped with a 2.30 GHz Intel Xeon Platinum 8260M CPU and 384 GiB of RAM, running Ubuntu 20.04.3. We compile our codebase using g++11.1 and the C++-20 language standard.

Datasets. We aimed at choosing very diverse datasets in terms of sources (such as the Web, bioinformatics, and databases) and features (such as the number n of strings, total number D of characters, alphabet size σ, and average/maximum length of the lcp between consecutive sorted strings) to depict a broader spectrum of the performances of the tested string dictionaries. We preprocess each dataset to keep, for each string s, the shortest prefix of s that distinguishes it from all the other strings in the dataset. It is well known that, for any trie-based structure, the remaining suffixes can be concatenated into a separate array and efficiently retrieved when needed [39, §8.5.3]. Table 1 shows the datasets and their characteristics after preprocessing.

Table 1. Datasets characteristics.

Name	Description	$n/10^6$	$D/10^6$	Avg lcp	Avg length	Max length	σ
url	URLs crawled from the web [13]	40.5	2 713.5	64.1	66.9	1 990	94
xml	Rows of an XML dump of dblp [3]	2.9	107.8	34.4	36.5	248	95
protein	Different sequences of amino acids [3]	2.9	155.6	36.7	53.3	16 191	26
dna	Unique 12-mer from a DNA seq. [3]	13.7	164.5	10.5	11.9	12	15
tpcds-id	Customers ids in TPC-DS-3TB [38]	30.0	446.4	13.4	14.8	15	16

State-of-the-art Competitors. We consider as competitors of our CoCo-trie the following static string dictionaries implementations because they are either the state of the art or offer efficient approaches to compact trie representations:

CART: a compact version of ART [32] obtained by constructing a plain ART and converting it to a static version [1, 49, 50].

PDT: the Centroid Path Decomposed Trie [24]. We experiment with both the vbyte version that encodes the labels of the edges with vbyte [47], and the csp version that adds another layer of compression on top of the edge labels.

FST: the Fast Succinct Trie [50]. We use a slightly-modified code [2] that solves lookup queries rather than range query filtering. We show the full space-time performance of FST by varying its parameter R as 2^i for $i = 0, \ldots, 10$.

Apart from the above static data structures, we also tested ART [32] and ctrie++ [46] as representative of the dynamic approaches. In our figures we do not show ctrie++ since its space usage on our datasets is from 2.8 to 6.2× larger than ART, which in turn uses up to one order of magnitude more space than the other tested solutions, and since it is faster than ART only on url (by 14%).

We do not experiment with Masstree [34] because [11] shows it uses from 1.8 to 3× more space than ART. We also do not experiment with HOT [11] because their implementation only supports strings shorter than 256, while our datasets contain much longer strings. Finally, we do not experiment with the implementation provided in [14,35] as we were unable to run its codebase in a fair environment due to some old software dependencies and incompatibilities with modern compilers.

Query Workloads. Given that our competitors do not implement rank (despite their design does support it), we decided to measure the performance of lookup. We can reasonably expect that rank would perform similarly to lookup, because of the way the former can be derived from the latter in trie-based (rather than hash-based) data structures, as the ones we experimentally test here.

Given a dataset of n strings, we measure the query time by averaging the performance of 3 repetitions of a batch of size n, where half of the strings are taken from the datasets and half are generated randomly. To generate each of these latter strings, we (i) extract a randomly-chosen string belonging to the dataset and truncate it to the average lcp of the entire dataset, and (ii) append a random string whose length matches the average length of the strings in the

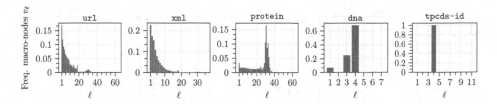

Fig. 3. Normalised frequency of macro-nodes collapsing subtries having ℓ levels.

dataset. This way, the queries we generate mimic a fair query workload that guarantees a balance between existent and not existent queried strings, and for the latter that the traversal does not stop at the very first steps because of a mismatch.

Experimental Results. Figure 3 shows our first experimental result: the number of macro-nodes collapsing a certain amount of levels forms a non-trivial distribution whose shape differs from dataset to dataset. This provides a clear answer to both questions **Q2** and **Q3** in Sect. 3: the number of levels to be collapsed in a subtrie greatly depends on the strings the trie is built on, and it must be chosen locally. Therefore, the data-aware approach to subtrie compaction implemented in our CoCo-trie optimiser is essential to attain the most from these features.

In particular, on `url` and `xml`, the CoCo-trie optimiser selects many times the lowest possible values of ℓ (each horizontal axis ranges from $\ell = 1$ to the largest ℓ over all macro-nodes v_ℓ). For `protein`, instead, the CoCo-trie optimiser selects high values of ℓ (very often $\ell \approx 30$) so that, in the end, the distribution resembles a Gaussian one. On `dna`, CoCo-trie optimiser collapses at most $\ell = 7$ levels at a time and selects $\ell = 4$ for 67% of the times. The results on `tpcds-id` are also of interest for their simplicity: due to the regularity of the dataset, the CoCo-trie optimiser here creates a macro-node for the root that collapses $\ell = 11$ levels, and each of its 4096 children collapses $\ell = 4$ levels.

Figure 4 shows the results about the space and time performance of CoCo-trie and the five competitors.

Firstly, we observe that ART and CART, though fast, are generally very space-demanding (note the vertical axis is logarithmic). On `url` they are also slower due to their large size and the high average lcp among the dictionary strings, which causes longer trie traversals and thus more cache misses.

FST is dominated in space and time performance by our CoCo-trie (and also by other data structures) on all the datasets. This is especially evident on `url`, `xml`, and `protein`. We argue that this is due to the high average lcp of these datasets that require FST to perform longer trie traversals that proceed one character at a time (indeed, FST does not compact unary paths).

PDT shows overall a good space-time performance, with the exception of `dna` and `tpcds-id`, for which the average height of the PDT nodes is 6.0 and 8.7, respectively. The average height of the macro-nodes in the CoCo-trie (see the $\alpha = 0\%$ configuration in Fig. 4) is instead 3.5 and 4.0, thus requiring nearly

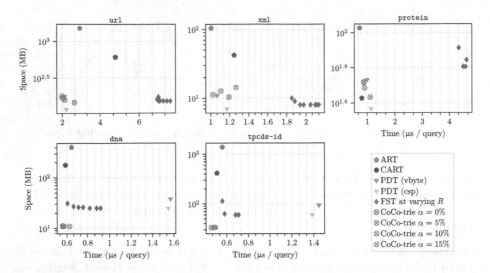

Fig. 4. Space-time performance of the lookup query for various tested approaches. Note that the y-axis is logarithmic and that the ranges of both axes change among the plots.

half accesses to nodes, on average. Indeed, CoCo-trie is 2.3× faster and 3× more succinct than PDT on `dna`, and it is 2.6× faster and 2.4× more succinct than PDT on `tpcds-id`. This means that, on these two datasets, the approach of exploiting the small local alphabet at each macro-node is particularly effective. On the rest of the datasets, CoCo-trie always has some configuration on the Pareto frontier of PDT, thus offering other competitive space-time trade-offs.

In summary, with respect to the highly-engineered competitors we test on the diverse five datasets, the CoCo-trie results space-time efficient, robust and flexible: in fact, it significantly dominates the space-time performance of all competitors on `dna` and `tpcds-id`; it is on the Pareto frontier of the best competitors over `url` and `xml`; and, lastly, it is very close to the Pareto frontier for `protein`.

5 Conclusions and Future Work

We have introduced a new design of compressed string dictionaries that collapses and succinctly encodes properly-chosen subtries via novel data-aware encoding of (possibly long) edge labels and a space optimisation procedure. The experimental results over a variety of the datasets and highly-engineered competitors suggest that our CoCo-trie does advance the state of the art of string dictionaries.

The novel design scheme on which CoCo-trie hinges, paves the way to further new approaches for the *multicriteria* optimisation of trie data structures that take into account possibly other encoding schemes and different multi-objective functions (e.g., over time, space, energy usage, etc.).

Acknowledgements. Supported by the Italian MUR PRIN project "Multicriteria data structures and algorithms" (Prot. 2017WR7SHH), and by the EU H2020 projects "SoBigData++: European Integrated Infrastructure for Social Mining and Big Data Analytics" (grant #871042) and "HumanE AI Network" (grant #952026).

A   Calculations for the Motivating Example of Sect. 2

We first recall that the γ-code of a positive integer x consists of a number of 0s equal to the number of bits minus one of the binary representation of x, followed by that binary representation, e.g. $\gamma(6) = 00\,110$. Thus, $\gamma(x)$ takes $2\lfloor \log_2 x \rfloor + 1$ bits.

The case of trie $\mathcal{A}$. The succinct representation of the edge labels in $\mathcal{A}$ takes $3 + 3 + 3 = 9$ bits. In fact, the encoding of the edge labels $\{A, C\}$ of the root is $enc(A)\,\gamma(C - A) = 00\,\gamma(1) = 001$, then the encoding of the edge labels $\{G, T\}$ of the first node at the second level is $enc(G)\,\gamma(G - T) = 10\,\gamma(1) = 101$, and finally the encoding of the edge labels $\{A, C\}$ (again) of the second node at the second level is $enc(A)\,\gamma(C - A) = 001$.

If, instead, we collapse the two levels of $\mathcal{A}$ in the root of $\mathcal{A}^c$, this root gets four children whose edge labels are $\{AG, AT, CA, CC\}$, and their succinct representation takes 7 bits. In fact, we encode the first branching macro-symbol $enc(AG) = 0010$ as it is, followed by the encoding of the other three branching macro-symbols as: $\gamma(AT - AG) = \gamma(0011 - 0010) = \gamma(1) = 1$, $\gamma(CA - AT) = \gamma(0100 - 0011) = \gamma(1) = 1$, and $\gamma(CC - CA) = \gamma(0101 - 0100) = \gamma(1) = 1$.

Thus in terms of space cost, $\mathcal{A}$ is worse than $\mathcal{A}^c$. This result is even more evident when accounting for the space cost for the topology, simply because $\mathcal{A}$ has more nodes than $\mathcal{A}^c$. We conclude that, under this setting, it is better to collapse the trie and keep $\mathcal{A}^c$.

The case of trie $\mathcal{B}$. Surprisingly, one comes to the opposite conclusion with $\mathcal{B}$, despite having the same topology of $\mathcal{A}$. Here, the larger alphabet together with the different distribution of the edge labels changes the optimal choice.

The succinct representation of the edge labels in $\mathcal{B}$ takes at most $5b + 1$ bits. We can indeed represent the edge labels $\{A, \xi\}$ of the root with $enc(A)\,\gamma(|\Sigma| - 1)$ which takes at most $3b - 1$ bits; the root gets followed by the encoding of the edge labels $\{A, \xi\}$ of the first node at the second level, namely $enc(A)\,\gamma(1)$ which takes $b + 1$ bits, and by the encoding of the edge labels $\{\xi', \xi\}$ of the second node at the second level, which is $enc(\xi')\,\gamma(1)$ which takes $b + 1$ bits.

Conversely, the succinct representation of $\mathcal{B}^c$ may take up to $6b - 1$ bits, since we encode AA with $2b$ bits set to 0, followed by $\gamma(AC - AA) = \gamma(1) = 1$, then by $\gamma(\xi\xi' - AC) = \gamma(01\ldots 101)$ (which takes $4b - 3$ bits, because the γ-encoded number consists of $2b - 1$ bits), and finally by $\gamma(\xi\xi - \xi\xi') = \gamma(1) = 1$.

Hence, differently from the example on $\mathcal{A}$, here it is better not to collapse $\mathcal{B}$ because its succinct encoding takes $b - 2$ bits less than the one of $\mathcal{B}^c$, and b can make this gap arbitrarily large, up to the point that the cost of representing their topology becomes negligible.

B   Proof of Theorem 1

Starting from a node v of height $h(v)$, we can compute $C(v_\ell)$ for any $\ell = 1, 2, \ldots, h(v)$ by obtaining incrementally all the optimisation parameters u_ℓ and m_ℓ from the already (inductively) known $u_{\ell-1}$ and $m_{\ell-1}$.

To compute the universe size u_ℓ for v_ℓ we need to determine the enc_ℓ-code of the leftmost and rightmost length-ℓ strings descending from v, and these can be computed by extending the respective $enc_{\ell-1}$-codes computed at the previous step with one character, in constant time. This costs overall $\mathcal{O}(Nh)$ time because, for each node, we have to visit the leftmost and rightmost branching strings that are of length at most h.

To compute m_ℓ (i.e. the number of children of the collapsed macro-node v_ℓ), we need to visit once the whole subtrie rooted at v. Knowing $m_{\ell-1}$, we add to it the number of leaves at the ℓ-th level. Performing for every node v a complete visit of its whole subtrie costs overall $\mathcal{O}(Nh)$ time: indeed, each of the N nodes has at most h different ancestors and thus belongs to at most h different subtries, thereby getting visited at most h times.

For every node v we maintain just the optimal C^*-cost, thus the required space amounts to $\mathcal{O}(N)$.

References

1. ART and CART implementations. https://github.com/efficient/fast-succinct-trie/tree/master/third-party/art. Accessed June 2022
2. FST implementation. https://github.com/kampersanda/fast_succinct_trie. Accessed June 2022
3. Pizza&Chili corpus. http://pizzachili.dcc.uchile.cl/texts.html. Accessed June 2022
4. Acharya, A., Zhu, H., Shen, K.: Adaptive algorithms for cache-efficient trie search. In: Goodrich, M.T., McGeoch, C.C. (eds.) ALENEX 1999. LNCS, vol. 1619, pp. 300–315. Springer, Heidelberg (1999). https://doi.org/10.1007/3-540-48518-X_18
5. Apostolico, A., Crochemore, M., Farach-Colton, M., Galil, Z., Muthukrishnan, S.: 40 years of suffix trees. Commun. ACM **59**(4), 66–73 (2016). https://doi.org/10.1145/2810036
6. Arz, J., Fischer, J.: Lempel–Ziv-78 compressed string dictionaries. Algorithmica **80**(7), 2012–2047 (2017). https://doi.org/10.1007/s00453-017-0348-7
7. Askitis, N., Sinha, R.: Engineering scalable, cache and space efficient tries for strings. VLDB J. **19**(5), 633–660 (2010). https://doi.org/10.1007/s00778-010-0183-9
8. Bender, M.A., Farach-Colton, M., Kuszmaul, B.C.: Cache-oblivious String B-trees. In: Proceedings of the 25th ACM Symposium on Principles of Database Systems (PODS), pp. 233–242 (2006). https://doi.org/10.1145/1142351.1142385
9. Bentley, J.L., Sedgewick, R.: Fast algorithms for sorting and searching strings. In: Proceedings of the 8th Annual ACM-SIAM Symposium on Discrete Algorithms (SODA), pp. 360–369 (1997). https://doi.org/10.5555/314161.314321
10. Bille, P., Gørtz, I.L., Skjoldjensen, F.R.: Deterministic indexing for packed strings. In: Proceedings of the 28th Annual Symposium on Combinatorial Pattern Matching (CPM), vol. 78, pp. 6:1–6:11 (2017). https://doi.org/10.4230/LIPIcs.CPM.2017.6

11. Binna, R., Zangerle, E., Pichl, M., Specht, G., Leis, V.: HOT: a height optimized trie index for main-memory database systems. In: Proceedings of the ACM International Conference on Management of Data (SIGMOD), pp. 521–534 (2018). https://doi.org/10.1145/3183713.3196896
12. Boffa, A., Ferragina, P., Vinciguerra, G.: A learned approach to design compressed rank/select data structures. ACM Trans. Algorithms (2022). https://doi.org/10.1145/3524060
13. Boldi, P., Vigna, S.: The WebGraph framework I: compression techniques. In: Proceedings of the 13th International World Wide Web Conference (WWW), pp. 595–601 (2004). https://law.di.unimi.it/webdata/it-2004/
14. Brisaboa, N.R., Cerdeira-Pena, A., de Bernardo, G., Navarro, G.: Improved compressed string dictionaries. In: Proceedings of the 28th ACM International Conference on Information and Knowledge Management (CIKM), pp. 29–38 (2019). https://doi.org/10.1145/3357384.3357972
15. Darragh, J.J., Cleary, J.G., Witten, I.H.: Bonsai: a compact representation of trees. Softw. Pract. Exp. **23**(3), 277–291 (1993). https://doi.org/10.1002/spe.4380230305
16. Elias, P.: Efficient storage and retrieval by content and address of static files. J. ACM **21**(2), 246–260 (1974). https://doi.org/10.1145/321812.321820
17. Fano, R.M.: On the number of bits required to implement an associative memory. Memo 61. Massachusetts Institute of Technology, Project MAC (1971)
18. Ferragina, P., Grossi, R.: The String B-tree: a new data structure for string search in external memory and its applications. J. ACM **46**(2), 236–280 (1999). https://doi.org/10.1145/301970.301973
19. Ferragina, P., Grossi, R., Gupta, A., Shah, R., Vitter, J.S.: On searching compressed string collections cache-obliviously. In: Proceedings of the 27th ACM Symposium on Principles of Database Systems (PODS), pp. 181–190 (2008). https://doi.org/10.1145/1376916.1376943
20. Ferragina, P., Luccio, F., Manzini, G., Muthukrishnan, S.: Compressing and indexing labeled trees, with applications. J. ACM **57**(1), 1–33 (2009). https://doi.org/10.1145/1613676.1613680
21. Ferragina, P., Venturini, R.: The compressed Permuterm index. ACM Trans. Algorithms **7**(1), 1–21 (2010). https://doi.org/10.1145/1868237.1868248
22. Ferragina, P., Venturini, R.: Compressed cache-oblivious String B-tree. ACM Trans. Algorithms **12**(4), 1–17 (2016). https://doi.org/10.1145/2903141
23. Gog, S., Pibiri, G.E., Venturini, R.: Efficient and effective query auto-completion. In: Proceedings of the 43rd ACM International Conference on Research and Development in Information Retrieval (SIGIR), pp. 2271–2280 (2020). https://doi.org/10.1145/3397271.3401432
24. Grossi, R., Ottaviano, G.: Fast compressed tries through path decompositions. ACM J. Exp. Algorithmics **19**(1), 1 (2014). https://doi.org/10.1145/2656332
25. Heinz, S., Zobel, J., Williams, H.E.: Burst tries: a fast, efficient data structure for string keys. ACM Trans. Inf. Syst. **20**(2), 192–223 (2002). https://doi.org/10.1145/506309.506312
26. Hsu, B.J.P., Ottaviano, G.: Space-efficient data structures for Top-k completion. In: Proceedings of the 22nd International Conference on World Wide Web (WWW), pp. 583–594 (2013). https://doi.org/10.1145/2488388.2488440
27. Huston, S.J., Moffat, A., Croft, W.B.: Efficient indexing of repeated n-grams. In: Proceedings of the 4th International Conference on Web Search and Web Data Mining (WSDM), pp. 127–136 (2011). https://doi.org/10.1145/1935826.1935857

28. Kanda, S., Köppl, D., Tabei, Y., Morita, K., Fuketa, M.: Dynamic path-decomposed tries. ACM J. Exp. Algorithmics **25**, 1–28 (2020). https://doi.org/10.1145/3418033

29. Kang, Y.M., Liu, W., Zhou, Y.: QueryBlazer: efficient query autocompletion framework. In: Proceedings of the 14th International Conference on Web Search and Data Mining (WSDM), pp. 1020–1028 (2021). https://doi.org/10.1145/3437963.3441725

30. Kärkkäinen, J., Kempa, D., Puglisi, S.J.: Hybrid compression of bitvectors for the FM-index. In: Proceedings of the 24th Data Compression Conference (DCC), pp. 302–311 (2014). https://doi.org/10.1109/DCC.2014.87

31. Knuth, D.E.: The art of computer programming, vol. 3. 2 edn. Addison-Wesley (1998)

32. Leis, V., Kemper, A., Neumann, T.: The adaptive radix tree: ARTful indexing for main-memory databases. In: Proceedings of the 29th IEEE International Conference on Data Engineering (ICDE), pp. 38–49 (2013). https://doi.org/10.1109/ICDE.2013.6544812

33. Mäkinen, V., Belazzougui, D., Cunial, F., Tomescu, A.I.: Genome-Scale Algorithm Design. Cambridge University Press, Cambridge (2015)

34. Mao, Y., Kohler, E., Morris, R.T.: Cache craftiness for fast multicore key-value storage. In: Proceedings of the 7th European Conference on Computer Systems (EuroSys), pp. 183–196 (2012). https://doi.org/10.1145/2168836.2168855

35. Martínez-Prieto, M.A., Brisaboa, N.R., Cánovas, R., Claude, F., Navarro, G.: Practical compressed string dictionaries. Inf. Syst. **56**, 73–108 (2016). https://doi.org/10.1016/j.is.2015.08.008

36. Mavlyutov, R., Wylot, M., Cudré-Mauroux, P.: A comparison of data structures to manage URIs on the web of data. In: Proceedings of the 12th European Semantic Web Conference (ESWC), pp. 137–151 (2015). https://doi.org/10.1007/978-3-319-18818-8_9

37. Morrison, D.R.: PATRICIA-practical algorithm to retrieve information coded in alphanumeric. J. ACM **15**(4), 514–534 (1968). https://doi.org/10.1145/321479.321481

38. Nambiar, R.O., Poess, M.: The making of TPC-DS. In: Proceedings of the 32nd International Conference on Very Large Data Bases (VLDB), pp. 1049–1058 (2006). http://www.tpc.org/tpcds/

39. Navarro, G.: Compact Data Structures: A Practical Approach. Cambridge University Press, NY (2016)

40. Nilsson, S., Tikkanen, M.: Implementing a dynamic compressed trie. In: Proceedings of the 2nd International Workshop on Algorithm Engineering (WAE), pp. 25–36 (1998)

41. Ottaviano, G., Venturini, R.: Partitioned Elias-Fano indexes. In: Proceedings of the 37th ACM International Conference on Research and Development in Information Retrieval (SIGIR), pp. 273–282 (2014). https://doi.org/10.1145/2600428.2609615

42. Pibiri, G.E., Venturini, R.: Efficient data structures for massive n-gram datasets. In: Proceedings of the 40th ACM International Conference on Research and Development in Information Retrieval (SIGIR), pp. 615–624 (2017). https://doi.org/10.1145/3077136.3080798

43. Poyias, A., Puglisi, S.J., Raman, R.: m-Bonsai: a practical compact dynamic trie. Int. J. Found. Comput. Sci. **29**(8), 1257–1278 (2018). https://doi.org/10.1142/S0129054118430025

44. Silvestri, F., Venturini, R.: VSEncoding: efficient coding and fast decoding of integer lists via dynamic programming. In: Proceedings of the 19th ACM International Conference on Information and Knowledge Management (CIKM), pp. 1219–1228 (2010). https://doi.org/10.1145/1871437.1871592
45. Takagi, T., Inenaga, S., Sadakane, K., Arimura, H.: Packed compact tries: a fast and efficient data structure for online string processing. IEICE Trans. Fundam. Electron. Commun. Comput. Sci. **100**(9), 1785–1793 (2017). https://doi.org/10.1587/transfun.E100.A.1785
46. Tsuruta, K., et al.: c-Trie++: a dynamic trie tailored for fast prefix searches. In: Proceedings of the 30th Data Compression Conference (DCC), pp. 243–252 (2020). https://doi.org/10.1109/DCC47342.2020.00032
47. Williams, H.E., Zobel, J.: Compressing integers for fast file access. Comput. J. **42**(3), 193–201 (1999). https://doi.org/10.1093/comjnl/42.3.193
48. Yata, S.: Dictionary compression by nesting prefix/patricia tries. In: Proceedings of the 17th Meeting of the Association for Natural Language (2011)
49. Zhang, H., Andersen, D.G., Pavlo, A., Kaminsky, M., Ma, L., Shen, R.: Reducing the storage overhead of main-memory OLTP databases with hybrid indexes. In: Proceedings of the ACM International Conference on Management of Data (SIGMOD), pp. 1567–1581 (2016). https://doi.org/10.1145/2882903.2915222
50. Zhang, H., et al.: SuRF: practical range query filtering with fast succinct tries. In: Proceedings of the ACM International Conference on Management of Data (SIGMOD), pp. 323–336 (2018). https://doi.org/10.1145/3183713.3196931

On Representing the Degree Sequences of Sublogarithmic-Degree Wheeler Graphs

Travis Gagie[(✉)]

Dalhousie University, Halifax, Canada
`travis.gagie@dal.ca`

Abstract. We show how to store a searchable partial-sums data structure with constant query time for a static sequence S of n positive integers in $o\left(\frac{\log n}{(\log \log n)^2}\right)$, in $nH_k(S) + o(n)$ bits for $k \in o\left(\frac{\log n}{(\log \log n)^2}\right)$. It follows that if a Wheeler graph on n vertices has maximum degree in $o\left(\frac{\log n}{(\log \log n)^2}\right)$, then we can store its in- and out-degree sequences D_{in} and D_{out} in $nH_k(D_{\mathsf{in}}) + o(n)$ and $nH_k(D_{\mathsf{out}}) + o(n)$ bits, for $k \in o\left(\frac{\log n}{(\log \log n)^2}\right)$, such that querying them for pattern matching in the graph takes constant time.

1 Introduction

A Wheeler graph [7] is a directed edge-labelled graph whose vertices can be ordered such that vertices with no in-edges come first; if u has an in-edge labelled a and v has an in-edge labelled b with $a \prec b$ then $u < v$; if edges (u, v) and (w, x) are both labelled a and $u < w$ then $v \leq x$. Wheeler graphs are interesting because graphs that arise in some important applications are Wheeler—such as collections of edge-labelled paths and cycles, tries, and de Bruijn graphs—and if a graph is Wheeler then we can build a small index for it such that, given a pattern, we can quickly tell which vertices can be reached by paths labelled with that pattern.

The index for a Wheeler graph consists of four components:

1. a data structure supporting sum queries on the list D_{out} of the vertices' out-degrees, with $D_{\mathsf{out}}.\mathsf{sum}(i)$ returning the ith partial sum of the out-degrees (that is, the sum of the out-degrees of the first i vertices in the Wheeler order);
2. a data structure supporting rank queries on the list L of edge labels sorted by the edges' origins, with $L.\mathsf{rank}_a(i)$ returning the frequency of a among the first i edge labels;
3. a data structure supporting sum queries on the list C of the edge labels' frequencies, with $C.\mathsf{sum}(a)$ returning the sum of the frequencies of the edge labels lexicographically strictly less than a;
4. a data structure supporting search queries on the list D_{in} of the vertices' in-degrees, with $D_{\mathsf{in}}.\mathsf{search}(j)$ returning the largest i such that the sum of the in-degrees of the first i vertices in the Wheeler order is at most j.

D. Arroyuelo and B. Poblete (Eds.): SPIRE 2022, LNCS 13617, pp. 250–256, 2022.
https://doi.org/10.1007/978-3-031-20643-6_18

To see how the index works, first notice that, by the definition of a Wheeler graph, the vertices reachable by paths labelled with a pattern P form a single interval in the Wheeler order. In particular, all the vertices are reachable by paths labelled with the empty string. Suppose we have already found the endpoints V_P.start and V_P.end of the interval V_P in the Wheeler order containing vertices reachable by paths labelled P, and we want to find the endpoints $V_{P\cdot a}$.start and $V_{P\cdot a}$.end of the interval $V_{P\cdot a}$ in the Wheeler order containing vertices reachable by paths labelled $P\cdot a$, where $\cdot$ denotes concatenation.

We use the first data structure to find the endpoints $E_{P,\text{out}}$.start $=$ D_{out}.sum($V_{P\cdot a}$.start) $+ 1$ and $E_{P,\text{out}}$.end $=$ D_{out}.sum(V_P.end) of the interval $E_{P,\text{out}}$ in L that contains the labels of the out-edges of the vertices in V_P. We then use the second and third data structures to find the endpoints $E_{P\cdot a}$.start $=$ $L.\text{rank}_a(E_{P,\text{out}}.\text{start} - 1) + 1 + C.\text{sum}(a)$ and $E_{P\cdot a}$.end $= L.\text{rank}_a(E_{P,\text{out}}.\text{end}) + C.\text{sum}(a)$ of the interval $E_{P\cdot a}$ in the list of edge labels sorted into lexicographic order with ties broken by origin, that contains the copies of a in $E_{P,\text{out}}$. Finally, we use the fourth data structure to find the endpoints $V_{P\cdot a}$.start $=$ $D_{\text{in}}.\text{search}(E_{P\cdot a}.\text{start})$ and $V_{P\cdot a}$.end $= D_{\text{in}}.\text{search}(E_{P\cdot a}.\text{end})$ of $V_{P\cdot a}$.

This works because, again by the definition of a Wheeler graph, the list of edge labels sorted into lexicographic order with ties broken by origin, is also sorted by the ranks in the Wheeler order of the edges' destinations. For the sake of brevity, however, we refer the reader to Gagie et al.'s [7] original paper on Wheeler graphs for a full proof of correctness, and offer here only the example in Fig. 1 (modified from [4]), which shows the BOSS [5] representation of a de Bruijn graph (with the out-edge leaving vertex ACT and labelled $ deleted).

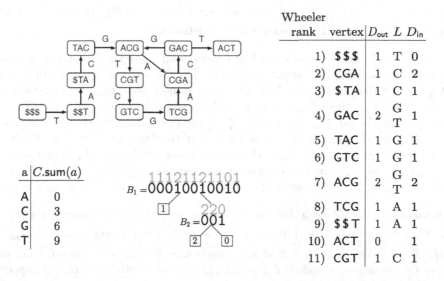

Fig. 1. A Wheeler graph (**upper left;** [4]); a table with D_{out}, L and D_{in} (**right**); C (**lower left**); and a degenerate wavelet tree supporting sum on D_{out} (**lower center**).

Suppose we have already found the endpoints V_C.start $= 4$ and V_C.end $= 6$ of the interval V_C of vertices reachable by paths labelled C, and we want to find the endpoints V_{CG}.start $= 7$ and V_{CG}.end $= 8$ of the interval V_{CG} of vertices reachable by paths labelled CG. We compute

$$D_{out}.\text{sum}(4-1) + 1 = 4$$
$$D_{out}.\text{sum}(6) = 7$$

$$L.\text{rank}_G(4-1) + 1 + C.\text{sum}(G) = 7$$
$$L.\text{rank}_G(7) + C.\text{sum}(G) = 9$$

$$D_{in}.\text{search}(7) = 7$$
$$D_{in}.\text{search}(9) = 8$$

and correctly conclude $V_{CG} = [7, 8]$ (containing vertices ACG and TCG).

There have been many papers on how to represent L compactly while supporting fast rank queries on it, and representing C compactly while supporting fast sum queries on it is trivial unless the alphabet of edge labels is unusually large, so in this paper we focus on how to represent D_{out} and D_{in} compactly while supporting fast sum and search queries on them. Specifically, we describe the first searchable partial-sums data structure for a static sequence S of sublogarithmic positive integers, with constant query time and space bounded in terms of the kth-order empirical entropy $H_k(S)$ of S:

Theorem 1. *Let $S[1..n]$ be a static sequence of positive integers. If $\max_i\{S[i]\}, k \in o\left(\frac{\log n}{(\log \log n)^2}\right)$ then we can store S in $nH_k(S) + o(n)$ bits and support* sum *and* search *queries on it in constant time.*

Theorem 1 may be of independent interest and it is easy to apply to support sum queries on D_{out} and search queries on D_{in}. To see how we can apply it to D_{out}, notice that if D'_{out} is the sequence obtained from D_{out} by incrementing each out-degree, then D'_{out} contains only positive integers, $|D'_{out}|H_k(D'_{out}) = |D_{out}|H_k(D_{out})$ and $D_{out}.\text{sum}(i) = D'_{out}.\text{sum}(i) - i$. To see how we can apply it to D_{in}, notice that all the 0s in D_{in} are at the beginning (by the definition of a Wheeler graph), so if D'_{in} is the sequence obtained from D_{in} by deleting its leading 0s, then D'_{in} contains only positive integers, $|D'_{in}|H_k(D'_{in}) \leq |D_{in}|H_k(D_{in})$ and $D_{in}.\text{search}(j) = D'_{in}.\text{search}(j) + |D_{in}| - |D'_{in}|$. This gives us our main result:

Theorem 2. *Let G be a Wheeler graph on n vertices with maximum degree Δ. If $\Delta, k \in o\left(\frac{\log n}{(\log \log n)^2}\right)$ then we can store G's out-degree sequence D_{out} in $nH_k(D_{out}) + o(n)$ bits such that it supports* sum *queries in constant time, and store G's in-degree sequence D_{in} in $nH_k(D_{in}) + o(n)$ bits such that it supports* search *queries in constant time.*

2 Intuition

The standard approach, proposed by Mäkinen and Navarro [8], to storing a compact searchable partial-sums data structure for a static sequence $S[1..n]$ of positive integers that sum to u, is as a bitvector B in which there are $S[1] - 1$ copies of 0 before the first 1 and, for $i > 1$, there are $S[i] - 1$ copies of 0 between the $(i - 1)$st and ith copies of 1. This takes $n \lg \frac{u}{n} + o(u)$ bits and supports $S.\mathsf{sum}(i) = B.\mathsf{select}_1(i)$ and $S.\mathsf{search}(j) = B.\mathsf{rank}_1(j)$ in constant time. If we use it to store the in- and out-degrees in a BOSS representation of a de Bruijn graph then we use about $\lg \sigma + 2$ bits per edge.

There are many other searchable partial-sums data structures (see, e.g., [3,9] and references therein) but, as far as we know, only a very recent one by Arroyuelo and Raman [1] achieves a space bound in terms of the empirical entropy of S and still answers queries in constant time. It takes $nH_0(S) +$ $O\left(\frac{u(\log \log u)^2}{\log u}\right)$ bits so, if $\max_i\{S[i]\} \in o\left(\frac{\log n}{(\log \log n)^2}\right)$, then $u \in o\left(\frac{n \log n}{(\log \log n)^2}\right)$ and it takes $nH_0(S) + o(n)$ bits. If we apply this instead of Theorem 1 then we obtain a slightly weaker form of Theorem 2, in which H_k is replaced by H_0.

To prove Theorem 1, our starting point is Ferragina and Venturini's [6] well-known result about storing a static string in nH_k-compressed space while supporting fast random access to it:

Theorem 3 (Ferragina and Venturini). *We can store S as a string of n characters from an alphabet of size σ in*

$$nH_k(S) + O\left(\frac{n \log \sigma}{\log n}(k \log \sigma + \log \log n)\right)$$

bits for $k \in o\left(\frac{\log n}{\log \sigma}\right)$ such that we can extract any substring of S of length ℓ in $O\left(1 + \frac{\ell \log \sigma}{\log n}\right)$ *time.*

Assuming S consists of positive integers with $\max_i\{S[i]\} \in o\left(\frac{\log n}{(\log \log n)^2}\right)$, we have $\sigma \in o\left(\frac{\log n}{(\log \log n)^2}\right)$ and the space bound in Theorem 3 is $nH_k(S) + o(n)$ bits for $k \in o\left(\frac{\log n}{(\log \log n)^2}\right)$. Notice the extraction time is constant for $\ell \in O\left(\frac{\log n}{\log \sigma}\right)$.

In order to support sum and search on S in constant time, we augment Ferragina and Venturini's representation of S with sublinear data structures similar to those Raman, Raman and Rao [10] used to support rank and select on their succinct bitvectors. Since these augmentations are fairly standard, we omit the details of the how we support sum and leave the details of how we support search to the next section.

Lemma 1. *We can add $o(n)$ bits to Ferragina and Venturini's representation of S and support sum in constant time.*

Lemma 2. *We can add $o(n)$ bits to Ferragina and Venturini's representation of S and support search in constant time.*

Combining Theorem 3 and Lemmas 1 and 2, we immediately obtain Theorem 1. We note that we need $\sigma \in o\left(\frac{\log n}{(\log \log n)^2}\right)$ only to prove Lemma 2. In the full version of this paper we will show how we can store S in $nH_k(S) + o(n)$ bits of space and support sum queries on it in constant time even when σ is polylogarithmic in n, for example—which could be of interest when storing Wheeler graphs with large maximum out-degree but small maximum in-degree, such as some tries.

3   Proof of Lemma 2

Proof. We first store $\mathsf{search}(c\sigma \lg^2 n)$ for each multiple $c\sigma \lg^2 n$ of $\sigma \lg^2 n$. Since $\mathsf{sum}(n) \leq \sigma n$, this takes a total of

$$O\left(\frac{\sigma n}{\sigma \lg^2 n} \cdot \lg n\right) \subset o(n)$$

bits. We then store the difference

$$\mathsf{search}\left(c \cdot \frac{\lg n}{2 \lg \sigma}\right) - \mathsf{search}\left(\sigma \lg^2(n) \cdot \left\lfloor \frac{c \cdot \frac{\lg n}{2 \lg \sigma}}{\sigma \lg^2 n} \right\rfloor\right)$$

for each multiple $c \cdot \frac{\lg n}{2 \lg \sigma}$ of $\frac{\lg n}{2 \lg \sigma}$ and the preceding multiple $\sigma \lg^2(n) \cdot \left\lfloor \frac{c \cdot \frac{\lg n}{2 \lg \sigma}}{\sigma \lg^2 n} \right\rfloor$ of $\sigma \lg^2 n$. Since each of these differences is at most $\sigma \lg^2 n$, this takes a total of

$$O\left(\frac{\sigma n \log \sigma}{\log n} \cdot \log(\sigma \log^2 n)\right) \subset o(n)$$

bits. Finally, we store a universal table that, for each possible $\frac{\lg n}{2}$-bit encoding of a substring of S consisting of $\frac{\lg n}{2 \lg \sigma}$ integers (each represented by $\lg \sigma$ bits) and each value q between 1 and the maximum possible sum $\sigma \cdot \frac{\lg n}{2 \lg \sigma}$ of such a substring, tells us how many of that substring's integers we can sum before exceeding q. This takes

$$2^{\frac{\lg n}{2} + \lg\left(\sigma \cdot \frac{\lg n}{2 \lg \sigma}\right)} \lg\left(\frac{\lg n}{2 \lg \sigma}\right) \in o(n)$$

bits.

To evaluate $\mathsf{search}(j)$ in constant time, we first look up $\mathsf{search}\left(\sigma \lg^2(n) \cdot \left\lfloor \frac{j}{\sigma \lg^2 n} \right\rfloor\right)$ and

$$\mathsf{search}\left(\frac{\lg n}{2 \lg \sigma} \cdot \left\lfloor \frac{j}{\frac{\lg n}{2 \lg \sigma}} \right\rfloor\right) - \mathsf{search}\left(\sigma \lg^2(n) \cdot \left\lfloor \frac{j}{\sigma \lg^2 n} \right\rfloor\right),$$

which tells us search $\left(\frac{\lg n}{2 \lg \sigma} \cdot \left\lfloor \frac{j}{\frac{\lg n}{2 \lg \sigma}} \right\rfloor \right)$. Since

$$ j - \frac{\lg n}{2 \lg \sigma} \cdot \left\lfloor \frac{j}{\frac{\lg n}{2 \lg \sigma}} \right\rfloor < \frac{\lg n}{2 \lg \sigma} $$

and the integers in S are positive,

$$ \mathsf{search}(j) - \mathsf{search} \left(\frac{\lg n}{2 \lg \sigma} \cdot \left\lfloor \frac{j}{\frac{\lg n}{2 \lg \sigma}} \right\rfloor \right) < \frac{\lg n}{2 \lg \sigma}. $$

It follows that we can find $\mathsf{search}(j)$ by extracting the substring of $\frac{\lg n}{2 \lg \sigma} \in$ $O \left(\frac{\log n}{\log \sigma} \right)$ characters starting at $S \left[\mathsf{search} \left(\frac{\lg n}{2 \lg \sigma} \cdot \left\lfloor \frac{j}{\frac{\lg n}{2 \lg \sigma}} \right\rfloor \right) \right]$ and using the universal table to learn how many of that substring's integers we can sum before exceeding

$$ j - \mathsf{sum} \left(\mathsf{search} \left(\frac{\lg n}{2 \lg \sigma} \cdot \left\lfloor \frac{j}{\frac{\lg n}{2 \lg \sigma}} \right\rfloor \right) - 1 \right). $$

4   Postscript

We have not implemented Theorems 1 or 2 because there are other approaches that perform poorly in the worst case but are likely unbeatable in practice. If we store S as a degenerate wavelet tree, then we can implement an $S.\mathsf{sum}$ query with σ rank queries on the wavelet trees bitvectors, together with σ multiplications and additions: for example, to find $D_{\mathsf{out}}.\mathsf{sum}(8)$ for the sequence $D_{\mathsf{out}} = 1, 1, 1, 2, 1, 1, 2, 1, 1, 0, 1$ with the degenerate wavelet tree shown in Fig. 1, we compute

$$ 1 \cdot B_1.\mathsf{rank}_0(8) + 2 \cdot B_2.\mathsf{rank}_0(8 - B_1.\mathsf{rank}_0(8)) = 1 \cdot 6 + 2 \cdot 2 = 10. $$

In practice σ is usually a small constant—often 4—and if the bitvectors in the wavelet tree are entropy-compressed, then it takes $nH_0(S) + o(n \log \sigma)$ bits. If we store a minimal monotone perfect hash function [2] mapping each value $S.\mathsf{sum}(i)$ to i, together with a small sample of those pairs, then we should also be able to support $S.\mathsf{search}$ queries by computing a few hash values and $S.\mathsf{sum}$ queries, quickly and in small space in practice. We leave the details for the full version of this paper.

Acknowledgments. Many thanks to Jarno Alanko for bringing the topic of this paper to our attention, to Rossano Venturini for pointing out Arroyuelo and Raman's result, and to Meng He, Gonzalo Navarro and Srinivasa Rao Satti for helpful discussions.

References

1. Arroyuelo, D., Raman, R.: Adaptive succinctness. Algorithmica **84**, 694–718 (2022)
2. Belazzougui, D., Boldi, P., Pagh, R., Vigna, S.: Theory and practice of monotone minimal perfect hashing. ACM J. Exp. Algorithmics **16**, 1–26 (2011)
3. Bille, P., Gørtz, I.L., Skjoldjensen, F.R.: Partial sums on the ultra-wide word RAM. Theor. Comput. Sci. **905**, 99–105 (2022)
4. Boucher, C., Bowe, A., Gagie, T., Puglisi, S.J., Sadakane, K.: Variable-order de Bruijn graphs. In: Proceedings of the Data Compression Conference (DCC), pp. 383–392 (2015)
5. Bowe, A., Onodera, T., Sadakane, K., Shibuya, T.: Succinct de Bruijn graphs. In: Raphael, B., Tang, J. (eds.) WABI 2012. LNCS, vol. 7534, pp. 225–235. Springer, Heidelberg (2012). https://doi.org/10.1007/978-3-642-33122-0_18
6. Ferragina, P., Venturini, R.: A simple storage scheme for strings achieving entropy bounds. Theoret. Comput. Sci. **372**, 115–121 (2007)
7. Gagie, T., Manzini, G., Sirén, J.: Wheeler graphs: a framework for BWT-based data structures. Theoret. Comput. Sci. **698**, 67–78 (2017)
8. Mäkinen, V., Navarro, G.: Rank and select revisited and extended. Theoret. Comput. Sci. **387**, 332–347 (2007)
9. Pibiri, G.E., Venturini, R.: Practical trade-offs for the prefix-sum problem. Softw. Pract. Exp. **51**, 921–949 (2021)
10. Raman, R., Raman, V., Satti, S.R.: Succinct indexable dictionaries with applications to encoding k-ary trees, prefix sums and multisets. ACM Trans. Algorithms **3**, 43 (2007)

Engineering Compact Data Structures for Rank and Select Queries on Bit Vectors

Florian Kurpicz[✉][iD]

Karlsruhe Institute of Technology, Karlsruhe, Germany
kurpicz@kit.edu

Abstract. Bit vectors are fundamental building blocks of succinct data structures used in compressed text indices, e.g., in the form of the wavelet trees. Here, two types of queries are of interest: rank and select queries. In practice, the smallest (uncompressed) rank and select data structure cs-poppy has a space overhead of $\approx 3.51\%$ [Zhou et al. SEA 2013] [26]. Using the same overhead, we present a data structure that can answer queries up to 8 % (rank) and 16.5 % (select) faster compared with cs-poppy.

Keywords: Rank and select · Bit vectors · SIMD · Succinct data structures

1 Introduction and Related Work

Given a bit vector B of length n and $\alpha \in \{0, 1\}$, *rank* and *select* are defined as:

rank: given $i \in [0, n)$, rank returns the number of ones (or zeros) in $B[0, i]$, i.e.,

$$B.rank_\alpha(i) = |\{j \in [0, i] \colon B[j] = \alpha\}|$$

select: given a rank i, select returns the leftmost position where the bit vector contains a one (or zero) with rank i, i.e.,

$$B.select_\alpha(i) = \min\{j \in [0, n) \colon B.rank_\alpha(j) = i\}$$

Bit vectors are building blocks of many important compact and succinct data structures like wavelet trees [10] that have applications in many compressed full-text indices (e.g., the FM-index [10] and r-index [11]; we point to the following surveys [6,9,17,18] for more information on wavelet trees), succinct graph representations (e.g., LOUDS [14]), and can also be used as a representation of monotonic sequences of integers (e.g., Elias-Fano coding [7,8]) that supports predecessor queries. It should be noted that all of the applications mentioned above require rank and/or select queries on bit vectors.

Given a length-n bit vector, it is known how to solve rank and select queries in constant time using only $n + o(n)$ bits of space [5,14]. Here, the bit vector

© The Author(s) 2022
D. Arroyuelo and B. Poblete (Eds.): SPIRE 2022, LNCS 13617, pp. 257–272, 2022.
https://doi.org/10.1007/978-3-031-20643-6_19

occupies n bits. The rank and select data structures require only $o(n)$ additional bits. There also exist more precise results, when considering a length-n bit vector containing k ones, focusing on applications where the number of ones is small. Currently, the best known result requires $\lg \binom{n}{k} + \frac{n}{(\lg n/t)^t} + \tilde{O}(n^{3/4})$ bits of space and can answer rank and select queries in $O(t)$ time [23] (by no explicitly storing the bit vector).

Related Work. In this paper, we focus on practical space-efficient *uncompressed* rank and select data structures that can handle bit vectors of arbitrary size. Prominent implementations can be found in the popular succinct data structure library (SDSL) [12]. Furthermore, there exist highly tuned select-only data structures by Vigna [25], which currently can answer select queries the fastest while being reasonably space-efficient. The currently most space-efficient rank and select data structure by Zhou et al. [26] requires only 3.51 % additional space. There exists more work on practical space-efficient rank and select data structures for bit vectors that require more space, answer queries slower, and/or can handle only bit vectors up to size 2^{32}, e.g., [13,15,19], which we, therefore, do not consider in this paper. There also exists work on *compressed* rank and select data structures, e.g., [1–3,24] and on rank and select data structures for *mutable* bit vectors, e.g., [21,22].

2   Preliminaries

Due to the simplicity of the problem, the notations in this paper are rather simple. We have a *bit vector* of length n, where we can access each bit in constant time. In the following descriptions, we make use of the notion of *blocks*. Here, a block is an interval within the bit vector that *covers* the bits within the interval. Given (a part of) a bit vector, the *population count* or *popcount* is the number of ones within (the part of) the bit vector. Popcount instructions are supported by most modern CPUs for up to 64 bit words. Since their first introduction, rank and select data structures that require sub-linear additional space and can answer rank and/or select queries in constant time have similar structures. In this section, we briefly describe the commonly used designs for rank and select data structures.

Almost all practical rank data structures follow the same layout. First, the bit vector is partitioned into consecutive *basic blocks*, which are the smallest unit for which any information is collected in an index. The rank of bits within a basic block is determined directly on the bit vector. Then, there exists a hierarchy of different (overlaying) blocks of different sizes. For each type of block, there exists an index that stores information about the number of ones. The scope in which the number of ones is considered can differ, e.g., the number of ones in the block or the number of ones up to the beginning of the block from the beginning of either the bit vector or another overlaying block. See Fig. 1 for an example. Queries are then answered using the information provided by the blocks and the popcount of the basic block up to the position. Depending on the sizes of the blocks, the pertinent information in the indices can be accessed very efficiently. It should be noted, that it suffices to store information about the number of ones in each block, as $rank_0(i) = i - rank_1(i)$.

Unlike rank data structures, select data structures come in two flavors: *rank-based* and *position*-based. Rank-based select data structures utilize a rank data structure that is enhanced by a small index containing sample positions for every k-th one (or zero). To answer a query, we first determine the closest block using the sampled positions. Then, we have to look at blocks until we have found the basic block that contains the correct bit with the requested rank. The position in the basic block can then easily be computed. While fast in practice (see Sect. 4), this type of select data structure usually cannot guarantee a constant query time.

Position-based select data structures on the other hand only store sample positions. Usually, they differentiate between different block sizes and densities, e.g., for very sparse blocks, the answer of every select query can be stored directly. One advantage of position-based select data structures is a constant worst-case query time can be achieved, e.g., [25]. However, there is also a disadvantage of position-based select data structures. Unlike when answering rank queries, we cannot use a $select_1$ query to answer a $select_0$ query. The significant difference between rank- and position-based select data structures is that we can easily use the information in a rank-based data structure to answer both $select_0$ and $select_1$ queries. We can simply transform the number of ones (or zeros) up to a position to the number of zeros (or ones) up to that position.[1] The same is not possible with position-based select data structures.

3 Space Efficient Rank and Select Data Structures

First, in Sect. 3.1, we describe the design of a space efficient rank and select data structure by Zhou et al. [26] named *cs-poppy* that makes use of fast popcount instructions. In Sect. 3.2, we present our main result—a significantly faster rank and select data structure—that requires the same space as cs-poppy and makes use of SIMD. Finally, in Sect. 3.3, we present a new and relatively simple rank data structure that provides better rank query times (in practice) by combining some techniques from our main result with a slightly higher space usage.

3.1 CS-Poppy: Rank-Based Rank and Select Data Structure

We describe the rank and select data structure *cs-poppy* top-down, starting with the largest blocks. For an overview, see Fig. 1. First, the bit vector is split up into consecutive L0-blocks of size 2^{32} bits. For each L0-block we store the number of ones occurring in the bit vector before the L0-block in an L0-index. To accommodate bit vectors of arbitrary size, each entry in the L0-index requires 64 bits. Note that the indices for the different block types are just plain arrays where the entry of the i-th block is stored at the i-th array entry. Given a position

---

[1] The sampled positions described above have to be stored for ones and zeros.

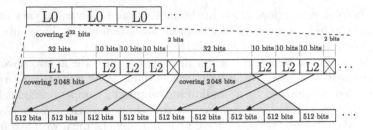

Fig. 1. L0-index and interleaved L1- and L2-index of cs-poppy. Arrows indicate that the popcount of the basic block is stored. Wasted bits are marked by red crosses. (Color figure online)

in the bit vector, we can identify all blocks that cover the position by dividing it by the block size (in bits). Next, each L0-block is split up into consecutive L1-blocks of size 2048 bits. This time, we are interested in the number of ones occurring in the L0-block before the L1-block. For each L1-block, we store this information in the L1-index. Since the number of ones in an L0-block can be at most 2^{32}, each entry in the L1-index requires only 32 bits. Finally, each L1-block is split up into four L2-blocks of size 512, i.e., the basic blocks of cs-poppy. We store the number of ones in the L2-blocks for the first three L2-blocks in each L1-block in the L2-index. The number of ones in each L1-block's last L2-block is not stored, as it does not provide any information that cannot be computed using the L1-index. (We can look at the number of one-bits occurring before the next L1-block and subtract the number of one-bits seen before the fourth L2-block.) A L2-index entry has to encode a number in $[0, 512]$ and thus requires 10 bits.

One important technique used by the authors of cs-poppy is the *interleaved L1- and L2-index*. Here, for each L1-block and the corresponding L2-blocks, a 64-bit word is used to store the entry of the L1- and L2-index. Since the L1-index contains 32-bit words and the L2-index contains 10-bit entries (of which three are pertinent to the L1-block), everything fits into 64 bits. While this approach wastes two bits for each L2-block (0.09 % additional space), it reduces the number of cache misses, as the required part of the L2-index should be loaded whenever the L1-index is accessed. Zhou et al. [26] introduced more practical improvements that can speed up answering rank and select queries using cs-poppy. We refer to their paper for a detailed description.

Answering Rank Queries. Now, we want to answer a rank query for position i. To this end, we first have to identify the L0- and L1-block the position is covered by. We obtain both blocks by dividing i with the bit size of an L0- and L1-block respectively. The corresponding entries in the L0- and L1-index contain the number of bits occurring from the beginning of the bit vector to the beginning of the L0-block and from the beginning of the L0-block to the beginning of the L1-block. This is the first part of the result. Next, we have to determine the number of ones within the L1-block up to the position. To this

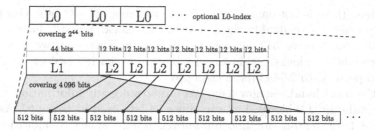

Fig. 2. L0-index and interleaved L1- and L2-index of flat-popcount. Boxes indicate that the number of ones in the L1-block up to this point is stored in the L2-index.

end, we scan the entries of the L2-index pertinent to the L1-block until we have reached the L2-block that covers position i. We add entries of the L2-index we have scanned to the result. Finally, we have to determine the number of ones in the final L2-block directly on the bit vector. This can be done using fast popcount instructions. Overall, rank queries can be answered in constant time using cs-poppy.

Answering Select Queries. To answer a select query for a rank i, we first identify the L0-block where the first position with rank i can occur. This can be done by scanning the L0-index until we see an entry greater or equal to i. The position has to be in the L0-block belonging to the previous entry in the L0-index. Now, we have to scan the L1-index, until we have identified the L1-block that contains the searched position. To speed up select queries, the position of every 8 192-th one is sampled. We can use these samples to skip some parts of the L1-index. When we have identified the correct L1-block, we continue scanning the L2-index until we find the L2-block that contains the position we are looking for. During each step, we subtract the number of ones in the L0-, L1-, and L2-entries from the rank i, because otherwise, we could not identify the block containing the result in the L1- and L2-index, resp. Finally, we have to identify the position within the L2-block and return it. The search in the L2-block has been further optimized by using SIMD by Pandey et al. [20]. Note that this query algorithm requires linear time in the worst-case, but is fast in practice, see Sect. 4.

3.2   Flat-Popcount: Storing More Information Wasting No Bits

Now, we present the main result of this paper, a rank and select data structure that requires the same space as cs-poppy but is faster in practice, see Sect. 4. We call this data structure *flat-popcount*. To achieve this, we have to store additional information about blocks without using any additional space. Here, we make use of the two bits that cs-poppy wastes for every L1-block.

As mentioned before, we want to store additional information. To be more precise, we store the number of ones that occur before each L2-block within each L1-block—similar to the L0- and L1-index—see Fig. 2 for an example. If we consider cs-poppy, we have to store numbers in $[0, 1536]$, which require 11 bits each. Thus, there is not enough space to do so using only 32 bits. Our solution to this problem is to double the size of an L1-block. We still have L2-blocks that cover consecutive 512 bits of the bit vector. Therefore, we now have to consider eight L2-blocks within one L1-block, i.e., each L1-block covers 4096 bits (compared with 2048 bits in cs-poppy).

On the other hand, we now have 128 bits to store any information regarding the L1- and eight L2-blocks, i.e., two times 64 bits used in cs-poppy. As before, we do not have to store any information regarding the last L2-block in the L1-block, as we can compute this information using the information stored for the next L1-block. We can therefore store the number of ones up to each L2-block within the L1-block using 12 bits each, as the number of previously set bits is in $[0, 3584]$. This requires $7 \cdot 12 = 84$ bits, leaving us with 44 bits for the entry in the L1-index. Similar to cs-poppy, we interleave the L1- and L2-index within one 128-bit word, which is supported by almost all modern CPUs. Not only can we now store more information for each L2-block, we can also increase the size of the L0-blocks to 2^{44} bits. This allows us to make the L0-index optional, as it would be required only for bit vectors of size greater than 2^{44} bits, which is significantly larger than the 2^{32} bits that cs-poppy supports without L0-index and would require 2 TiB space for the bit vector alone.

The space requirements of both cs-poppy and flat-popcount are nearly the same. However, there is one big advantage of our approach: in flat-popcount, we have random access to the L2-index. When using cs-poppy, we have to scan the entries in the L2-index, as they are delta encoded. Using flat-popcount, we store the number of ones occurring to the left of the L2-block within the L1-block *directly*, as for the L1- and L0-index. This allows us to answer both rank queries slightly and select queries significantly more efficiently (in practice). We now take a look at the changes in the query algorithms.

Answering Rank Queries with Flat-Popcount. Answering rank queries for a position i in this data structure is similar to answering rank queries using cs-poppy, which we describe in Sect. 3.1. At least for identifying the entries in the L0- and L1-index. Here, nothing has changed, as we only improved on the L2-index. As mentioned earlier, the main advantage of our new design is that we do not have to scan all L2-entries to compute the number of one-bits up the final L2-block. Thus, we can now determine the number of bits occurring in the L1-block before the L2-block that contains the position i accessing only one entry of the L2-index. This entry can be computed the same way we compute the entries of the L0- and L1-index. In our experiments (see Sect. 4), we can see that scanning up to three L2-entries and adding up their values contributes significantly to the running time of a query.

Answering Select Queries with Flat-Popcount. Answering a select query for a rank i is also very similar to cs-poppy. We still sample the position of every 8192-th one in the bit vector and use the samples to speed up identifying the L1-block that contains the first position that has rank i. Thus, the first part of

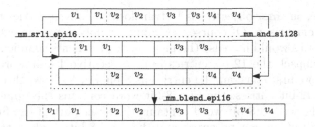

Fig. 3. Simplified transformation of four packed 12-bit integers to 16-bit integers. Each block represents 8 bit. The split blocks contain 4 bit of two different entries.

the query algorithm that changes is the identification of the L2-block, where the first position with rank i occurs. Due to us storing the number of ones occurring (within the L1-block) before each L2-block, we have a monotonic increasing sequence of integers, allowing us to search for the L2-block more efficiently.

Finding the Correct L2-Entry using Linear or Binary Search. When we use a linear search, we scan the L2-index until we have found an entry that contains the rank that we are looking for. Therefore, answering queries is not much different than before, when using cs-poppy. However, since we do store the number of ones before the L2-block, we save some additions (of L2-entries) when answering queries, which results in a small improvement. Since the number of ones in the L1-block occurring before each L2-block is a monotonic increasing sequence, we can also use a binary search on the entries of the L2-index. As we store seven entries in the L2-index for each L1-block, we can use a *uniform* binary search [16, p. 414f], which always requires three iterations to identify the correct block. Most importantly, we can pre-compute the whole decision tree and reuse entries of the L2-index. Both of these approaches are rather simple and do not make use of the fact that all entries of the L2-block that we are interested in are contained in a single 128-bit word. Modern CPUs are able to compare multiple values contained in such a word at the same time. However, there are some limitations, which we describe in the following section.

Finding the Correct L2-Entry Using SIMD. In addition to these more obvious approaches described above, we can also search for the L2-block using the *streaming SIMD extension* (SSE), which allows us to divide a 128-bit computer word into consecutive blocks and apply operations on each of the blocks at the same time. One limitation of these instructions is that all blocks must have the same size and that the sizes of the block must be either 8, 16, 32, or 64 bits. Our main goal is to use the compare operation _mm_cmpgt_epi16 to compare all seven entries of the L2-index that are covered by the L1-block at the same time. Unfortunately, we cannot simply use the compare operation, because we have to store the L2-entries using only 12 bits and there is no compare operation working on 12-bit blocks.

Therefore, our first objective is to transform the seven entries to use 16 bits during the comparison. Of course, we cannot simply store the entries using 16 bits each, as this would increase the memory requirements significantly. Instead, we have to unpack the 12-bit entries into consecutive 16-bit words, see Fig. 3. To this end, we first consider the entries as 8-bit words. Now, three 8-bit words contain two 12-bit entries: the first 8-bit word contains the upper part of the first entry, the second 8-bit word contains the lower part of the first entry and the upper part of the second entry, and the third 8-bit word contains the lower part of the second entry. This pattern repeats for all entries.

We now can shift the words that have their upper part in a whole 8-bit word four bits to the right (_mm_srli_epi16) and mask the lower 12 bit of the other entries (_mm_set1_epi16) to obtain 16-bit words containing the entries as results. Then, we can take one 16-bit word of each result alternately to obtain our final result, where each 12-bit entry is stored in consecutive 16-bit words. Using this final result, we can compare (_mm_cmpgt_epi16) with the remaining rank minus one (because there does not exist a greater-or-equal comparison). As the compare operation does not return the word, where the first match occurs, we have to transform the result to the position of the block. The compare operation returns a word, where the result of the compare operation is marked by all-ones (true) or all-zeros (false). Therefore, we can take the most significant bit of each 8-bit word (_mm_cmpgt_epi16) and apply a simple popcount operation on the result, because we are only interested in the first result where the entry is greater or equal. Now, we can continue the select query as before.

3.3   Wide-Popcount: Faster Rank

As mentioned before, using 16 bits to store an entry of the L2-index would allow us to directly use the SIMD instructions on the L2-index without transforming it first. We do so in our final rank data structure that we call *wide-popcount*. Since we now have 16 bits available for each entry in the L2-index, we also have to make the L1-block bigger, because otherwise, the additional space required by the index would be too much. Therefore, we do now let each L1-block cover 128 L2-blocks, or 65 536 bits. As before, we are only interested in the first 127 L2-blocks within each L1-block. Thus, each entry in the L2-index is in $[0, 65\,024]$ and can be stored using 16 bits. For the L1-index, we use 64-bit words, because then we do not need an L0-index. This increases the required space of the data structure for rank queries to 3.198 % additional space. On the other hand, we have only two levels of indices instead of three. We also do not interleave the L1- and L2-index, as there is no advantage because not all L2-entries can be loaded directly into the cache together with the L1-entry, due to their size and number.

Answering rank and select queries works the same as with flat-popcount (without an L0-index). In Sect. 4, we will see that this approach works very well for rank queries, but it is not well suited for select queries. This is mostly because we have to search through a lot of L2-entries during a select query. Even when speeding up the search using a uniform binary search or SIMD instructions, answering select queries requires significantly more time than using flat-popcount.

4   Experimental Evaluation

Our implementations are available at https://github.com/pasta-toolbox/bit-vector as open-source (GPLv3, header-only C++-library). In addition to the source code of our implementations, we also provide scripts to easily reproduce all results presented in this paper (https://github.com/pasta-toolbox/bit-vector-experiments). Our experiments were conducted on a workstation equipped with an AMD Ryzen 9 3950X (16 cores with frequencies up to 3.5 GHz, 32 KiB L1d and L1i cache and 512 MiB L2 cache per core, and 4 times 16 MiB L3 cache) and 64 GiB DDR4 RAM running Ubuntu 20.04.2 LTS. Since our experiments are sequential, only one core was used at a time. We compiled the code using GCC 10.2 with flags -O3, -march=native, and -DNDEBUG and created the makefile using CMake version 3.22.1.

In our experiments, we use two types of random inputs with different densities of ones in the bit vector (10 %, 50 %, and 90 % of all bits are ones). For the first type of input, the ones are uniformly distributed. This type of input should be the easiest one of the two. The second type of input is an adversarial input similar to the one used by Vigna [25]. Assume that k % of the bits in the bit vector should be ones. Then, we set 99 % of the ones in the last k % of the bit vector and the remaining one percent in the first $100 - k$ % of the bit vector. Here, the first part of the bit vector is very sparse while the second part of the bit vector is very dense. Overall, these two types of distribution are the extreme ends of distributions that can occur. All data structures are tested on the same bit vectors and the same queries. The reported running times are the average of three runs (each with a new bit vector and queries). During each run, we constructed the data structure and then asked 100 million queries of each query type supported by the data structure. We compare the following rank and select data structures:

- *cs-poppy* is the space-efficient rank-based rank and select data structure described in Sect. 3.1 by Zhou et al. [26],
- *cs-poppy-fs* is the same as cs-poppy but with the supposedly faster select algorithm used for the final 64-bit word by Pandey et al. [20],
- *simple-select$_x$* is a position-based select data structure by Vigna [25] that allows for tuning parameter x that determines the size of additional space the data structure is allowed to allocate,
- *simple-select$_h$* is a version of simple-select by Vigna [25] that is highly tuned for bit vectors that contain the same amount of ones and zeros,
- *rank9select* is a rank and select data structure by Vigna [25] that stores 9-bit values to answer rank queries and positions to answer select queries,
- *sdsl-mcl* Clark's select data structure [4] contained in the SDSL [12],
- *sdsl-v* is a simple rank data structure that requires 25 % additional memory and is contained in the SDSL, and
- *sdsl-v5* is a more space-efficient variant of the rank data structure above (only 6.25 % additional space) and also contained in the SDSL

Table 1. Average additional space in percent used by all evaluated data structures on bit vectors of different sizes over the uniform and adversarial distribution.

Name	$n = 1 \cdot 10^9$	$2 \cdot 10^9$	$4 \cdot 10^9$	$8 \cdot 10^9$	$16 \cdot 10^9$	$32 \cdot 10^9$
cs-poppy	3.32	3.32	3.32			
cs-poppy-fs	3.32	3.32	3.32			
pasta-poppy	3.58	3.58	3.58	3.58	3.58	3.58
pasta-flat$_{\text{SIMD}}$	3.58	3.58	3.58	3.58	3.58	3.58
pasta-wide	10.16	10.17	10.16	10.16	10.16	10.16
sdsl-v	25.00	25.00	25.00	25.00	25.00	25.00
sdsl-v5	6.25	6.25	6.25	6.25	6.25	6.25
sdsl-mcl	18.51	18.52	18.53	18.54	18.55	18.56
simple-select$_0$	8.72	8.72	8.72	8.72	8.72	8.72
simple-select$_1$	9.88	9.88	9.88	9.88	9.88	9.88
simple-select$_2$	12.21	12.20	12.20	12.20	12.20	12.20
simple-select$_3$	16.85	16.85	16.84	16.84	16.84	16.84
simple-select$_h$	15.62	15.63	15.63	15.63	15.63	15.63
rank9select	56.25	56.25	56.25	56.25	56.25	56.25

with our implementations that we describe in Sects. 3.1 to 3.3:

- *pasta-poppy* is our implementation of cs-poppy,
- *pasta-flat$_t$* is the rank-based rank and select data structure that we describe in Sect. 3.2 with $t \in \{\texttt{linear}, \texttt{binary}, \texttt{SIMD}\}$, and
- *pasta-wide* is the rank-data structure that we describe in Sect. 3.3.

Unfortunately, two competitors cs-poppy [26] and cs-poppy-fs [20] were not able to compute the select queries on all inputs in a reasonable time (3 h for all queries on a single bit vector). We were not able to find the error causing this problem, but want to highlight that all queries asked were feasible queries.

We only include *pasta-flat$_{\text{SIMD}}$* in the plots as it is overall the fastest variant. For a comparison of the select query times of the three *pasta-flat* query versions, see Fig. 6. The rank query times are identical, as the same data structure and query algorithm is used. Note that we did not include the rank and select data structures that have already been shown to be slower (and to require more additional memory) than the ones included in the experimental evaluation, e.g., *combined-sampling* [19], which is slower than cs-poppy and only works for bit vectors up to size 2^{32} bits, *BitRankF* [13], which is slower than simple-select and also requires more space, and the data structures described by Kim et al. [15].

Space Requirements and Select $_0$ & Select $_1$ Queries. First, we discuss the space requirements of all data structures evaluated in this paper. For an overview, see Table 1. We measured the additional space by overwriting *malloc* to see all allocations on the heap of the different data structures. This is also the

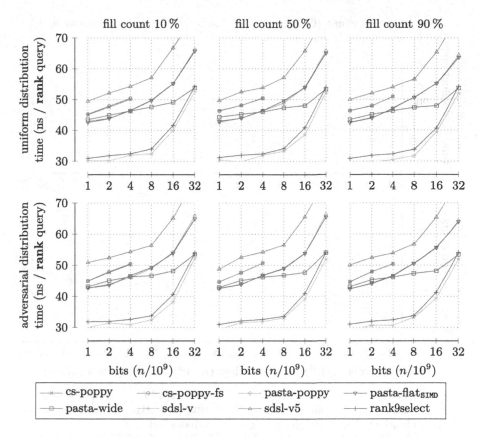

Fig. 4. Average rank query time in nanoseconds on all tested inputs.

reason why it looks like cs-poppy and cs-poppy-fs require less space than pasta-poppy and pasta-flat, because the former allocates memory on the stack to store the samples for the select queries (which we did not modify to not change the results of the running time experiments). This makes our new data structures, cs-poppy, and cs-poppy-fs the most space-efficient rank and select data structures, requiring roughly half the space that the smallest rank- and select-only data structures (sdsl-v5 and simple-select$_0$) require.

Rank Queries. Let us take a look at all data structures that support rank queries. In Fig. 4, we report the average query time on all tested inputs. Here, we can see that sdsl-v and sdsl-v5 provide the fastest query times. For large inputs, pasta-wide has query times similar to sdsl-v and sdsl-v5. All these data structures can only answer rank queries and require more space than our new data structures (1.75–6.98 times as much). While rank9select also requires more memory it is also significantly slower. Both pasta-poppy and pasta-flat have similar query times and get slower for larger inputs. Nevertheless, they are roughly 8 % faster than cs-poppy and cs-poppy-fs.

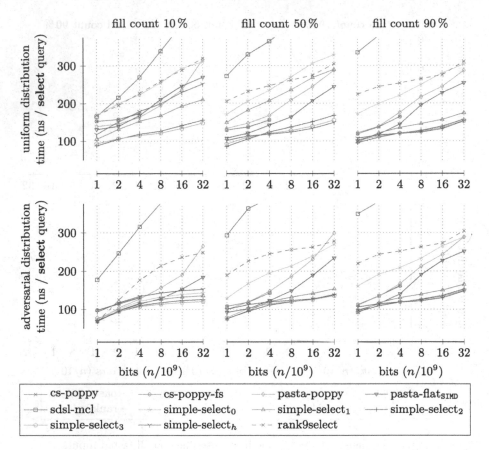

Fig. 5. Average select query time in nanoseconds on all tested inputs.

Select Queries. We report select query times in Fig. 5. Here, we can see that sdsl-mcl, rank9select, and simple-select$_0$ are among the slowest approaches. Depending on the input, either simple-select$_1$ or simple-select$_2$ is always faster than our new data structures. All other evaluated select data structures are somewhere between pasta-flat and simple-select$_2$ when it comes to select query times. This comes without surprise, simple-select are highly tuned select-only data structures that also use at least 2.43 times as much memory as pasta-flat. However, pasta-flat$_{SIMD}$ is 16.5 % faster than cs-poppy and cs-poppy-fs, making it the fastest and most space-efficient uncompressed rank *and* select data structure.

5   Conclusion

With pasta-flat, we present a space-efficient rank and select data structure that is fast in practice. It requires the same space as the previously most space-efficient rank and select data structure cs-poppy and is between 8 % (rank) and 16.5 % (select) faster than cs-poppy. While there exist faster rank- and select-only data structures, they require significantly more memory and cannot easily answer both $select_0$ and $select_1$ queries, a necessity for many applications, e.g., wavelet trees [10] or succinct tree representations [14]. Pasta-flat can answer both (with a slowdown of up to 1.5 for one of the queries) without requiring additional memory.

Acknowledgements. This project has received funding from the European Research Council (ERC) under the European Union's Horizon 2020 research and innovation programme (grant agreement No. 882500).

A   Additional Experimental Results

A.1   Comparison of Our Implementations Only

For better readability, we only show our fastest select algorithm in the main part of this paper. Here, in Fig. 6, we compare our implementations with each other.

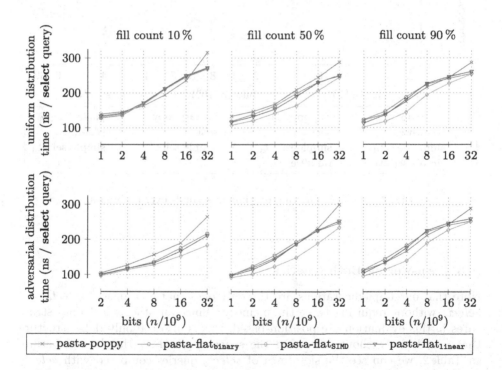

Fig. 6. Average select query time on all tested inputs of our implementations.

A.2   Construction Times

Finally, we want to report the construction times for the different data structures, see Fig. 7. While the query times of the data structures are definitely more
important, as we usually ask multiple queries but construct the data structure
only once, we discovered that there are huge differences. Overall, pasta-poppy,
pasta-flat$_{\text{SIMD}}$, and pasta-wide are the fastest to construct, followed by the data
structures contained in the SDSL. All other data structures (all variants of
simple-select and cs-poppy) are orders of magnitude slower to construct. The
difference in construction time is so big, that we can answer more than $2\,000\,000$
select queries using pasta-flat$_{\text{SIMD}}$ and simple-select$_2$ requires the same amount
of time when we include the construction time.

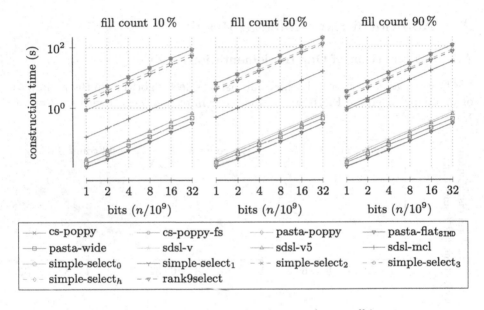

Fig. 7. Average construction time in seconds over all inputs.

A.3   Answering Both Queries

Note that our implementations are the only ones that can answer $select_0$ and
$select_1$ without requiring twice the memory.[2] Since all other select data structures in this evaluation are position-based, this is also the only data structure
that can do so without increasing the space required by the data structure.
In Table 2, we can see the slowdown of $select_0$ queries compared with $select_1$
queries, when the data structure is optimized for $select_1$ queries. Surprisingly,
the binary search approach has the least slowdown. Here, we have a trade-off:

---

[2] Per design, every rank-based select data structure can do so.

we can either double the required memory to answer both types of select queries or we can use a data structure that answers one of the queries 1.5 times slower.

Table 2. Slowdown of $select_0$ queries compared with $select_1$ queries, when the ranks of ones are stored for each block. Note that our implementations also support optimized $select_0$ queries.

Name	Slowdown (uniform)	Slowdown (adversarial)
pasta-poppy	1.632	1.559
pasta-flat$_{binary}$	**1.526**	**1.402**
pasta-flat$_{SIMD}$	1.817	1.813
pasta-flat$_{linear}$	1.651	1.604

References

1. Arroyuelo, D., Weitzman, M.: A hybrid compressed data structure supporting rank and select on bit sequences. In: SCCC, pp. 1–8. IEEE (2020). https://doi.org/10.1109/SCCC51225.2020.9281244
2. Beskers, K., Fischer, J.: High-order entropy compressed bit vectors with rank/select. Algorithms **7**(4), 608–620 (2014). https://doi.org/10.3390/a7040608
3. Boffa, A., Ferragina, P., Vinciguerra, G.: A learned approach to design compressed rank/select data structures. ACM Trans. Algorithms **18**(3), 1–28 (2022). https://doi.org/10.1145/3524060
4. Clark, D.R.: Compact pat trees. Ph.D. thesis, University of Waterloo (1997)
5. Clark, D.R., Munro, J.I.: Efficient suffix trees on secondary storage (extended abstract). In: SODA, pp. 383–391. ACM/SIAM (1996)
6. Dinklage, P., Ellert, J., Fischer, J., Kurpicz, F., Löbel, M.: Practical wavelet tree construction. ACM J. Exp. Algorithmics **26**, 1–67 (2021). https://doi.org/10.1145/3457197
7. Elias, P.: Efficient storage and retrieval by content and address of static files. J. ACM **21**(2), 246–260 (1974). https://doi.org/10.1145/321812.321820
8. Fano, R.M.: On the number of bits required to implement an associative memory (1971)
9. Ferragina, P., Giancarlo, R., Manzini, G.: The myriad virtues of wavelet trees. Inf. Comput. **207**(8), 849–866 (2009). https://doi.org/10.1016/j.ic.2008.12.010
10. Ferragina, P., Manzini, G.: Opportunistic data structures with applications. In: FOCS, pp. 390–398. IEEE Computer Society (2000). https://doi.org/10.1109/SFCS.2000.892127
11. Gagie, T., Navarro, G., Prezza, N.: Fully functional suffix trees and optimal text searching in BWT-runs bounded space. J. ACM **67**(1), 1–54 (2020). https://doi.org/10.1145/3375890
12. Gog, S., Beller, T., Moffat, A., Petri, M.: From theory to practice: plug and play with succinct data structures. In: Gudmundsson, J., Katajainen, J. (eds.) SEA 2014. LNCS, vol. 8504, pp. 326–337. Springer, Cham (2014). https://doi.org/10.1007/978-3-319-07959-2_28
13. González, R., Grabowski, S., Mäkinen, V., Navarro, G.: Practical implementation of rank and select queries. In: WEA, pp. 27–38 (2005)

14. Jacobson, G.: Space-efficient static trees and graphs. In: FOCS, pp. 549–554. IEEE Computer Society (1989). https://doi.org/10.1109/SFCS.1989.63533

15. Kim, D.K., Na, J.C., Kim, J.E., Park, K.: Efficient implementation of rank and select functions for succinct representation. In: Nikoletseas, S.E. (ed.) WEA 2005. LNCS, vol. 3503, pp. 315–327. Springer, Heidelberg (2005). https://doi.org/10.1007/11427186_28

16. Knuth, D.E.: The Art of Computer Programming, vol. III, 2nd Edn. Addison-Wesley (1998)

17. Makris, C.: Wavelet trees: a survey. Comput. Sci. Inf. Syst. **9**(2), 585–625 (2012). https://doi.org/10.2298/CSIS110606004M

18. Navarro, G.: Wavelet trees for all. J. Discrete Algorithms **25**, 2–20 (2014). https://doi.org/10.1016/j.jda.2013.07.004

19. Navarro, G., Providel, E.: Fast, small, simple rank/select on bitmaps. In: Klasing, R. (ed.) SEA 2012. LNCS, vol. 7276, pp. 295–306. Springer, Heidelberg (2012). https://doi.org/10.1007/978-3-642-30850-5_26

20. Pandey, P., Bender, M.A., Johnson, R.: A fast x86 implementation of select. arXiv preprint arXiv:1706.00990 (2017)

21. Pibiri, G.E., Kanda, S.: Rank/select queries over mutable bitmaps. Inf. Syst. **99**, 101756 (2021). https://doi.org/10.1016/j.is.2021.101756

22. Prezza, N.: A framework of dynamic data structures for string processing. In: SEA. LIPIcs, vol. 75, pp. 11:1–11:15. Schloss Dagstuhl - Leibniz-Zentrum für Informatik (2017). https://doi.org/10.4230/LIPIcs.SEA.2017.11

23. Patrascu, M.: Succincter. In: FOCS, pp. 305–313. IEEE Computer Society (2008). https://doi.org/10.1109/FOCS.2008.83

24. Raman, R., Raman, V., Satti, S.R.: Succinct indexable dictionaries with applications to encoding k-ary trees, prefix sums and multisets. ACM Trans. Algorithms **3**(4), 43 (2007). https://doi.org/10.1145/1290672.1290680

25. Vigna, S.: Broadword implementation of rank/select queries. In: McGeoch, C.C. (ed.) WEA 2008. LNCS, vol. 5038, pp. 154–168. Springer, Heidelberg (2008). https://doi.org/10.1007/978-3-540-68552-4_12

26. Zhou, D., Andersen, D.G., Kaminsky, M.: Space-efficient, high-performance rank and select structures on uncompressed bit sequences. In: Bonifaci, V., Demetrescu, C., Marchetti-Spaccamela, A. (eds.) SEA 2013. LNCS, vol. 7933, pp. 151–163. Springer, Heidelberg (2013). https://doi.org/10.1007/978-3-642-38527-8_15

Pattern Matching on Strings, Graphs, and Trees

Matching Patterns with Variables Under Edit Distance

Paweł Gawrychowski[1], Florin Manea[2], and Stefan Siemer[2]($\boxtimes$)

[1] Faculty of Mathematics and Computer Science,
University of Wrocław, Wrocław, Poland
gawry@cs.uni.wroc.pl
[2] Computer Science Department and CIDAS, Göttingen University,
Göttingen, Germany
{florin.manea,stefan.siemer}@cs.uni-goettingen.de

Abstract. A pattern α is a string of variables and terminal letters. We say that α matches a word w, consisting only of terminal letters, if w can be obtained by replacing the variables of α by terminal words. The matching problem, i.e., deciding whether a given pattern matches a given word, was heavily investigated: it is NP-complete in general, but can be solved efficiently for classes of patterns with restricted structure. If we are interested in what is the minimum Hamming distance between w and any word u obtained by replacing the variables of α by terminal words (so matching under Hamming distance), one can devise efficient algorithms and matching conditional lower bounds for the class of regular patterns (in which no variable occurs twice), as well as for classes of patterns where we allow unbounded repetitions of variables, but restrict the structure of the pattern, i.e., the way the occurrences of different variables can be interleaved. Moreover, under Hamming distance, if a variable occurs more than once and its occurrences can be interleaved arbitrarily with those of other variables, even if each of these occurs just once, the matching problem is intractable. In this paper, we consider the problem of matching patterns with variables under edit distance. We still obtain efficient algorithms and matching conditional lower bounds for the class of regular patterns, but show that the problem becomes, in this case, intractable already for unary patterns, consisting of repeated occurrences of a single variable interleaved with terminals.

Keywords: Pattern with variables · Matching · Edit distance

1 Introduction

A *pattern with variables* is a string consisting of *constant* or *terminal letters* from a finite alphabet Σ (e.g., a, b, c), and *variables* (e.g., x, y, x_1, x_2) from a potentially infinite set $\mathcal{X}$, with $\Sigma \cap \mathcal{X} = \emptyset$. In other words, a pattern α is an element of $PAT_\Sigma = (\mathcal{X} \cup \Sigma)^+$. A pattern α is mapped (by a function h called substitution) to a word by substituting the variables by arbitrary strings of

D. Arroyuelo and B. Poblete (Eds.): SPIRE 2022, LNCS 13617, pp. 275–289, 2022.
https://doi.org/10.1007/978-3-031-20643-6_20

terminal letters; as such, h simply maps the variables occurring in α to words over Σ. For example, xxbbbyy can be mapped to aaaabbbbb by the substitution h defined by $(x \to$ aa, $y \to$ b). In this framework, $h(\alpha)$ denotes the word obtained by substituting every occurrence of a variable x in α by $h(x)$ and leaving all the terminals unchanged. If a pattern α can be mapped to a string of terminals w, we say that α matches w; the problem of deciding whether there exists a substitution which maps a given pattern α to a given word w is called the *(exact) matching problem*.

Exact Matching Problem: Match
Input: A pattern α, with $|\alpha| = m$, a word w, with $|w| = n$.
Question: Is there a substitution h with $h(\alpha) = w$?

Match appears frequently in various areas of theoretical computer science, such as combinatorics on words (e.g., unavoidable patterns [30], string solving and the theory of word equations [29]), stringology (e.g., generalized function matching [1]), language theory (e.g., pattern languages [2], the theory of extended regular expressions with backreferences [6,17,20,21]), database theory (e.g., the theory of document spanners [12,18,19,25,38,39]), or algorithmic learning theory (e.g., the theory of descriptive patterns for finite sets of words [2,13,41]).

Match is NP-complete [2], in general. In fact, a detailed analysis [14–16,35,37,40] of the matching problem has provided a better understanding of the parameterized complexity of this problem, highlighting, in particular, several subclasses of patterns for which the matching problem is polynomial, when various structural parameters of patterns are bounded by constants. Prominent examples in this direction are patterns with a bounded number of repeated variables, patterns with bounded scope coincidence degree [35], patterns with bounded locality [10], or patterns with a bounded treewidth [35]. See [10,14,35] for efficient algorithms solving Match$_P$ restricted to (or, in other words, parameterized by) to such classes P of patterns. In general, each of the structural parameters defining such classes P is a number k characterizing in some way the structure of the patterns of the class P and the matching algorithms for the respective class of patterns runs in $O(n^{ck})$ for some constant c. Moreover, these restricted matching problems are usually shown to be $W[1]$-hard [11] w.r.t. the respective parameters.

In [22], the study of efficient matching algorithms for patterns with variables was extended to an approximate setting. More precisely, the problem of deciding, for a pattern α from a class of patterns P (defined by structural restrictions), a word w, and a non-negative integer Δ, whether there exists a substitution h such that the Hamming distance $d_{\mathrm{HAM}}(h(\alpha), w)$ between $h(\alpha)$ and w is at most Δ was investigated. The corresponding minimization problem of computing $d_{\mathrm{HAM}}(\alpha, w) = \min\{d_{\mathrm{HAM}}(h(\alpha), w) \mid h$ is a substitution of the variables of $\alpha\}$ was also considered. The main results of [22] were rectangular time algorithms and matching conditional lower bounds for the class of regular patterns Reg (which contain at most one occurrence of any variable). Moreover, polynomial time algorithms were obtained for unary patterns (also known as one-variable patterns, which consist in one or more occurrences of a single variable, potentially interleaved with terminal strings) or non-cross patterns (which consist in concatenations of unary patterns, whose variables are pairwise distinct). However, as soon as the patterns may contain multiple variables, whose occurrences are interleaved, the problems became NP-hard, even if only one of the variables occurs more than once. As such, unlike the case of exact matching, the approximate matching problem under Hamming distance is NP-hard even if some of the aforementioned parameters (number of repeated variables, scope coincidence degree, treewidth, but, interestingly, not locality) were upper bounded by small constants.

Our Contribution. In this paper, inspired by, e.g., [4,5,7–9,26,32,33] where various stringology patterns are considered in an approximate setting under *edit distance* [27,28], and as a natural extension of the results of [22], we consider the aforementioned approximate matching problems (parameterized by a class of patterns P) for the edit distance $d_{ED}(\cdot, \cdot)$, instead of Hamming Distance:

Approximate Matching Decision Problem for `MisMatch`$_P$
Input: A pattern $\alpha \in P$, with $|\alpha| = m$, a word w, with $|w| = n$, an integer $\Delta \le m$.
Question: Is $d_{ED}(\alpha, w) \le \Delta$?

Approximate Matching Minimisation Problem for `MinMisMatch`$_P$
Input: A pattern $\alpha \in P$, with $|\alpha| = m$, a word w, with $|w| = n$.
Question: Compute $d_{ED}(\alpha, w)$.

Our paper presents two main results, which allow us to paint a rather comprehensive picture of the approximate matching problem under edit distance.

Firstly, we consider the class of regular patterns, and show that `MisMatch`$_{Reg}$ and `MinMisMatch`$_{Reg}$ can be solved in $O(n\Delta)$ time (where, for `MinMisMatch`, Δ is the computed result); a matching conditional lower bound follows from the literature [3]. This is particularly interesting because the problem of computing $d_{ED}(\alpha, w)$ for $\alpha = w_0 x_1 w_1 \ldots x_k w_k$ can be seen as the problem of computing the minimal edit distance between any string in which $w_1, \ldots, w_k$ occur, without overlaps, in this exact order and the word w.

Secondly, we show that, unlike the case of matching under Hamming distance, `MisMatch`$_P$ becomes $W[1]$-hard already for P being the class of unary patterns, with respect to the number of occurrences of the single variable. So, interestingly, the problem of matching patterns with variables under edit distance is computationally hard for all the classes (that we are aware of) of structurally restricted patterns with polynomial exact matching problem, as soon as at least one variable is allowed to occur an unbounded number of times.

To complement the results presented in this paper, we note that, for the classes of patterns considered in [10,14,22,35], which admit polynomial-time exact matching algorithms, one can straightforwardly adapt those algorithms to work in polynomial time in the case of matching under edit distance, when a constant upper bound k_1 on the number of occurrences of each variable exists. The complexity of these algorithms is usually $O(n^{f(k_1,k_2)})$, for a polynomial function f and for k_2 being a constant upper bound for the value of the structural parameter considered when defining these classes (locality, scope coincidence degree, treewidth, etc.). If no restriction is imposed on the structure of the pattern, Match (and, as such, the matching under both Hamming and edit distances) is NP-hard even if there are at most two occurrences of each variable [15].

2   Preliminaries

Some basic notations and definitions regarding strings and patterns with variables were already given in the introduction and, for more details, we also refer to [22,31]. We only recall here some further notations. The set of all patterns, over all terminal-alphabets Σ, is denoted $PAT = \bigcup_\Sigma PAT_\Sigma$. Given a word or pattern γ, we denote by $\mathtt{alph}(\gamma) = B$ the smallest set (w.r.t. inclusion) $B \subseteq \Sigma$ and by $\mathtt{var}(\gamma) = Y$ the smallest set $Y \subseteq \mathcal{X}$ such that $\gamma \in (B \cup Y)^\star$. For any symbol $t \in \Sigma \cup \mathcal{X}$ and $\alpha \in PAT_\Sigma$, $|\alpha|_t$ denotes the total number of occurrences of the symbol t in α. For a pattern $\alpha = w_0 x_1 w_1 \dots x_k w_k$, we denote by $\mathtt{term}(\alpha) = w_0 w_1 \dots w_k$ the projection of α on the terminal alphabet Σ.

For words $u, w \in \Sigma^\star$, the *edit distance* [27,28] between u and w is defined as the minimal number $d_{\mathtt{ED}}(u, w)$ of letter insertions, letter deletions, and letter to letter substitutions which one has to apply to u to obtain w.

We recall some basic facts about the edit distance. Assume that u is transformed into w by a sequence of edits γ (i.e., u is aligned to w by γ). We can assume without losing generality that the edits in γ are ordered left to right with respect to the position of u where they are applied. Then, for each factorization $u = u_1 \dots u_k$ of u, there exists a factorization $w = w_1 \dots w_k$ of w such that w_i is obtained from u_i when applying the edits of γ which correspond to the positions of u_i, for $i \in \{1, \dots, k\}$. Note that this factorization of w is not unique: we assume that the insertions applied at the beginning of u correspond to positions of u_1, the insertions applied at the end of u correspond to positions of u_k, but the insertions applied between u_{i-1} and u_i can be split arbitrarily in two parts: when considering them in the order in which they occur in γ (so left to right w.r.t. the positions of u where they are applied) we assume to first have a (possibly empty) set of insertions which correspond to positions of u_{i-1} and then a (possibly empty) set of insertions which correspond to positions of u_i. On the other hand, if $w = w_1 w_2$, we can uniquely identify the shortest prefix u_1 (respectively, the longest prefix u_1') of u from which, when applying the edits of γ we obtain the prefix w_1 of w.

Now, for a pattern α and a word w, we can define the edit distance between α and w as $d_{\mathtt{ED}}(\alpha, w) = \min\{d_{\mathtt{ED}}(h(\alpha), w) \mid h$ is a substitution of the variables of $\alpha\}$. It is worth noting that $d_{\mathtt{ED}}(\alpha, w) \leq |w| + |\mathtt{term}(\alpha)|$.

With these definitions, we can consider the two pattern matching problems for families of patterns $P \subseteq PAT$, as already defined in the introduction. In the first problem MisMatch$_P$, which extends Match$_P$, we allow for a certain edit distance Δ between the image $h(\alpha)$ of α under a substitution h and the target word w instead of searching for an exact matching. In the second problem, MinMisMatch_P, we are interested in finding the substitution h for which the edit distance between $h(\alpha)$ and the word w is minimal, over all possible choices of h.

As a remark, based on our general comments regarding the edit distance, the following theorem follows.

Theorem 1. MisMatch_{PAT} and MinMisMatch_{PAT} can be solved in $O(n^{2k_2+k_1})$ time, where k_1 is the maximum number of occurrences of any variable in the input pattern α and k_2 is the total number of occurrences of variables in α.

As mentioned in the Introduction, the result of the previous Theorem can be improved if we consider the two problems for classes of patterns with restricted structure, where we obtain algorithms whose complexity depends on the structural parameter associated to that class, rather than the total number of occurrences of variables.

3 Our Results

The first main result of our paper is about the class of regular patterns. A pattern α over the terminal alphabet Σ is regular if $\alpha = w_0(\Pi_{i=1}^k x_i w_i)$ where, for $i \in \{1, \ldots, k\}$, $w_i \in \Sigma^*$ and x_i is a variable, and $x_i \neq x_j$ for all $i \neq j$. The class of regular patterns is denoted by Reg. We can show the following theorem.

Theorem 2. $\text{MisMatch}_{\text{Reg}}$ can be solved in $O(n\Delta)$ time. For an accepted instance w, α, Δ of $\text{MisMatch}_{\text{Reg}}$ we also compute $d_{\text{ED}}(\alpha, w)$ (which is at most Δ).

Proof. **Preliminaries and setting.** We begin with an observation. For $\alpha = w_0(\Pi_{i=1}^k x_i w_i)$, we can assume w.l.o.g. that $w_i \in \Sigma^+$ for all $i \leq k$ as otherwise we would have neighboring variables that could be replaced by a single variable; thus, $k \leq |\text{term}(\alpha)|$. To avoid some corner cases, we can assume w.l.o.g. that α and w start with the same terminal symbol (this can be achieved by adding a fresh letter \$ in front of both α and w). While not fundamental, these simplifications make the exposure of the following algorithm easier to follow.

Before starting the presentation of the algorithm, we note that a solution for $\text{MisMatch}_{\text{Reg}}$ with distance $\Delta = 0$ is a solution to $\text{Match}_{\text{Reg}}$ and can be solved in $\mathcal{O}(n)$ by a greedy approach (as shown, for instance, in [14]). Further, the special case $x_1 w_1 x_2$ can be solved by an algorithm due to Landau and Vishkin [26] in $\mathcal{O}(n\Delta)$ time. In the following, we are going to achieve the same complexity for the general case of $\text{MisMatch}_{\text{Reg}}$ by extending the ideas of this algorithm to accommodate the existence of an unbounded number of pairwise-distinct variables.

One important idea which we use in the context of computing the edit distance between an arbitrary regular pattern and a word is to interpret each regular variable as an arbitrary amount of "free" insertions on that position, where "free" means that they will not be counted as part of the actual distance (in other words, they do not increase this distance). Indeed, we can see that the factor which substitutes a variable should always be equal to the factor to which it is aligned (after all the edits are performed) from the target word, hence does not add anything to the overall distance (and, therefore, it is "free"). As such, this factor can be seen as being obtained via an arbitrary amount of letter insertions. Now, using this observation, it is easier to design an $O(nm)$-time algorithm which computes the edit distance between the terminal words $\beta = \texttt{term}(\alpha)$ (instead of the pattern α) and w with the additional property that, for the positions $F_g = |(\Pi_{i=0}^{g} w_i)|$ for $0 \leq g \leq k-1$, we have that the insertions done between positions $\beta[F_g]$ and $\beta[F_g + 1]$ when editing β to obtain w do not count towards the total edit distance between β and w. For simplicity, we denote the set $\{F_g | 0 \leq g \leq k-1\}$ by F, we set $F_k = +\infty$, and note that $|\beta| = m - k$ (so $|\beta| \in \Theta(m)$).

The description of our algorithm is done in two phases. We first explain how $\texttt{MisMatch}_{\texttt{Reg}}$ can be solved by dynamic programming in $O(nm)$ time. Then, we refine this approach to an algorithm which fulfills the statement of the theorem.

When presenting our algorithms, we refer to an alignment of prefixes $\beta[1:j]$ of β and $w[1:\ell]$ of w, which simply means editing $\beta[1:j]$ to obtain $w[1:\ell]$.

First phase: a classical dynamic programming solution. We define the $(|\beta| + 1) \times (n+1)$ matrix $D[\cdot][\cdot]$, where $D[j][\ell]$ is the edit distance between the prefixes $\beta[1:j]$, with $0 \leq j \leq |\beta|$, and $w[1:\ell]$, with $0 \leq \ell \leq n$, with the additional important property that the insertions done between positions $\beta[F_g]$ and $\beta[F_g + 1]$, for $F_g \leq j$, are not counted in this distance (they correspond to variables in the pattern α). As soon as this matrix is computed, we can retrieve the edit distance between α and w from the element $D[m-k][n]$. Clearly, now the instance (α, w, Δ) of $\texttt{MisMatch}_{\texttt{Reg}}$ is answered positively if and only if $D[m-k][n] \leq \Delta$. So, let us focus on an algorithm computing this matrix.

The elements of the matrix $D[\cdot][\cdot]$ can be computed by dynamic programming in $O(mn)$ time (see full version [23]). Moreover, by tracing back the computation of $D[m-k][n]$, we obtain a path consisting in elements of the matrix, leading from $D[0][0]$ to $D[m-k][n]$, which encodes the edits needed to transform β into w. An edge between $D[j-1][\ell]$ and $D[j][\ell]$ corresponds to the deletion of $\beta[j]$; and edge between $D[j-1][\ell-1]$ and $D[j][\ell]$ corresponds to a substitution of $\beta[j]$ by $w[\ell]$, or to the case where $\beta[j]$ and $w[\ell]$ are left unchanged, and will be aligned in the end. Moreover, an edge between $D[j][\ell-1]$ and $D[j][\ell]$ corresponds to an insertion of $w[\ell]$ after position j in β; this can be a free insertion too (and part of the image of a variable of α), but only when $j \in F$. This concludes the first phase of our proof.

Second phase: a succinct representation and more efficient computation of the dynamic programming table. In the second phase of our proof, we will focus on how to solve $\texttt{MisMatch}_{\texttt{Reg}}$ more efficiently. The idea is to avoid computing all the elements of the matrix $D[\cdot][\cdot]$, and compute, instead, only the relevant elements of this matrix, following the ideas of the algorithm by Landau and Vishkin [26]. The main difference between the setting of that algorithm (which can be directly used to compute the edit distance between two terminal words or between a word w and a pattern α of the form xuy, xu, or uy, where

x and y are variables and u is a terminal word) and ours is that, in our case, the diagonals of the matrix $D[\cdot][\cdot]$ are not non-decreasing (when traversed in increasing order of the rows intersected by the respective diagonal), as we now also have free insertions which may occur at various positions in β (not only at the beginning and end). This is a significant complication, which we will address next.

The main idea of the optimization done in this second phase is that we could actually compute and represent the matrix $D[\cdot][\cdot]$ more succinctly, by only computing and keeping track of at most Δ relevant elements on each diagonal of this matrix, where relevant means that we cannot explicitly rule out the existence of a path leading from $D[0][0]$ to $D[m-k][n]$ which goes through that element.

For the clarity of exposure, we recall that the diagonal d of the matrix $D[\cdot][\cdot]$ is defined as the array of elements $D[j][\ell]$ where $\ell - j = d$ (ordered in increasing order w.r.t. the first component j), where $-|\beta| + 1 \leq d \leq n$. Very importantly, for a diagonal d, we have that if $D[j][j+d] \leq D[j+1][j+1+d]$ then $D[j+1][j+1+d] - D[j][j+d] \leq 1$; however, it might also be the case that $D[j][j+d] > D[j+1][j+1+d]$, when $D[j+1][j+1+d]$ is obtained from $D[j+1][j+d]$ by a free insertion.

Analysis of the diagonals, definition of $M_d[\delta]$ and its usage. Now, for each diagonal d, with $-|\beta| + 1 \leq d \leq n$, and $\delta \leq \Delta$, we define $M_d[\delta] = \max\{j \mid D[j][j+d] = \delta$, and $D[j'][j'+d] > \delta$ for all $j' > j\}$ (by convention, $M_d[\delta] = -\infty$, if $\{j \mid D[j][j+d] = \delta$, and $D[j'][j'+d] > \delta$ for all $j' > j\} = \emptyset$). That is, $M_d[\delta]$ is the greatest row where we find the value δ on the diagonal d and, moreover, all the elements appearing on greater rows on that diagonal are strictly greater than δ (or $M_d[\delta] = -\infty$ if such a row does not exist).

Note that if a value δ appears on diagonal d and there exists some j' such that $D[j][j+d] \geq \delta$ for all $j \geq j'$, then, due to the only relations which may occur between two consecutive elements of d, we have that $M_d[\delta] \neq -\infty$. In particular, if a value δ appears on diagonal d then $M_d[\delta] \neq -\infty$ if and only if $D[|\beta|][|\beta|+d] \geq \delta$. Consequently, if there exists $k > 0$ such that $M_d[\delta - k] = |\beta|$ then $M_d[\delta] = -\infty$.

In general, all values $M_d[\delta]$ which are equal to $-\infty$ are not relevant to our computation. To understand which other values $M_d[\delta]$ are not relevant for our algorithm, we note that if there exist some $k > 0$ and $s \geq 0$ such that $M_{d+s}[\delta - k] = |\beta|$ then it is not needed to compute $M_{d-g}[\delta + h]$, for any $g, h \geq 0$, at all, as any path going from $D[0][0]$ to $D[|\beta|][n]$, which corresponds to an optimal sequence of edits, does not go through $D[M_{d-g}[\delta + h]][M_d[\delta + h] + d]$. If $s = 0$, then it is already clear that $M_d[\delta] = -\infty$, and we do not need to compute it. If $s \geq 1$, it is enough to show our claim for $h = 0$ and $g = 0$. Indeed, assume that the optimal sequence of edits transforming β into w corresponds to a path from $D[0][0]$ to $D[|\beta|][n]$ going through $D[M_d[\delta]][M_d[\delta] + d]$. By the fact that $M_d[\delta]$ is the largest j for which $D[j][j+d] \leq \delta$, we get that this path would have to intersect, after going through $D[M_d[\delta]][M_d[\delta] + d]$, the path from $D[0][0]$ to $D[M_{d+s}[\delta - k]][M_{d+s}[\delta - k] + d + s] = D[|\beta|][|\beta| + d + s]$ (which goes only through elements $\leq \delta - k$). As $k > 0$, this is a contradiction, as the path from $D[0][0]$

to $D[|\beta|][n]$ going through $D[M_d[\delta]][M_d[\delta] + d]$ goes only through elements $\geq \delta$ after going through $D[M_d[\delta]][M_d[\delta] + d]$. So, $M_d[\delta]$ is not relevant if there exist $k > 0$ and $s > 0$ such that $M_{d+s}[\delta - k] = m$.

Once all relevant values $M_d[\delta]$ are computed, for d diagonal and $\delta \leq \Delta$, we simply have to check if $M_{n-|\beta|}[\delta] = |\beta|$ (i.e., $D[|\beta|][n] = \delta$) for some $\delta \leq \Delta$. So, we can focus, from now on, on how to compute the relevant elements $M_d[\delta]$ efficiently. In particular, all these elements are not equal to $-\infty$.

Towards an algorithm: understanding the relations between elements on consecutive diagonals. Let us now understand under which conditions $D[j][\ell] = \delta$ holds, as this is useful to compute $M_d[\delta]$. In general, this means that there exists a path leading from $D[0][0]$ to $D[j][\ell]$ consisting only in elements with value $\leq \delta$, and which ends with a series of edges belonging to the diagonal $d = \ell - j$, that correspond to substitutions or to letters being left unchanged. In particular, if all the edges connecting $D[j'][j' + d]$ and $D[j][\ell]$ on this path correspond to unchanged letters, then $\beta[j' : j]$ is a common prefix of $\beta[j' : |\beta|]$ and $w[j' + d : n]$. Looking more into details, there are several cases when $D[j][\ell] = \delta$.

If $j \notin F$ and $\beta[j] \neq w[\ell]$, then $D[j - 1][\ell - 1] \geq \delta - 1$ and $D[j - 1][\ell] \geq \delta - 1$ and $D[j][\ell - 1] \geq \delta - 1$ and at least one of the previous inequalities is an equality (i.e., one of the following must hold: $D[j][\ell - 1] = \delta - 1$ or $D[j - 1][\ell - 1] = \delta - 1$ or $D[j - 1][\ell] = \delta - 1$). If $j \notin F$ and $\beta[j] = w[\ell]$, then $D[j - 1][\ell - 1] \geq \delta$ and $D[j - 1][\ell] \geq \delta - 1$ and $D[j][\ell - 1] \geq \delta - 1$ and at least one of the previous inequalities is an equality.

If $j \in F$ and $\beta[j] \neq w[\ell]$, then $D[j - 1][\ell - 1] \geq \delta - 1$ and $D[j - 1][\ell] \geq \delta - 1$ and $D[j][\ell - 1] \geq \delta$ and at least one of the previous inequalities is an equality. If $j \in F$ and $\beta[j] = w[\ell]$ then $D[j - 1][\ell - 1] \geq \delta$ and $D[j - 1][\ell] \geq \delta - 1$ and $D[j][\ell - 1] \geq \delta$ and at least one of the previous inequalities is an equality.

Moving forward, assume now that $M_d[\delta] = j \neq -\infty$. This means that $D[j][\ell] = \delta$, and $D[j''][j'' + d] > \delta$ for all $j'' > j$. By the observations above, there exists $j' \leq j$ such that $D[j'][j' + d] = \delta$ and the longest common prefix of $\beta[j' : |\beta|]$ and $w[j' + d : n]$ has length $j - j' + 1$, i.e., it equals $\beta[j' : j]$. The last part of this statement means that once we have aligned $\beta[1 : j']$ to $w[1 : j' + d]$, we can extend this alignment to an alignment of $\beta[1 : j]$ to $w[1 : j + d]$ by simply leaving the symbols of $\beta[j' + 1 : j]$ unchanged.

Let us see now what this means for the elements of diagonals d, $d + 1$, and $d - 1$.

Firstly, we consider the diagonal d. Here we have that $j' \geq M_d[\delta - 1] + 1$. Note that if $\delta - 1$ appears on diagonal d then $M_d[\delta - 1] \neq -\infty$.

Secondly, we consider the diagonal $d + 1$. Here, for all rows ℓ with $j' \leq \ell \leq j$, we have that $D[\ell - 1][\ell + d] \geq \delta - 1$ and $D[j'' - 1][j'' + d] > \delta - 1$, for all j'' with $|\beta| \geq j'' > j$. Therefore, if $\delta - 1$ appears on diagonal $d + 1$, either $D[m][m + d + 1] \leq d - 1$ or $M_{d+1}[\delta - 1] \neq -\infty$ and $M_d[\delta - 1] + 1 \leq j$.

Finally, we consider the diagonal $d - 1$. Here, for all rows ℓ with $j' \leq \ell \leq j$, we have that $D[\ell][\ell + d - 1] \geq \delta - 1$ and $D[j''][j'' + d - 1] \geq \delta$, for all j'' with $m \geq j'' > j$. Thus, either all elements on the diagonal $d - 1$ are $\geq \delta$, or $\delta - 1$

occurs on diagonal $d-1$ and $M_{d-1}[\delta-1] \neq -\infty$. In the second case, when $M_{d-1}[\delta-1] \neq -\infty$, we have that $j \geq M_{d-1}[\delta-1]$ as, otherwise, we would have that $D[M_{d-1}[\delta-1]][M_{d-1}[\delta-1]+d] \leq \delta$ and $M_{d-1}[\delta-1] > j$, a contradiction.

Still on diagonal $d-1$, if δ occurs on it, then $M_d[\delta] \neq -\infty$ holds. So, for $g \leq k-1$ with $F_g \leq M_{d-1}[\delta] < F_{g+1}$, we have that $F_g \leq M_d[\delta]$. Indeed, otherwise we would have two possibilities. If the path connecting $D[0][0]$ to $D[M_{d-1}[\delta]][M_{d-1}[\delta]+d-1]$ via elements $\leq d$ intersects row F_g on $D[F_g][F_g+d']$ for some $d' \leq d$, then $D[F_g][F_g+d] \leq D[F_g][F_g+d'] \leq \delta$ and $F_g > j$, a contradiction. If the path connecting $D[0][0]$ to $D[M_{d-1}[\delta]][M_{d-1}[\delta]+d-1]$ via elements $\leq d$ intersects row F_g on $D[F_g][F_g+d']$ for some $d' > d$, then the respective path will also intersect diagonal d on a row $> j$ before reaching $M_{d-1}[\delta]$, a contradiction with the fact that j is the last row on diagonal d where we have an element $\leq \delta$.

So, for $M_d[\delta]$ to be relevant, we must have $D[|\beta|][|\beta|+d+1] \geq \delta$ (so there exists no $k > 0$ such that $M_{d+1}[\delta-k] = |\beta|$). In this case, if $M_d[\delta] = j$, then the following holds. The path (via elements $\leq d$) from $D[0][0]$ to $D[j][j+d]$ goes through an element $D[g][g+d'] = \delta-1$. If the last such element on the respective path is on diagonal d, then it must be $M_d[\delta-1]$. If it is on diagonal $d-1$, then either $g = M_{d-1}[\delta-1]$ (and then the path moves on diagonal d via an edge corresponding to an insertion) or $g < M_{d-1}[\delta-1]$ (and then the path moves on diagonal d via an edge corresponding to an insertion); in this second case, we could replace the considered path by a path connecting $D[0][0]$ to $D[M_{d-1}[\delta-1]][M_{d-1}[\delta-1]+d-1]$ (via elements $\leq \delta-1$), which then moves on diagonal d via an edge corresponding to an insertion, and continues along that diagonal (with edges corresponding to letters left unchanged). If $D[g][g+d']$ is on diagonal $d+1$ (i.e., $d' = d+1$) then, just like in the previous case, we can simply consider the path connecting $D[0][0]$ to $D[M_{d+1}[\delta-1]][M_{d+1}[\delta-1]+d+1]$ (via elements $\leq \delta-1$), which then moves on diagonal d via an edge corresponding to a deletion, and then continues along diagonal d (with edges corresponding to letters left unchanged). If $D[g][g+d']$ is on none of the diagonals $d-1, d, d+1$ then we reach diagonal d by edges corresponding to free insertions from some diagonal $d'' < d$. The respective path also intersects diagonal $d-1$ (when coming from d'' to d by free insertions), so diagonal $d-1$ contains δ and $M_{d-1}[\delta] \neq \infty$, and we might simply consider as path between $D[0][0]$ and $D[j][j+d]$ the path reaching diagonal $d-1$ on position $D[F_g][F_g+d-1]$ (via elements $\leq \delta$), where $F_g \leq M_{d-1}[\delta] < F_{g+1}$, which then moves on diagonal d by an edge corresponding to a free insertion, and then continues along d (with edges corresponding to letters left unchanged, as F_g is greater or equal to the row where the initial path intersected diagonal d). This analysis covers all possible cases.

Computing $M_d[\delta]$. Therefore, if $M_d[\delta]$ is relevant (and, as such, $M_d[\delta] \neq -\infty$), then $M_d[\delta]$ can be computed as follows. Let g be such that $F_g \leq M_{d-1}[\delta] < F_{g+1}$ (and $g = -1$ and $F_g = -\infty$ if $M_{d-1}[\delta] = -\infty$). Let $H = \max\{M_{d-1}[\delta-1], F_g, M_d[\delta-1]+1, M_{d+1}[\delta-1]+1\}$ (as explained, in the case we are discussing, at least one of these values is not $-\infty$). Then we have that $j \geq H$ and the longest common prefix of $\beta[H+1:|\beta|]$ and $w[H+d+1:n]$ is exactly $\beta[H+1:j]$ (or we could increase j). So, to compute $j = M_d[\delta]$, we compute H and then we compute the longest common prefix $\beta[H + 1 : j]$ of $\beta[H + 1 : |\beta|]$ and $w[H + d + 1 : n]$.

In general, $M_d[\delta]$ is not relevant either because there exists some $s \geq 0$ and $\delta' < \delta$ such that $M_{d+s}[\delta'] = |\beta|$ or because all elements of diagonal d are strictly greater than δ. In the second case, we note that all values $M_{d-1}[\delta - 1]$, F_g, $M_d[\delta - 1]$, and $M_{d+1}[\delta - 1]$ must be $-\infty$ (as otherwise the diagonal d would contain an element equal to δ), so our computation of $M_d[\delta]$ returns $-\infty$ (which is correct).

Now, based on these observations, we can see a way to compute the relevant values $M_d[\delta]$, for $-|\beta| \leq d \leq n$ and $\delta \leq \Delta$ (without computing the matrix D).

We first construct the word β and longest common prefix data structures for the word βw, allowing us to compute $\mathtt{LCP}(\beta[h : |\beta|], w[h + d : n])$, the length of the longest common prefix of $\beta[h : |\beta|]$ and $w[h + d : n]$ for all h and d.

Then, we will compute the values of $M_d[0]$ for all diagonals d. Basically, we need to identify, if it exists, a path from $D[0][0]$ to $D[M_d[0]][M_d[0] + d]$ which consists only of edges corresponding to letters left unchanged, or to free insertions. By an analysis similar to the one done above, we can easily show that $M_0[0]$ is $\mathtt{LCP}(\beta[1 : |\beta|], w[1 : n])$ (which is ≥ 1, by our assumptions). Further, $M[d][0] = -\infty$ for $d < 0$ and, for $d \geq 0$, $M_d[0] = F_g + \mathtt{LCP}(\beta[F_g + 1 : |\beta|], w[F_g + 1 + d : n])$, where $F_g \in F$ is such that $F_g \leq M_{d-1}[0] < F_{g+1}$ ($M_d[0] = -\infty$ if such an element F_g does not exist).

Further, for δ from 1 to Δ we compute all the values $M_d[\delta]$, in order for d from $-|\beta| + 1$ to n. We first compute the largest diagonal d' such that $M_{d'}[\delta - k] = |\beta|$, for some $k > 0$. We will only compute $M_d[\delta]$, for d from $d' + 1$ to n. For each such diagonal d, we compute g such that $F_g \leq M_{d-1}[\delta] < F_{g+1}$ and $H = \max\{M_{d-1}[\delta - 1], F_g, M_d[\delta - 1] + 1, M_{d+1}[\delta - 1] + 1\}$. Then we set $M_d[\delta]$ to be $H + \mathtt{LCP}(\beta[H + 1 : |\beta|], w[H + d + 1 : n]) - 1$.

Conclusions. This algorithm, which computes all relevant values $M_d[\delta]$, can be implemented in $O((n+m)\Delta)$ time, as discussed in the full version [23] (where also its pseudocode is given). As explained before, this allows us to solve $\mathtt{MisMatch_{Reg}}$ for the input (α, w, Δ). Moreover, if the instance can be answered positively, the value δ for which $M_{n-|\beta|}[\delta] = |\beta|$ equals $d_{\mathrm{ED}}(\alpha, w)$. $\qquad\square$

The following result now follows.

Theorem 3. $\mathtt{MinMisMatch_{Reg}}$ *can be solved in* $O(n\Phi)$ *time, where* $\Phi = d_{\mathrm{ED}}(\alpha, w)$.

The upper bounds reported in Theorems 2 and 3 are complemented by the following conditional lower bound, known from the literature [3, Thm. 3] (see full version [23]).

Theorem 4. $\mathtt{MisMatch_{Reg}}$ *can not be solved in time* $\mathcal{O}(|w|^h \Delta^g)$ *(or* $\mathcal{O}(|w|^h |\alpha|^g)$*) where* $h + g = 2 - \epsilon$ *with* $\epsilon > 0$, *unless the Orthogonal Vectors Conjecture fails.*

It is worth noting that the lower bound from Theorem 4 already holds for very restricted regular patterns, i.e., for $\alpha = xuy$, where u is a string of terminals and

x and y are variables. Interestingly, a similar lower bound (for such restricted patterns) does not hold in the case of the Hamming distance, covered in [22].

Our second main result addresses another class of restricted patterns. To this end, we consider the class of unary (or one-variable) patterns 1Var, which is defined as follows: $\alpha \in$ 1Var if there exists $x \in X$ such that $\text{var}(\alpha) = \{x\}$. An example of unary pattern is $\alpha_1 = \text{ab}x\text{ab}x\text{xbaab}$.

We can show the following theorem.

Theorem 5. MisMatch$_{1\text{Var}}$ *is* $W[1]$-*hard w.r.t. the number of occurrences of the single variable* x *of the input pattern* α.

Proof (Sketch). We begin by recalling the following problem:

Median String: MS
Input: k strings $w_1, \ldots, w_k \in \sigma^*$ and an integer Δ.
Question: Does there exist a string s such that $\sum_{i=1}^{k} d_{\text{ED}}(w_i, s) \leq \Delta$?
            (The string s for which $\sum_{i=1}^{k} d_{\text{ED}}(w_i, s)$ is minimum is called the
            median string of the strings $\{w_1, \ldots, w_k\}$.)

Without loss of generality, we can assume that $\Delta \leq \sum_{i=1}^{k} |w_i|$ as, otherwise, the answer is clearly yes (for instance, for $s = \varepsilon$ we have that $\sum_{i=1}^{k} d_{\text{ED}}(w_i, \varepsilon) \leq \sum_{i=1}^{k} |w_i|$). Similarly, we can assume that $|s| \leq \Delta + \max\{|w_i| \mid i \in \{1, \ldots, k\}\}$. In [34] it was shown that MS is NP-complete even for binary input strings and $W[1]$-hard with respect to the parameter k, the number of input strings.

We will reduce now MS to MisMatch$_{1\text{Var}}$, such that an instance of MS with k input strings is mapped to an instance of MisMatch$_{1\text{Var}}$ with exactly k occurrences of the variable x (the single variable occurring in the pattern).

Thus, we consider an instance of MS which consists in the k binary strings $w_1, \ldots, w_k \in \{0,1\}^*$ and the integer $\Delta \leq \sum_{i=1}^{k} |w_i|$. The instance of MisMatch$_{1\text{Var}}$ which we construct consists of a word w and a pattern α, such that α contains exactly k occurrences of a variable x, and both strings are of polynomial size w.r.t. the size of the MS-instance. Moreover, the bound on the $d_{\text{ED}}(\alpha, w)$ defined in this instance of MisMatch$_{1\text{Var}}$ equals Δ. That is, if there exists a solution for the MS-instance such that $\sum_{i=1}^{k} d_{\text{ED}}(w_i, s) \leq \Delta$, then, and only then, we should be able to find a solution of the MisMatch$_{1\text{Var}}$-instance with $d_{\text{ED}}(\alpha, w) \leq \Delta$. The construction of the MisMatch$_{1\text{Var}}$ instance is realized in such a way that the word w encodes the k input strings for MS, conveniently separated by some long strings over two fresh symbols \$, #, while α can be obtained from w by simply replacing each of the words w_i by a single occurrence of the variable x. Intuitively, in this way, for $d_{\text{ED}}(\alpha, w)$ to be minimal, x should be mapped to the median string of $\{w_1, \ldots, w_k\}$. In this proof sketch, we just define the reduction. The proof of its correctness is given in the long version of this paper [23].

For the strings $w_1, \ldots w_k \in \{0,1\}^*$, let $S = 6(\sum_{i=1}^{k} |w_i|)$; clearly, $S \geq 6\Delta$. Let $w = w_1(\$^S \#^S)^S w_2(\$^S \#^S)^S \ldots w_k(\$^S \#^S)^S$ and $\alpha = \left(x(\$^S \#^S)^S\right)^k$.

The constructed instance of MisMatch$_{1\text{Var}}$ (i.e., α, w, Δ) is of polynomial size w.r.t. the size of the MS-instance (i.e., $\{w_1, \ldots, w_k\}, \Delta$). Therefore, it (and our entire reduction) can be computed in polynomial time. Moreover, we can show that the instance (w, α, Δ) of $\texttt{MisMatch}_{\texttt{1Var}}$ is answered positively if and only if the original instance of MS is answered positively. Finally, as the number of occurrences of the variable x blocks in α is k, where k is the number of input strings in the instance of MS, and MS is $W[1]$-hard with respect to this parameter, it follows that $\texttt{MisMatch}_{\texttt{1Var}}$ is also $W[1]$-hard when the number of occurrences of the variable x in α is considered as parameter. The statement follows.   □

A simple corollary of Theorem 1 is the following:

Theorem 6. $\texttt{MisMatch}_{\texttt{1Var}}$ *and* $\texttt{MinMisMatch}_{\texttt{1Var}}$ *can be solved in* $O(n^{3|\alpha|_x})$ *time, where x is the single variable occurring in α.*

Clearly, finding a polynomial time algorithm for $\texttt{MisMatch}_{\texttt{1Var}}$, for which the degree of the polynomial does not depend on $|\alpha|_x$, would be ideal. Such an algorithm would be, however, an FPT-algorithm for $\texttt{MisMatch}_{\texttt{1Var}}$, parameterized by $|\alpha|_x$, and, by Theorem 5 and common parameterized complexity assumptions, the existence of such an algorithm is unlikely. This makes the straightforward result reported in Theorem 6 relevant, to a certain extent.

4   Conclusion

Our results regarding the problem MisMatch for various classes of patterns are summarized in Table 1, which highlights the differences to the case of exact matching and to the case of approximate matching under Hamming distance.

Note that the results reported in the first row of the rightmost column of this table are based on Theorem 2 (the upper bound) and Theorem 4 (the lower bound). The rest of the cells of that rightmost column are all consequences of the result of Theorem 5. Indeed, the classes of patterns covered in this table, which are presented in detail in [22], are defined based on a common idea. In the pattern α, we identify for each variable x the x-blocks: maximal factors of α (w.r.t. length) which contain only the variable x and terminals, and start and end with x. Then, classes of patterns are defined based on the way the blocks defined for all variables occurring in α are interleaved. However, in the patterns of all these classes, there may exist at least one variable which occurs an unbounded number of times, i.e., they all include the class of unary patterns. Therefore, the hardness result proved for unary patterns carries over and, as the structural parameters used to define those classes do not take into account the overall number of occurrences of a variable, but rather the number of blocks for the variables (or the way they are interleaved), we obtain NP-hardness for MisMatch for that class, even if the structural parameters are trivial.

While our results, together with those reported in [22], seem to completely characterize the complexity of MisMatch and MinMisMatch under both Hamming and edit distances, there are still some directions for future work. Firstly, in [24] the fine-grained complexity of computing the median string under edit distance for k input strings is discussed. Their main result, a lower bound, was only

Table 1. Our new results are listed in column 4. The results overviewed in column 3 were all shown in [22]. We assume $|w| = n$ and $|\alpha| = m$.

Class	Match(w,α)	MisMatch(w,α,Δ) for $d_{\text{HAM}}(\cdot,\cdot)$	MisMatch(w,α,Δ) for $d_{\text{ED}}(\cdot,\cdot)$
Reg	$O(n)$ [folklore]	$O(n\Delta)$, matching cond. lower bound	$O(n\Delta)$, matching cond. lower bound
1Var ($\text{var}(\alpha) = \{x\}$)	$O(n)$ [folklore]	$O(n)$	$O(n^{3\|\alpha\|_x})$ W[1]-hard w.r.t. $\|\alpha\|_x$
NonCross	$O(nm\log n)$ [14]	$O(n^3p)$	NP-hard
1RepVar $k=\#$ x-blocks	$O(n^2)$ [14]	$O(n^{k+2}m)$ W[1]-hard w.r.t. k	NP-hard for $k \geq 1$
kLOC	$O(mkn^{2k+1})$ [10] W[1]-hard w.r.t. k	$O(n^{2k+2}m)$ W[1]-hard w.r.t. k	NP-hard for $k \geq 1$
kSCD	$O(m^2n^{2k})$ [14] W[1]-hard w.r.t. k	NP-hard for $k \geq 2$	NP-hard for $k \geq 1$
kRepVar	$O(n^{2k})$ [14] W[1]-hard w.r.t. k	NP-hard for $k \geq 1$	NP-hard for $k \geq 1$
k-bounded treewidth	$O(n^{2k+4})$ [35] W[1]-hard w.r.t. k	NP-hard for $k \geq 3$	NP-hard for $k \geq 1$

shown for inputs over unbounded alphabets; it would be interesting to see if it still holds for alphabets of constant size. Moreover, it would be interesting to obtain similar lower bounds for MisMatch$_{\text{1Var}}$, as the two problem seem strongly related. To that end, it would be interesting if the upper bound of Theorem 6 can be improved, and brought closer to the one reported for median string in [36]. Secondly, another interesting problem is related to Theorem 4. The lower bound we reported in that theorem holds for regular patterns with a constant number of variables (e.g., two variables). It is still open what is the complexity of MisMatch for regular patterns with a constant number of variables under Hamming distance.

References

1. Amir, A., Nor, I.: Generalized function matching. J. Discrete Algorithms **5**, 514–523 (2007). https://doi.org/10.1016/j.jda.2006.10.001
2. Angluin, D.: Finding patterns common to a set of strings. J. Comput. Syst. Sci. **21**(1), 46–62 (1980)
3. Backurs, A., Indyk, P.: Edit distance cannot be computed in strongly subquadratic time (unless SETH is false). SIAM J. Comput. **47**(3), 1087–1097 (2018)
4. Bernardini, G., et al.: String sanitization under edit distance. In: 31st Annual Symposium on Combinatorial Pattern Matching (CPM 2020). LIPIcs, vol. 161, pp. 7:1–7:14 (2020). https://doi.org/10.4230/LIPIcs.CPM.2020.7

5. Bernardini, G., Pisanti, N., Pissis, S.P., Rosone, G.: Approximate pattern matching on elastic-degenerate text. Theor. Comput. Sci. **812**, 109–122 (2020)
6. Câmpeanu, C., Salomaa, K., Yu, S.: A formal study of practical regular expressions. Int. J. Found. Comput. Sci. **14**, 1007–1018 (2003). https://doi.org/10.1142/S012905410300214X
7. Charalampopoulos, P., Kociumaka, T., Mozes, S.: Dynamic string alignment. In: 31st Annual Symposium on Combinatorial Pattern Matching (CPM 2020). LIPIcs, vol. 161, pp. 9:1–9:13 (2020). https://doi.org/10.4230/LIPIcs.CPM.2020.9
8. Charalampopoulos, P., Kociumaka, T., Wellnitz, P.: Faster approximate pattern matching: a unified approach. In: Irani, S. (ed.) 61st IEEE Annual Symposium on Foundations of Computer Science, FOCS 2020, pp. 978–989. IEEE (2020). https://doi.org/10.1109/FOCS46700.2020.00095
9. Charalampopoulos, P., Kociumaka, T., Wellnitz, P.: Faster pattern matching under edit distance. arXiv preprint arXiv:2204.03087 (2022)
10. Day, J.D., Fleischmann, P., Manea, F., Nowotka, D.: Local patterns. In: Proceedings of the 37th IARCS Annual Conference on Foundations of Software Technology and Theoretical Computer Science (FSTTCS 2017). LIPIcs, vol. 93, pp. 24:1–24:14 (2017). https://doi.org/10.4230/LIPIcs.FSTTCS.2017.24
11. Downey, R.G., Fellows, M.R.: Parameterized complexity. In: Monographs in Computer Science, Springer, NY (1999). https://doi.org/10.1007/978-1-4612-0515-9
12. Fagin, R., Kimelfeld, B., Reiss, F., Vansummeren, S.: Document spanners: a formal approach to information extraction. J. ACM **62**(2), 12:1–12:51 (2015). https://doi.org/10.1145/2699442
13. Fernau, H., Manea, F., Mercas, R., Schmid, M.L.: Revisiting Shinohara's algorithm for computing descriptive patterns. Theor. Comput. Sci. **733**, 44–54 (2018)
14. Fernau, H., Manea, F., Mercas, R., Schmid, M.L.: Pattern matching with variables: efficient algorithms and complexity results. ACM Trans. Comput. Theory **12**(1), 6:1–6:37 (2020). https://doi.org/10.1145/3369935
15. Fernau, H., Schmid, M.L.: Pattern matching with variables: a multivariate complexity analysis. Inf. Comput. **242**, 287–305 (2015). https://doi.org/10.1016/j.ic.2015.03.006
16. Fernau, H., Schmid, M.L., Villanger, Y.: On the parameterised complexity of string morphism problems. Theory Comput. Syst. **59**(1), 24–51 (2016)
17. Freydenberger, D.D.: Extended regular expressions: succinctness and decidability. Theory Comput. Syst. **53**, 159–193 (2013). https://doi.org/10.1007/s00224-012-9389-0
18. Freydenberger, D.D.: A logic for document spanners. Theory Comput. Syst. **63**(7), 1679–1754 (2019)
19. Freydenberger, D.D., Holldack, M.: Document spanners: from expressive power to decision problems. Theory Comput. Syst. **62**(4), 854–898 (2018)
20. Freydenberger, D.D., Schmid, M.L.: Deterministic regular expressions with backreferences. J. Comput. Syst. Sci. **105**, 1–39 (2019)
21. Friedl, J.E.F.: Mastering Regular Expressions, 3rd edn. O'Reilly, Sebastopol, CA (2006)
22. Gawrychowski, P., Manea, F., Siemer, S.: Matching patterns with variables under hamming distance. In: 46th International Symposium on Mathematical Foundations of Computer Science, MFCS 2021. LIPIcs, vol. 202, pp. 48:1–48:24 (2021). https://doi.org/10.4230/LIPIcs.MFCS.2021.48
23. Gawrychowski, P., Manea, F., Siemer, S.: Matching patterns with variables under edit distance (2022). https://doi.org/10.48550/ARXIV.2207.07477

24. Hoppenworth, G., Bentley, J.W., Gibney, D., Thankachan, S.V.: The fine-grained complexity of median and center string problems under edit distance. In: 28th Annual European Symposium on Algorithms (ESA 2020). LIPIcs, vol. 173, pp. 61:1–61:19 (2020). https://doi.org/10.4230/LIPIcs.ESA.2020.61
25. Kleest-Meißner, S., Sattler, R., Schmid, M.L., Schweikardt, N., Weidlich, M.: Discovering event queries from traces: laying foundations for subsequence-queries with wildcards and gap-size constraints. In: 25th International Conference on Database Theory, ICDT 2022. LIPIcs, vol. 220, pp. 18:1–18:21 (2022). https://doi.org/10.4230/LIPIcs.ICDT.2022.18
26. Landau, G.M., Vishkin, U.: Fast parallel and serial approximate string matching. J. Algorithms **10**(2), 157–169 (1989)
27. Levenshtein, V.: Binary codes capable of correcting spurious insertions and deletions of ones. Probl. Inf. Transm. **1**, 8–17 (1965)
28. Levenshtein, V.I.: Binary codes capable of correcting deletions, insertions, and reversals. Sov. Phys. Dokl. **10**, 707 (1966)
29. Lothaire, M.: Combinatorics on Words. Cambridge University Press, Cambridge (1997). https://doi.org/10.1017/CBO9780511566097
30. Lothaire, M.: Algebraic Combinatorics on Words. Cambridge University Press, Cambridge (2002). https://doi.org/10.1017/CBO9781107326019
31. Manea, F., Schmid, M.L.: Matching patterns with variables. In: Mercaş, R., Reidenbach, D. (eds.) WORDS 2019. LNCS, vol. 11682, pp. 1–27. Springer, Cham (2019). https://doi.org/10.1007/978-3-030-28796-2_1
32. Mieno, T., Pissis, S.P., Stougie, L., Sweering, M.: String sanitization under edit distance: improved and generalized. In: 32nd Annual Symposium on Combinatorial Pattern Matching, CPM 2021. LIPIcs, vol. 191, pp. 19:1–19:18 (2021)
33. Navarro, G.: A guided tour to approximate string matching. ACM Comput. Surv. **33**(1), 31–88 (2001)
34. Nicolas, F., Rivals, E.: Hardness results for the center and median string problems under the weighted and unweighted edit distances. J. Discrete Algorithms **3**(2–4), 390–415 (2005)
35. Reidenbach, D., Schmid, M.L.: Patterns with bounded treewidth. Inf. Comput. **239**, 87–99 (2014)
36. Sankoff, D.: Minimal mutation trees of sequences. SIAM J. Appl. Math. **28**(1), 35–42 (1975)
37. Schmid, M.L.: A note on the complexity of matching patterns with variables. Inf. Process. Lett. **113**(19), 729–733 (2013). https://doi.org/10.1016/j.ipl.2013.06.011
38. Schmid, M.L., Schweikardt, N.: A purely regular approach to non-regular core spanners. In: Proceedings of the 24th International Conference on Database Theory, ICDT 2021. LIPIcs, vol. 186, pp. 4:1–4:19 (2021). https://doi.org/10.4230/LIPIcs.ICDT.2021.4
39. Schmid, M.L., Schweikardt, N.: Document spanners - a brief overview of concepts, results, and recent developments. In: PODS 2022: International Conference on Management of Data, pp. 139–150. ACM (2022). https://doi.org/10.1145/3517804.3526069
40. Shinohara, T.: Polynomial time inference of pattern languages and its application. In: Proceedings of the 7th IBM Symposium on Mathematical Foundations of Computer Science, MFCS, pp. 191–209 (1982)
41. Shinohara, T., Arikawa, S.: Pattern inference. In: Jantke, K.P., Lange, S. (eds.) Algorithmic Learning for Knowledge-Based Systems. LNCS, vol. 961, pp. 259–291. Springer, Heidelberg (1995). https://doi.org/10.1007/3-540-60217-8_13

On the Hardness of Computing the Edit Distance of Shallow Trees

Panagiotis Charalampopoulos[1]([✉]) [ID], Paweł Gawrychowski[2] [ID], Shay Mozes[3] [ID], and Oren Weimann[4] [ID]

[1] Birkbeck, University of London, London, UK
p.charalampopoulos@bbk.ac.uk
[2] University of Wrocław, Wrocław, Poland
gawry@cs.uni.wroc.pl
[3] Reichman University, Herzliya, Israel
smozes@idc.ac.il
[4] University of Haifa, Haifa, Israel
oren@cs.haifa.ac.il

Abstract. We consider the edit distance problem on rooted ordered trees parameterized by the trees' depth. For two trees of size at most n and depth at most d, the state-of-the-art solutions of Zhang and Shasha [SICOMP 1989] and Demaine et al. [TALG 2009] have runtimes $\mathcal{O}(n^2 d^2)$ and $\mathcal{O}(n^3)$, respectively, and are based on so-called *decomposition algorithms*. It has been recently shown by Bringmann et al. [TALG 2020] that, when $d = \Theta(n)$, one cannot compute the edit distance of two trees in $\mathcal{O}(n^{3-\epsilon})$ time (for any constant $\epsilon > 0$) under the APSP hypothesis. However, for small values of d, it is not known whether the $\mathcal{O}(n^2 d^2)$ upper bound of Zhang and Shasha is optimal. We make the following twofold contribution. First, we show that under the APSP hypothesis there is no algorithm with runtime $\mathcal{O}(n^2 d^{1-\epsilon})$ (for any constant $\epsilon > 0$) when $d = \text{poly}(n)$. Second, we show that there is no *decomposition algorithm* that runs in time $o(\min\{n^2 d^2, n^3\})$.

1 Introduction

Let F and G be two rooted and ordered trees of size n where each node is assigned a label from an alphabet Σ. The edit distance between trees F and G is the minimum cost of transforming F into G by a sequence of elementary *edit operations*: changing the label of a node v, deleting a node v and setting the children of v as the children of v's parent (in the place of v in the left-to-right order), and inserting a node v (defined as the inverse of a deletion); see Fig. 1. The cost of these elementary operations is given by two *cost functions*: $c_{del}(x)$ is the cost of deleting or inserting a node with label x, and $c_{match}(x, y)$ is the cost of changing the label of a node from x to y.

Tree edit distance is the most common similarity measure between labeled trees. It is instrumental in computational biology [7,17,25,31], structured text

---

S. Mozes and O. Weimann—Supported by Israel Science Foundation grant 810/21.

D. Arroyuelo and B. Poblete (Eds.): SPIRE 2022, LNCS 13617, pp. 290–302, 2022.
https://doi.org/10.1007/978-3-031-20643-6_21

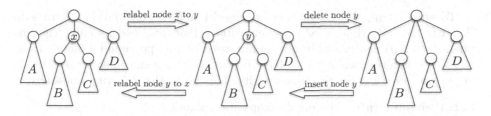

Fig. 1. The three edit operations on a node-labeled tree.

processing [10,11,16], programming languages [18], computer vision [6,19], character recognition [23], automatic grading [3], answer extraction [33], and many more (see the popular survey of Bille [7] and the books of Apostolico and Galil [4] and Valiente [29]).

The tree edit distance (TED) problem was introduced by Tai [27] as a generalization of the well known string edit distance problem [30]. Zhang and Shasha [35] showed that the classical dynamic-programming algorithm for string edit distance naturally extends to tree edit distance. Namely, to compute the edit distance of two forests F and G, consider the rightmost roots of F and G: they are either matched or (at least) one of them is deleted. Checking all these options generates a constant number of (smaller) recursive subproblems. Zhang and Shasha [35] showed that when the depths of F and G are bounded by d the total number of generated recursive subproblems (and hence the algorithm's running time) is $\mathcal{O}(n^2 d^2)$. This is appealing for shallow trees, but can be as high as $\Omega(n^4)$ for trees of large depth.

Obviously, the choice of recursing on the rightmost root (and not the leftmost) is arbitrary. Klein [20] observed that if we carefully alternate between recursing on the rightmost and the leftmost roots the running time improves to $\mathcal{O}(n^3 \log n)$ (regardless of d). Dulucq and Touzet [14] called such algorithms (i.e., that are based on the same dynamic programming but only differ in their choices of rightmost and leftmost) *decomposition algorithms* and showed that there is no $o(n^2 \log^2 n)$-time decomposition algorithm. Demaine et al. [12] gave an $\mathcal{O}(n^3)$-time decomposition algorithm and showed that for trees of depth $d = \Omega(n)$ there is no $o(n^3)$-time decomposition algorithm. Of course there may be a faster TED algorithm that is not a decomposition algorithm. This however is probably not the case for trees of depth $d = \Omega(n)$. For such trees, it was shown in [9] that: (1) assuming the APSP hypothesis, there is no $\mathcal{O}(n^{3-\epsilon})$ algorithm for TED with alphabet-size $|\Sigma| = \Omega(n)$, and (2) assuming the stronger k-Clique hypothesis, there is no $\mathcal{O}(n^{3-\epsilon})$ algorithm for TED with alphabet-size $|\Sigma| = \mathcal{O}(1)$. An important exception is the special case of *unweighted* TED where $c_{del}(a) = c_{match}(a,b) = 1$ and $c_{match}(a,a) = 0$ (aka the Levenshtein distance) for which very recently a non-decomposition strongly subcubic algorithm (for any d) was devised by Mao [21]; the exponent of n was further reduced to 2.9149 by Dürr [15].

To sum it up, the fastest known algorithm for (weighted) TED runs in $\mathcal{O}(\min\{n^3, n^2d^2\})$ time, it is a decomposition algorithm, and its cubic runtime when $d = \Omega(n)$ can probably not be improved by any polynomial factor. However, for smaller values of $d = \text{poly(n)}$ (i.e., $d = n^\delta$ for some constant $0 < \delta < 1$) we do not yet know the right complexity. This raises the following questions:

1. Is there an $\mathcal{O}(n^2d^{2-\epsilon})$-time decomposition algorithm?
   (The lower bound of [12] does not rule this out.)
2. Is there an $\mathcal{O}(n^2d^{2-\epsilon})$-time algorithm that is not a decomposition algorithm?
   (The lower bound of [9] does not rule this out.)

We show that the answer to the first question is no. As for the second question, we do not yet know if an $\mathcal{O}(n^2d^{2-\epsilon})$-time algorithm exists, but we show that an $\mathcal{O}(n^2d^{1-\epsilon})$-time algorithm does not (assuming the APSP hypothesis).

Related Work. Pawlik and Augsten [22] developed a decomposition algorithm whose performance on any input is not worse (and possibly better) than that of any of the existing decomposition algorithms. Other attempts achieved better running times by restricting the edit operations or the scoring schemes [1,11,21,24,26,28,34], or by resorting to approximation [2,5,8]. However, in the worst case no algorithm currently beats $\mathcal{O}(\min\{n^3, n^2d^2\})$ (not even by a logarithmic factor). Finally, the edit distance between edge-labeled unrooted trees, first studied by Klein [20], can be computed in $\mathcal{O}(n^3)$ time as shown by Dudek and Gawrychowski [13]. In addition, Dudek and Gawrychowski [13] presented a simple $\mathcal{O}(|\text{input}|)$-time reduction from TED on node-labeled rooted trees to TED on edge-labeled unrooted trees. This reduction replaces the two rooted trees by unrooted trees with the same size and diameter asymptotically, and hence our lower bounds also apply to the TED problem on edge-labeled unrooted trees parameterized by the trees' diameter.

2   Preliminaries

We denote the tree edit distance of two trees F and G by $\text{TED}(F, G)$. The alphabet is denoted by Σ.

Now, let us look more closely at the allowed edit operations. First, observe that the insertion operation is redundant: an insertion to one of the trees is equivalent to a deletion in the other. We can thus consider the problem of computing a minimum-cost sequence of deletions and relabelings to both F and G that yields identical trees. Further, as we argue next, without loss of generality, we can assume that $c_{del}(x) = 0$ for all $x \in \Sigma$. Starting from general cost functions, we can define new cost functions $c'_{match}(x, y) := c_{match}(x, y) - c_{del}(x) - c_{del}(y)$, for all $x, y \in \Sigma$, and $c'_{del}(x) := 0$, for all $x \in \Sigma$, that preserve the tree edit distance up to a linear-time computable additive constant equal to the cost of deleting both trees with the original c_{del} function. Intuitively, we pay for the deletion of all nodes up front and get refunded for nodes that are not deleted.

A *left comb* (resp. *right comb*) of depth n is a tree with $2n - 1$ nodes that consists of a path P of length n, with one endpoint of the path being the root of the tree and the other one being a leaf, and $n - 1$ more leaf nodes, each being the right (resp. left) child of a distinct node of P. We call a pair of left and a right combs *opposing combs*. See Fig. 2 for an illustration.

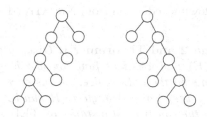

Fig. 2. Two opposing combs of depth 5; the left comb is shown in the left.

3   Lower Bound Conditioned on the APSP Hypothesis

In this section we present a hardness proof of TED on trees of depth $d = \text{poly}(n)$, conditioned on the All-Pairs Shortest Paths (APSP) hypothesis.

Conjecture 1 (APSP hypothesis). For any $\epsilon > 0$, there exists $c > 0$ such that APSP on n-vertex graphs with edge weights in $\{1, \ldots, n^c\}$ cannot be solved in time $\mathcal{O}(n^{3-\epsilon})$.

Instead of reducing APSP to TED, we will reduce from the equivalent (see [32]) NEGATIVETRIANGLE problem:

NEGATIVETRIANGLE

Input: A complete tripartite graph $H = (V, E)$ with three parts I, J, and K, each of size at most n, and a weight function $w : E \to \{-n^c, \ldots, n^c\}$.

Output: Yes if and only if there exist vertices $i \in I$, $j \in J$, and $k \in K$ such that $w(i, j) + w(j, k) + w(k, i) < 0$.

Lemma 1. *Consider an instance of* NEGATIVETRIANGLE *comprised of a complete tripartite graph* $H = (V, E)$ *with parts of size at most* n *and a weight-function* $w : E \to \{-n^c, \ldots, n^c\}$. *For any integer* $d \le n$, *this instance can be reduced to deciding whether any of* $\mathcal{O}((n/d)^3)$ *complete graphs on* $3d$ *vertices contains a negative triangle. The time required for this reduction is* $\mathcal{O}(n^2 + n^3/d)$.

Proof. Let us split each of the three parts I, J, and K into $\lceil n/d \rceil$ subsets, each of size at most d. Then, it suffices to solve separately, for each of the $\mathcal{O}((n/d)^3)$ triplets of subsets $A \subseteq I$, $B \subseteq J$, and $C \subseteq K$, an instance of the NEGATIVETRIANGLE problem for the subgraph of G induced by $A \cup B \cup C$. We consider each such induced subgraph and pad it with dummy vertices and edges so that it is a complete graph on $3d$ vertices, ensuring that we do not introduce any negative triangles. The latter can be achieved by setting the weights of dummy edges to be twice as large as the largest absolute value of an edge-weight in H. This reduction requires time linear in the total size of the input and the output and hence the stated bound follows.   □

We will use the following reduction from NEGATIVETRIANGLE to TED that was presented in [9].

Lemma 2 ([9, **Lemma 2 and Theorem 2**]). *Given a complete undirected n-vertex graph $H = (V, E)$ and a weight function $w : E \to \{1, \dots, n^c\}$, we can construct, in linear time in the output size, an instance of TED of size $\mathcal{O}(n)$ such that the minimum weight of a triangle in H can be extracted from the edit distance. In particular, the constructed instance of TED satisfies the following:*[1]

- *c_{del} is an all-zeroes function;*
- *the trees are two opposing combs of depth $2n + 1$;*
- *the edit distance of the two trees is equal to $-3M^2$ plus the minimum weight of a triangle in H, where M is a (sufficiently large) integer parameter that is used to define c_{match}.*

In particular, the fact that the (shapes of the) trees in the above lemma are fixed means that information about the NEGATIVETRIANGLE instance (H, w) is only encoded in the assignment of letters to nodes and the cost function $c_{match}(\cdot, \cdot)$. Hence, given t instances of NEGATIVETRIANGLE of the same size, one can construct $\sqrt{t}$ left combs and $\sqrt{t}$ right combs, such that each of the t pairs of left and right combs corresponds to one of the NEGATIVETRIANGLE instances. We can then assign a distinct letter to each node in each of the combs and define the cost function so that its restriction to any particular pair of left and right combs coincides with the cost function that Lemma 2 would yield for the NEGATIVETRIANGLE instance corresponding to this pair.

In the following lemma, we combine the above idea with Lemma 1 into a subcubic-time reduction from NEGATIVETRIANGLE to TED.

Lemma 3. NEGATIVETRIANGLE *reduces in $\mathcal{O}(n^2 + n^3/d)$ time to an instance of TED over $\mathcal{O}(n^{1.5}/\sqrt{d})$-size and $\mathcal{O}(d)$-depth trees.*

Proof. We first apply Lemma 1 to reduce our NEGATIVETRIANGLE instance to the problem of deciding whether any of $t = \mathcal{O}((n/d)^3)$ complete graphs on $3d$ vertices contains a negative triangle. Our goal is to efficiently reduce the latter problem to a TED instance with trees of depth $\mathcal{O}(d)$.

Let us denote the obtained graphs by $H_1, H_2, \dots, H_t$. We construct a TED instance as follows. Let $s = \sqrt{t}$ and assume that it is an integer in order to avoid clutter. F consists of a root with $3s$ children, which we call *available nodes*. Each of these available nodes has a left comb of depth $2 \cdot 3d + 1 = 6d + 1$ attached to it. G consists of a root with s children, which we call *decider nodes*. Each of the decider nodes has a right comb of depth $6d + 1$ attached to it; see Fig. 3. Observe that the sizes and the depths of these trees are as desired.

---

[1] Not all of these properties are explicitly stated in [9, Lemma 2 and Theorem 2], but they are evident from their proofs.

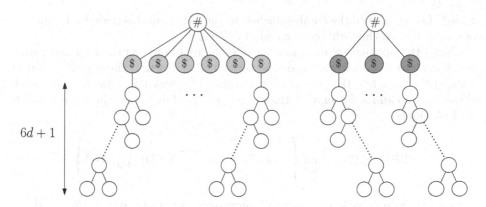

Fig. 3. A depiction of the instance of TED from the proof of Lemma 3 with $t = 4$ (and $s = 2$). Available nodes (in F) and selector nodes (in G) are colored yellow and purple, respectively. A comb of depth $6d+1$ is attached to *each* of the available/selector nodes—we only show a few of them for clarity. (Color figure online)

Our goal is to define a cost function $c_{match}(\cdot, \cdot)$ that ensures the following:

- the roots of the two trees are matched;
- each decider node is matched with an available node;
- the restriction of the cost function to the pair that consists of the $(s + i)$-th left comb in F and the j-th comb in G, for $i = 1, \ldots, s$ and $j = 1, \ldots, s$ is identical to the one yielded by Lemma 2 for H_m, where $m = s \cdot (i - 1) + j$;
- the restriction of the cost function to each other pair that consists of a left comb in F and a right comb in G is identical to the one yielded by Lemma 2 for some undirected graph on $3d$ vertices in which the minimum weight of a triangle is zero, where all such applications of Lemma 2 use the same (sufficiently large) integer parameter M.

To ensure the above properties, let us label the roots of both trees by $\#$ and all decider/available nodes by $\$$. In addition, let us label each other node with a unique letter from an alphabet Σ that is disjoint from $\{\#, \$\}$. We populate a table that corresponds to the c_{match} function as follows:

- For all $(x, y) \in \Sigma^2$ with x being a label of a node in F and y being a label of a node in G, i.e., pairs of labels of nodes from opposing combs, the cost $c_{match}(x, y)$ is given by an application of Lemma 2. Let us denote the sum of the absolute values of all these costs by ψ.
- $c_{match}(\#, \#) = c_{match}(\$, \$) = -2\psi$;
- $c_{match}(\#, \$) = \infty$;
- $c_{match}(x, y) = \infty$ for all pairs $(x, y) \in \{\#, \$\} \times \Sigma$.

Claim. There is a negative triangle in at least one of $H_1, H_2, \ldots, H_t$ if and only if $\text{TED}(F, G) < \eta := -2\psi \cdot (s + 1) - 3M^2 \cdot s$.

Proof. Let us denote the comb attached to the i-th available node by F_i and the one attached to the j-th decider node by G_j.

Now, the optimal solution must match the roots of the two trees and match every decider node with an available node. This is because these yield a cost of $-2\psi \cdot (s+1)$, while the cost of any sequence of operations that do not match these nodes cannot be smaller than $-2\psi \cdot s - \psi$. Thus, the edit distance of F and G equals

$$\text{TED}(F, G) = \min_p \left\{ -2\psi \cdot (s+1) + \sum_{j=1}^{s} \text{TED}(F_{p(j)}, G_j) \right\}$$

where the minimization is over all increasing functions $p : \{1, 2, \ldots, s\} \to \{1, 2, \ldots, 3s\}$. We therefore have two cases:

- If none of the graphs H_i contains a negative triangle, we might as well set every $p(j)$ to be j since the pairs (F_j, G_j) correspond to a graph in which the minimum triangle is of zero weight. So in this case we have

$$\text{TED}(F, G) = -2\psi \cdot (s+1) + \sum_{j=1}^{s} \text{TED}(F_j, G_j) = \eta.$$

- Else, some graph H_i contains a negative triangle of weight $-w$. Let $q = \lceil i/s \rceil$ and r be an integer in $\{1, \ldots, s\}$ satisfying $r \equiv i \pmod s$. Notice that matching the pair (F_{s+q}, G_r) is cheaper than matching a pair corresponding to a minimum weight triangle of value zero. We therefore have

$$\text{TED}(F, G) \leq -2\psi \cdot (s+1) + \sum_{j=1}^{r-1} \text{TED}(F_j, G_j) + \text{TED}(F_{s+q}, G_r)$$

$$+ \sum_{j=r+1}^{s} \text{TED}(F_{j+2s}, G_j) = \eta - w < \eta.$$

This completes the proof of the claim.   □

The lemma follows.   □

Theorem 1. *There exists no algorithm that solves* TED *for trees of size at most* n *and depth at most* $d = \text{poly}(n)$, *with node labels from an alphabet of size* $\Omega(n)$, *in* $\mathcal{O}(n^2 d^{1-\epsilon})$ *time, for any constant* $\epsilon > 0$, *unless the APSP hypothesis fails.*

Proof. To the contrary, suppose that there is such an algorithm with $\epsilon < 1$. Let N denote the size of an APSP instance. Using Lemma 3 with $d = \text{poly}(N)$, we obtain an algorithm for APSP with runtime $\mathcal{O}(N^2 + N^3/d + (N^{1.5}/\sqrt{d})^2 \cdot d^{1-\epsilon}) = \mathcal{O}(N^3/d^\epsilon)$, contradicting the APSP hypothesis.   □

4 Lower Bound for Decomposition Algorithms

Let us recall that the *decomposition algorithm* paradigm for the computation of tree edit distance is based on the following observation: given two forests F and G, the rightmost (or leftmost) roots of F and G are either matched or (at least) one of them is deleted. This observation leads to a dynamic programming approach: consider all three such options and recurse. The algorithm of Zhang and Shasha [35] proceeds by always considering the rightmost roots of the forests, while the algorithms of Klein [20] and Demaine et al. [12] use more intricate strategies (based on heavy-path decompositions) to decide whether to consider the rightmost or the leftmost roots of the forests in each step. In general, we call a mapping S from pairs of forests to the set {left, right} a *strategy*. Previous lower bounds on decomposition algorithms were established by proving a lower bound on the number of different pairs of forests F' of F and G' of G that a decomposition algorithm will consider irrespective of the strategy S that it follows; we do not deviate from this approach.

Let us introduce some more terminology and notation. For a node v in a tree T, we denote by T_v the subtree of T rooted at v. Further, we call v's child u such that T_u is largest *heavy*, resolving ties arbitrarily; all other children of v are called *light*. If a node v in T has two children, this allows us to naturally refer to the two subtrees of the children of v as v's heavy and light subtrees. Further, for a tree T, we denote by T° the forest obtained by deleting the root of T.

We next specify our hard instance of TED. It consists of trees of size $\Theta(n)$ and depth $\Theta(d)$ for any parameters $d, n \in \mathbb{Z}_+$ with $n > 100d \geq 300$. For simplicity, we assume that d divides n. Each of the trees that we will consider in this lower bound consists of a path (also called *spine*) P of length $d + 1$, with one endpoint of the path being the root of the tree and the other one being a leaf, and d trees of size $\lceil n/d \rceil$ and depth $\mathcal{O}(d)$, each attached to a distinct non-leaf node of P, in an alternating fashion with regards to being left/right subtrees. (The exact shape of these trees is not important.) See Fig. 4 for an illustration. We call a non-leaf spine node u of F and a non-leaf spine node v of G *opposing* if and only if their children that lie on the corresponding spine are in different directions (i.e., left and right); two such opposing nodes are indicated in Fig. 4.

Let us fix an arbitrary strategy S and denote by $\mathcal{U}(\mathsf{S})$ the set of subproblems that a decomposition algorithm with strategy S will encounter. Our goal is to give an $\Omega(\min\{n^2 d^2, n^3\})$ lower bound on $|\mathcal{U}(\mathsf{S})|$.

Consider a subproblem $(F', G') \in \mathcal{U}(\mathsf{S})$. Suppose that the strategy S for this subproblem is right. In this case, the subproblems obtained by (i) deleting the root of the rightmost tree in F', (ii) deleting the root of the rightmost tree in G', and (iii) matching the roots of the rightmost trees in F' and G', all belong to $\mathcal{U}(\mathsf{S})$. In what follows, when we say that in such a scenario we *delete from F* (resp. G), we mean that we concentrate our attention on the subproblem created in option (i) (resp. (ii)) above. We stress that in reality both these subproblems (as well as those created in option (iii) above) belong to $\mathcal{U}(\mathsf{S})$, but that for our purposes it suffices to focus on a particular subproblem.

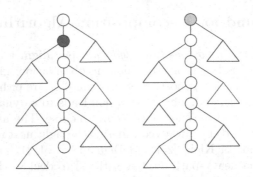

Fig. 4. An illustration of our hard instance for decomposition algorithms over shallow trees. Here, $d = 6$ and each of the subtrees attached to a spine node is of size $\mathcal{O}(\lceil n/6 \rceil) = \mathcal{O}(n)$ and depth $\mathcal{O}(1)$. A pair of opposing spine nodes is colored.

We rely on the following lemma from [12], stating that any strategy must consider matching every pair of nodes.

Lemma 4 ([12, **Lemma 2.3**]). *For any strategy* S*, for all* $u \in F$*,* $v \in G$*,* $(F_u^\circ, G_v^\circ) \in \mathcal{U}(\mathsf{S})$.

Our plan is to start from such subproblems $(F_u^\circ, G_v^\circ) \in \mathcal{U}(\mathsf{S})$, and, by choosing an appropriate sequence of deletions from F and from G, to obtain sufficiently many new subproblems. To make sure we do not double count subproblems obtained in this way, we will charge a subproblem consisting of a pair of forests F' and G' to the pair of spine nodes $p \in F$ and $q \in G$ that are the lowest common ancestors (LCAs) of the nodes of F' in F and of the nodes of G' in G, respectively. For a node v in a tree T, and for a positive integer x (resp. y) smaller than the size of the left (resp. right) subtree of T_v, we denote by $\mathsf{L}^x \mathsf{R}^y(T_v)$ the forest obtained from T_v° by x deletions of the leftmost root and y deletions of the rightmost root—we stress that the root of T_v has already been deleted prior to these $x + y$ deletions. Observe that v is the LCA of the nodes in $\mathsf{L}^x \mathsf{R}^y(T_v)$. In what follows, we refer to deletions of the leftmost and rightmost root as left deletions and right deletions, respectively.

Lemma 5. *Consider a spine node* p *in* F *of depth at most* $d/2$ *whose right child* c *is heavy. Let* q *be an opposing spine node in* G *of depth at most* $d/2$ *whose heavy child is* w*. Then, the total number of subproblems charged to* (p,q)*,* (c,q)*, and* (p,w) *is* $\Omega(\min\{n^2, n^3/d^2\})$.

Proof. Observe that the heavy subtree of each of p, c, q, and w has at least $n/3$ nodes. Let $\Delta := \{1, \ldots, \lceil n/4d \rceil\}$. We distinguish between two cases.

Case I: For every $(k, \ell) \in \Delta^2$, there is a subproblem in $\mathcal{U}(\mathsf{S})$ with exactly k deletions in the left subtree of F_p and exactly ℓ deletions in the right subtree of G_q, at least $\lfloor n/8 \rfloor$ nodes in the heavy (i.e., right) subtree of F_p, and at least $\lfloor n/8 \rfloor$ nodes in the heavy (i.e., left) subtree of G_q. In this case, starting from each of these $\lceil n/4d \rceil^2 = \Omega(n^2/d^2)$ pairs of forests, we consider the subproblems generated by always performing deletions only in the heavy subtrees of both F_p and G_q, i.e., deleting from F if and only if $\mathsf{S} = \mathsf{right}$. As each of these subtrees has at least $\lfloor n/8 \rfloor$ nodes, we obtain $\Omega(n)$ distinct subproblems of the form $(\mathsf{L}^k \mathsf{R}^x(F_p), \mathsf{L}^y \mathsf{R}^\ell(G_q))$ for each fixed k, ℓ. Since $k, \ell < \lceil n/4d \rceil$, $\mathsf{L}^k \mathsf{R}^x(F_p)$ (resp. $\mathsf{L}^y \mathsf{R}^\ell(G_q)$) contains nodes from both the left and right subtrees of F_p (resp. G_q). Hence the LCA of the nodes of $\mathsf{L}^k \mathsf{R}^x(F_p)$ (resp. $\mathsf{L}^y \mathsf{R}^\ell(G_q)$) is p (resp. q). We thus obtain a total of $\Omega(n^3/d^2)$ subproblems that are charged to the pair (p, q) of spine nodes. We are therefore done in this case.

For the remainder of the proof we can thus focus on the complementary case.

Case II: There is some pair $(k, \ell) \in \Delta^2$ for which we do not have a subproblem with k deletions in the left subtree of F_p and ℓ deletions in the right subtree of G_q, and at least $\lfloor n/8 \rfloor$ nodes in both heavy subtrees of F_p and G_q.

Claim. One of the following holds:

– for every integer $y \in [n/16, \lfloor n/8 \rfloor)$, there exists $\ell' < \ell$ such that

$$(\mathsf{L}^k(F_p), \mathsf{L}^y \mathsf{R}^{\ell'}(G_q)) \in \mathcal{U}(\mathsf{S}),$$

– for every integer $y \in [n/16, \lfloor n/8 \rfloor)$, there exists $k' < k$ such that

$$(\mathsf{L}^{k'} \mathsf{R}^y(F_p), \mathsf{R}^\ell(G_q)) \in \mathcal{U}(\mathsf{S}).$$

Proof. Consider applying the following sequence of operations starting from the pair (F_p°, G_q°), which is in $\mathcal{U}(\mathsf{S})$ by Lemma 4: delete from F whenever $\mathsf{S} = \mathsf{left}$ until we have reached $\mathsf{L}^k(F_p)$ and delete from G whenever $\mathsf{S} = \mathsf{right}$ until we have reached $\mathsf{R}^\ell(G_q)$. Let us only consider the case where the strategy says left k times before it says right ℓ times as the other case is symmetric.

Let A denote the subproblem obtained by performing k left deletions from F_p and some number $\ell'' < \ell$ right deletions from G_q. That is, $A = (\mathsf{L}^k(F_p), \mathsf{R}^{\ell''}(G_q))$. Since we are in Case II, for every integer $y \in [n/16, \lfloor n/8 \rfloor)$ we can, starting from A, only make deletions in G_q, making y left deletions from G_q and making less than $\ell - \ell''$ right deletions from G_q, thus obtaining the subproblem $(\mathsf{L}^k(F_p), \mathsf{L}^y \mathsf{R}^{\ell'}(G_q))$ for some ℓ' satisfying $\ell'' \leq \ell' < \ell$. $\qquad\square$

Let us assume without loss of generality that we are in the first case of the above claim, as the other case is symmetric. Let us fix some value of $y \in [n/16, \lfloor n/8 \rfloor)$ and the corresponding subproblem $(\mathsf{L}^k(F_p), \mathsf{L}^y \mathsf{R}^{\ell'}(G_q))$, where $\ell' < \ell$. We prove the following claim.

Claim. There exist $\Omega(\min\{n, n^2/d^2\})$ quadruples (x, m, s, v) where x, m, and s are integers, and $v \in \{p, c\}$, such that

$$(\mathsf{L}^x \mathsf{R}^m(F_v), \mathsf{L}^y \mathsf{R}^s(G_q)) \in \mathcal{U}(\mathsf{S}).$$

Proof. Starting from our fixed subproblem $(\mathsf{L}^k(F_p), \mathsf{L}^y \mathsf{R}^{\ell'}(G_q))$, we consider making all left deletions in F. However, for each $t \in \Delta$, we consider making the first t right deletions in G and the remaining ones in F. We distinguish between two cases depending on whether, for each pair $(m, t) \in \Delta^2$, we obtain a subproblem of the form:

$$(\mathsf{L}^x \mathsf{R}^m(F_v), \mathsf{L}^y \mathsf{R}^{\ell'+t}(G_q)), \text{ for some integer } x \text{ and } v \in \{p, c\}.$$

1. If this is the case, we obtain $\Omega(n^2/d^2)$ of the desired quadruples, and we are thus done.
2. Else, let $(m', t') \in \Delta^2$ be a pair for which there is no integer x and node $v \in \{p, c\}$ such that $(\mathsf{L}^x \mathsf{R}^{m'}(F_v), \mathsf{L}^y \mathsf{R}^{\ell'+t'}(G_q)) \in \mathcal{U}(\mathsf{S})$. This can only be the case if we eliminate the entire heavy (i.e., left) subtree of T_c prior to making the intended $m' + t'$ right deletions. In other words, this can only happen if along this computational path of the recursion, S says left $\Omega(n)$ times before it says right $m' + t'$ times. In this case, for each $x \in [n/16, \lfloor n/8 \rfloor)$, there exist $m'', t'' \in \Delta^2$ such that $(\mathsf{L}^x \mathsf{R}^{m''}(F_c), \mathsf{L}^y \mathsf{R}^{\ell'+t''}(G_q)) \in \mathcal{U}(\mathsf{S})$. In this case, we thus obtain $\Omega(n)$ quadruples of the desired form.

This completes the proof of the claim.   □

Thus, for each of $\Omega(n)$ values of y, we obtain $\Omega(\min\{n, n^2/d^2\})$ subproblems. Over all such y, we thus obtain $\Omega(\min\{n^2, n^3/d^2\})$ subproblems charged to (p, q) and (c, q), thus completing the analysis of Case II.   □

As our instance of TED has $\Omega(d^2)$ pairs of spine nodes p and q that satisfy the conditions of Lemma 5, we obtain the main result of this section:

Theorem 2. *Any decomposition algorithm for tree edit distance on trees of size at most n and depth at most d requires $\Omega(\min\{n^3, n^2 d^2\})$ time.*

References

1. Akmal, S., Jin, C.: Faster algorithms for bounded tree edit distance. In: 48th ICALP, pp. 12:1–12:15 (2021). https://doi.org/10.4230/LIPIcs.ICALP.2021.12
2. Akutsu, T., Fukagawa, D., Takasu, A.: Approximating tree edit distance through string edit distance. In: 17th ISAAC, pp. 90–99 (2006). https://doi.org/10.1007/11940128_11
3. Alur, R., D'Antoni, L., Gulwani, S., Kini, D., Viswanathan, M.: Automated grading of DFA constructions. In: 23rd IJCAI, pp. 1976–1982 (2013). http://dl.acm.org/citation.cfm?id=2540128.2540412
4. Apostolico, A., Galil, Z. (eds.): Pattern Matching Algorithms. Oxford University Press, Oxford, UK (1997)
5. Aratsu, T., Hirata, K., Kuboyama, T.: Approximating tree edit distance through string edit distance for binary tree codes. Fundam. Inform. **101**(3), 157–171 (2010). https://doi.org/10.3233/FI-2010-282
6. Bellando, J., Kothari, R.: Region-based modeling and tree edit distance as a basis for gesture recognition. In: 10th International Conference on Image Analysis and Processing, ISIAP 1999, pp. 698–703 (1999). https://doi.org/10.1109/ICIAP.1999.797676

7. Bille, P.: A survey on tree edit distance and related problems. Theoret. Comput. Sci. **337**(1–3), 217–239 (2005). https://doi.org/10.1016/j.tcs.2004.12.030
8. Boroujeni, M., Ghodsi, M., Hajiaghayi, M., Seddighin, S.: 1+ε approximation of tree edit distance in quadratic time. In: 51st STOC, pp. 709–720. ACM (2019). https://doi.org/10.1145/3313276.3316388
9. Bringmann, K., Gawrychowski, P., Mozes, S., Weimann, O.: Tree edit distance cannot be computed in strongly subcubic time (unless APSP can). ACM Trans. Algorithms **16**(4), 48:1-48:22 (2020). https://doi.org/10.1145/3381878
10. Buneman, P., Grohe, M., Koch, C.: Path queries on compressed XML. In: VLDB, pp. 141–152 (2003). https://doi.org/10.1016/B978-012722442-8/50021-5
11. Chawathe, S.: Comparing hierarchical data in external memory. In: VLDB, pp. 90–101 (1999). http://www.vldb.org/conf/1999/P8.pdf
12. Demaine, E.D., Mozes, S., Rossman, B., Weimann, O.: An optimal decomposition algorithm for tree edit distance. ACM Trans. Algorithms **6**(1), 2:1-2:19 (2009). https://doi.org/10.1145/1644015.1644017
13. Dudek, B., Gawrychowski, P.: Edit distance between unrooted trees in cubic time. In: 45th ICALP, pp. 45:1–45:14 (2018). https://doi.org/10.4230/LIPIcs.ICALP.2018.45
14. Dulucq, S., Touzet, H.: Decomposition algorithms for the tree edit distance problem. J. Discrete Algorithms **3**(2–4), 448–471 (2005). https://doi.org/10.1016/j.jda.2004.08.018
15. Dürr, A.: Improved bounds for rectangular monotone min-plus product and applications. Arxiv 2208.02862v1 (2022)
16. Ferragina, P., Luccio, F., Manzini, G., Muthukrishnan, S.: Compressing and indexing labeled trees, with applications. J. ACM **57**, 1–33 (2009). https://doi.org/10.1145/1613676.1613680
17. Gusfield, D.: Algorithms on Strings, Trees and Sequences: Computer Science and Computational Biology. Cambridge University Press, Cambridge (1997)
18. Hoffmann, C.M., O'Donnell, M.J.: Pattern matching in trees. J. ACM **29**(1), 68–95 (1982). https://doi.org/10.1145/322290.322295
19. Klein, P.N., Tirthapura, S., Sharvit, D., Kimia, B.B.: A tree-edit-distance algorithm for comparing simple, closed shapes. In: 11th SODA, pp. 696–704 (2000). http://dl.acm.org/citation.cfm?id=338219.338628
20. Klein, P.N.: Computing the edit-distance between unrooted ordered trees. In: 6th ESA, pp. 91–102 (1998). https://doi.org/10.1007/3-540-68530-8_8
21. Mao, X.: Breaking the cubic barrier for (unweighted) tree edit distance. In: 62nd FOCS, pp. 792–803 (2021). https://doi.org/10.1109/FOCS52979.2021.00082
22. Pawlik, M., Augsten, N.: Efficient computation of the tree edit distance. ACM Trans. Database Syst. **40**(1), 3:1-3:40 (2015). https://doi.org/10.1145/2699485
23. Rico-Juan, J.R., Micó, L.: Comparison of AESA and LAESA search algorithms using string and tree-edit-distances. Pattern Recogn. Lett. **24**(9–10), 1417–1426 (2003). https://doi.org/10.1016/S0167-8655(02)00382-3
24. Selkow, S.: The tree-to-tree editing problem. Inf. Process. Lett. **6**(6), 184–186 (1977). https://doi.org/10.1016/0020-0190(77)90064-3
25. Shapiro, B.A., Zhang, K.: Comparing multiple RNA secondary structures using tree comparisons. Comput. Appl. Biosci. **6**(4), 309–318 (1990). https://doi.org/10.1093/bioinformatics/6.4.309
26. Shasha, D., Zhang, K.: Fast algorithms for the unit cost editing distance between trees. J. Algorithms **11**(4), 581–621 (1990). https://doi.org/10.1016/0196-6774(90)90011-3

27. Tai, K.: The tree-to-tree correction problem. J. ACM **26**(3), 422–433 (1979). https://doi.org/10.1145/322139.322143
28. Touzet, H.: Comparing similar ordered trees in linear-time. J. Discrete Algorithms **5**(4), 696–705 (2007). https://doi.org/10.1016/j.jda.2006.07.002
29. Valiente, G.: Algorithms on Trees and Graphs. Springer, Cham (2002). https://doi.org/10.1007/978-3-030-81885-2
30. Wagner, R.A., Fischer, M.J.: The string-to-string correction problem. J. ACM **21**(1), 168–173 (1974). https://doi.org/10.1145/321796.321811
31. Waterman, M.: Introduction to Computational Biology: Maps, Sequences and Genomes, Chapters 13, 14. Chapman and Hall (1995)
32. Williams, V.V., Williams, R.R.: Subcubic equivalences between path, matrix, and triangle problems. J. ACM **65**(5), 27:1-27:38 (2018). https://doi.org/10.1145/3186893
33. Yao, X., Durme, B.V., Callison-Burch, C., Clark, P.: Answer extraction as sequence tagging with tree edit distance. In: HLT-NAACL 2013, pp. 858–867 (2013). http://aclweb.org/anthology/N/N13/N13-1106.pdf
34. Zhang, K.: Algorithms for the constrained editing distance between ordered labeled trees and related problems. Pattern Recogn. **28**(3), 463–474 (1995). https://doi.org/10.1016/0031-3203(94)00109-Y
35. Zhang, K., Shasha, D.E.: Simple fast algorithms for the editing distance between trees and related problems. SIAM J. Comput. **18**(6), 1245–1262 (1989). https://doi.org/10.1137/0218082

Quantum Time Complexity
and Algorithms for Pattern Matching
on Labeled Graphs

Parisa Darbari[1], Daniel Gibney[2(✉)], and Sharma V. Thankachan[1]

[1] University of Central Florida, Orlando, Fl 32816, USA
[2] Georgia Institute of Technology, Atlanta, GA 30332, USA
daniel.j.gibney@gmail.com

Abstract. The problem of matching (exactly or approximately) a pattern P to a walk in an edge labeled graph $G = (V, E)$, denoted PMLG, has received increased attention in recent years. Here we consider conditional lower bounds on the time complexity of quantum algorithms for PMLG as well as a new quantum algorithm. We first provide a conditional lower bound based on a reduction from the Longest Common Subsequence problem (LCS) and the recently proposed NC-QSETH. For PMLG under substitutions to the pattern, our results demonstrate (i) that a quantum algorithm running in time $O(|E|m^{1-\varepsilon} + |E|^{1-\varepsilon}m)$ for any constant $\varepsilon > 0$ would provide an algorithm for LCS on two strings X and Y running in time $\tilde{O}(|X||Y|^{1-\varepsilon} + |X|^{1-\varepsilon}|Y|)$, which is better than any known quantum algorithm for LCS, and (ii) that a quantum algorithm running in time $O(|E|m^{\frac{1}{2}-\varepsilon} + |E|^{\frac{1}{2}-\varepsilon}m)$ would violate NC-QSETH. Results (i) and (ii) hold even when restricted to binary alphabets for P and the edge labels in G. We then provide a quantum algorithm for all versions of PMLG (exact, only substitutions, and substitutions/insertions/deletions) that runs in time $\tilde{O}(\sqrt{|V||E|} \cdot m)$. This is an improvement over the classical $O(|E|m)$ time algorithm when the graph is non-sparse.

Keywords: Pattern matching · Labeled graphs · Quantum algorithms

1 Introduction

We consider an approximate version of the Pattern Matching on Labeled Graphs problem (PMLG) under substitutions to the pattern, defined as follows: Given a directed edge-labeled graph $G = (V, E)$ with alphabet Σ, a string P of length m also over alphabet Σ (which we call a pattern), and an integer $\delta \geq 0$, determine if there exists a walk in G that matches a string P' such that $d_H(P, P') \leq \delta$. Here, $d_H(P, P')$ denotes the Hamming distance between P and P' and a walk is a ordered list of edges in E, i.e., $e_1, ..., e_m$ where e_i and e_{i+1} are incident to the same vertex for $1 \leq i < m$. Edges are allowed to be repeated in a walk.

© The Author(s), under exclusive license to Springer Nature Switzerland AG 2022
D. Arroyuelo and B. Poblete (Eds.): SPIRE 2022, LNCS 13617, pp. 303–314, 2022.
https://doi.org/10.1007/978-3-031-20643-6_22

Letting $label(e)$ denote the edge label for an edge $e \in E$, we say a length m string $P'[1, m]$ matches a walk $e_1, ..., e_m$ if $P'[i] = label(e_i)$ for $1 \le i \le m$.

PMLG was first considered in the context of pattern matching in hypertext [5, 30, 32, 33]. It has become increasingly important in Computational Biology where labeled graphs are used as multi-genomic references and sets of reads obtained through sequencing must be mapped to the reference [1, 10, 13, 17, 27, 34]. PMLG is also used in variant calling [9, 12, 25] and read error correction [28, 31].

The theoretical aspects of PMLG have also received significant study. The classical algorithm for PMLG is a dynamic programming solution that runs in time $O(|E|m)$ [5, 32]. It was first shown in [15] that a PMLG algorithm running in time $O(|E|^{1-\varepsilon}m + |E|m^{1-\varepsilon})$ for any constant $\varepsilon > 0$ would contradict the Strong Exponential Time Hypothesis (SETH) even for directed acyclic graphs (DAGs) and $\delta = 0$. These results were later strengthened in [18] to show that the same lower bounds hold based on likely weaker assumptions in circuit complexity. Nevertheless, there exist classes of graphs where the exact matching problem can be solved in near-linear time, e.g., Wheeler graphs [16]. However, recognizing whether a given graph has these properties is a hard problem [4, 19, 20]. The version of the problem where modifications are allowed to labels in the graph rather than the pattern has also been considered, which is NP-hard even when restricted to only substitutions over binary alphabets on special classes of graphs [5, 21, 26].

Despite the extent of the applied and theoretical work, there has been sparse research on utilizing quantum computing to solve PMLG. Equi et al. recently considered the problem for leveled DAGs, where they presented an algorithm running in $O(|E| + \sqrt{m})$ [14]. Several closely related problems have been studied as well. In [2], quantum algorithms for the problem of determining whether a string is contained in a regular language were considered. However, these regular languages were represented as monoids rather than NFAs, meaning the input representation could differ drastically from the labeled graphs used here. For finding exact matches in a single string (which could be viewed as a path) there exists a quantum algorithm running in $\tilde{O}(\sqrt{n} + \sqrt{m})$ time[1] on a string of length n and pattern of length m [24, 36].

We provide the first hardness result for PMLG in the quantum computing setting based on a reduction from the Longest Common Subsequence problem (LCS) and the conjectured hardness NC-QSETH [8], along with a new algorithm yielding a quantum speedup for non-sparse graphs.

1.1 Quantum Computing and Input Model

Quantum algorithms typically have their problem instance expressed as an *oracle*, or a function that allows one to query the problem instance. On a quantum computer, these queries can be made with an input that is the superposition of multiple inputs, allowing for a type of parallelism. We refer the interested reader to [23]. These oracles are often treated as black boxes, but they can also

---

[1] $\tilde{O}(\cdot)$ suppresses poly-logarithmic factors.

be provided as a Boolean circuit, or through an algorithmic description (under the quantum random access assumption discussed more below). The query complexity of a quantum algorithm is defined as the number of times that the oracle gets queried by the algorithm, and the quantum time complexity is the number of elementary gates[2] needed to implement the quantum algorithm, in addition to the number of queries.

A lower-level description of a quantum algorithm in terms of unitary operators acting on a state vector (a quantum circuit) is necessary for many of the algorithms that are the fundamental building blocks of quantum computing. These include, for example, quantum random walk algorithms for finding marked vertices in a graph [29], period finding algorithms [35], and Grover's search [22]. However, it is often possible to utilize these fundamental algorithms on a higher level of abstraction. This accommodates algorithm descriptions more similar to those used in imperative programming. Examples of this approach include the graph algorithms presented in [11], the $\tilde{O}(\sqrt{n}+\sqrt{m})$ pattern matching algorithm mentioned in the introduction [24], and a recent algorithm for finding the longest common substring of two strings [3]. One useful assumption is quantum random access, described in [3,7]. Using quantum random access, a classical T-time algorithm can be invoked by a quantum search algorithm, like Grover's search, in $O(T)$ time. We assume quantum random access here as well and provide our algorithm description at a high level. In fact, our solution in Sect. 3 can be seen as an algorithm (or even implemented as a small Boolean circuit) that utilizes the oracles of the original PMLG instance to create new oracles that are then used as input for a pre-existing quantum algorithm for shortest st-path in a directed graph.

For PMLG, we assume that our oracles allow us to query the indegree/outdegree and adjacency list of any vertex and the label of any edge. Any symbol in the pattern P can also be queried by specifying an index.

1.2   NC-QSETH

When establishing computational complexity results for quantum algorithms, using known lower bounds on query complexity has an immediate limitation for proving super-linear lower bounds on quantum time complexity. For problems where the input represents something such as a graph, once a linear number of queries have been made, the entire input is obtained by the algorithm. An alternative approach taken by the authors of [8] is to establish conditional lower bounds on quantum time complexity using the hypothesized hardness NC-QSETH. Although the details of NC-QSETH, which is based on a conjectured hardness of determining properties of circuits in a subset of the circuit class NC, are too complex to be covered here, we can use the result below.

Lemma 1 (LCS lower bounds based on NC-QSETH [8]). *Under NC-QSETH the Longest Common Subsequence problem (LCS) on two strings of length n cannot be solved in quantum time $\tilde{O}(n^{1.5-\varepsilon})$ for any constant $\varepsilon > 0$.*

---

[2] Elementary gates are defined in [6].

1.3 Our Results

We prove the following theorem in Sect. 2.

Theorem 1. *There exists a reduction from LCS with strings X and Y over alphabet Σ to PMLG with substitutions over a binary alphabet. This requires $O((|X|+|Y|)\log(|X|+|Y|)\cdot\log^2|\Sigma|)$ time (on a classical computer) and outputs a graph $G = (V,E)$ where $|V|,|E| = O(|X|\log(|X|+|Y|)\cdot\log^2|\Sigma|)$ and pattern $P[1,m]$ where $m = O(|Y|\log(|X|+|Y|)\cdot\log^2|\Sigma|)$.*

Theorem 1 gives us the following Corollaries.

Corollary 1. *An algorithm for PMLG with substitutions to the pattern over a binary alphabet running in quantum time $\tilde{O}(|E|^{1-\varepsilon}m+|E|m^{1-\varepsilon})$ for any constant $\varepsilon > 0$ would provide an algorithm running in quantum time $\tilde{O}(|X||Y|^{1-\varepsilon} + |X|^{1-\varepsilon}|Y|)$ for LCS.*

It should be noted that no strongly sub-quadratic quantum algorithms for LCS are known.

Corollary 2. *An algorithm for PMLG with substitutions to the pattern over a binary alphabet running in quantum time $\tilde{O}(|E|^{\frac{1}{2}-\varepsilon}m + |E|m^{\frac{1}{2}-\varepsilon})$ for any constant $\varepsilon > 0$, would provide an algorithm running in quantum time $\tilde{O}(|X|^{\frac{1}{2}-\varepsilon}|Y|+ |X||Y|^{\frac{1}{2}-\varepsilon})$ for LCS, violating NC-QSETH.*

In Sect. 3, we provide an algorithm running in quantum time $\tilde{O}(m\sqrt{|V||E|})$ for PMLG based on Durr et al.'s quantum algorithm for shortest path [11], implying a quantum speedup over the classical algorithm when the graph is not sparse, i.e., $|E| = \Omega(|V|^{1+\varepsilon})$. This algorithm also works when insertions and deletions are allowed to the pattern in addition to substitutions.

2 Reduction from LCS to PMLG

We first present a simplified version of the reduction to PMLG with a larger alphabet and then show how to modify it to obtain the result for PMLG on binary alphabets. In the decision version of LCS, we are given two strings X, Y and $k \geq 0$ and have to decide where there exists a common subsequence of X and Y having a length at least k. Suppose $|Y| \geq |X|$ and let $n = |Y|$.

We construct our graph G based on the string X. We start by making two sets of vertices $u_1, u_2, ..., u_{|X|}$ and $v_1, v_2, ..., v_{|X|}$. We add directed edges (v_i, u_i) with labels $X[i]$ for $1 \leq i \leq |X|$. All remaining edges are labeled with a new symbol $\#$ that is not found in either X or Y. We then create edges (u_i, v_{i+1}) for $1 \leq i \leq |X| - 1$. Next, for v_i, $1 \leq i \leq |X|$ we create edges (v_i, v_i), (v_i, v_{i+1}), (v_i, v_{i+2}), (v_i, v_{i+4}), ..., (v_i, v_{i+2^c}) for the largest c such that $i + 2^c \leq |X|$ and the edge $(u_{|X|}, u_{|X|})$. See Fig. 1. Let $\delta = n - k$ and

$$P = \#^{\lceil\log n\rceil+1}\ Y[1]\ \#^{\lceil\log n\rceil+1}\ Y[2]\ \#^{\lceil\log n\rceil+1}\ ...\ \#^{\lceil\log n\rceil+1}\ Y[n].$$

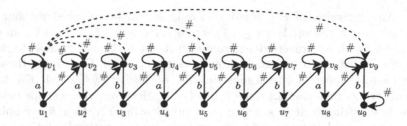

Fig. 1. Reduction from LCS to PMLG for $X = aababbbab$. The dashed edges are only shown from v_1 but similar edges are present from every v_i, $1 \leq i \leq |X|$. If $Y = baabbabaa$ then $P = \#^5 b \#^5 a \#^5 a \#^5 b \#^5 b \#^5 a \#^5 b \#^5 a \#^5 a$.

Lemma 2. *The graph distance from v_i to v_j for any $j > i$ is at most $\lceil \log n \rceil$.*

Proof. Let i' be the largest value such that $i \leq i' \leq j$ and there exists edge $(v_i, v_{i'}) \in E$. By construction $i' = i + 2^x$ for some $x \geq 0$. We claim $i' > \frac{j-i}{2} + i$. Otherwise $i' = i + 2^x \leq \frac{j-i}{2} + i$ implies $i + 2^{x+1} \leq j$, contradicting that index i' was the largest possible. Since the distance between the current index and j can always be at least halved, by repeatedly apply the same process, we need at most $\lceil \log n \rceil$ additional edges before reaching j.

The correctness of the reduction is established by the following lemma.

Lemma 3. *There exists an LCS of length at least k for strings X and Y iff there exists a walk in G that matches P after at most $\delta = n - k$ substitutions to P.*

Proof. First assume there exists an LCS of length $k' \geq k$, with $X[i_1]$, $X[i_2]$, ..., $X[i_{k'}]$ matching $Y[j_1]$, $Y[j_2]$, ..., $Y[j_{k'}]$. We obtain a walk on G as follows: starting at vertex v_{i_1}, we traverse the self-loop (v_{i_1}, v_{i_1}) until we reach the $Y[j_1]$ in P, substituting symbols in P to $\#$ as necessary. Then we follow edge (v_{i_1}, u_{i_1}) matching $Y[j_1]$ in P. We now traverse the edge (u_i, v_{i+1}) and the shortest path from v_{i+1} to v_{i_2}, which by Lemma 2 has at most $\lceil \log n \rceil$ edges. We next traverse the self-loop (v_{i_2}, v_{i_2}) until reaching the symbol $Y[j_2]$ in P, at which point we match $Y[j_2]$ with the edge (v_{j_2}, u_{j_2}). This process is repeated until $Y[n]$ is matched. Exactly $n - k' \leq n - k = \delta$ symbols in P are substituted to $\#$.

Next suppose there exists a walk in G that matches P with $\delta' \leq \delta$ substitutions. This implies that $n - \delta'$ of the non-$\#$-symbols in P are not substituted and instead matched with symbols on edges (v_i, u_i). By construction, once the edge (v_i, u_i) is traversed, the next edge with a non-$\#$-label traversed is an edge $(v_{i'}, u_{i'})$ where $i' > i$. Hence, the non-$\#$ symbols in P matched with edges in G correspond to a common subsequence of X and Y of length $n - \delta' \geq n - \delta = k$.

It can be easily shown that statement of Lemma 3 holds when deletions and insertions are also allowed to P. Lemma 4 proves this result.

Lemma 4. *Given graph G and path P as in our reduction, if a walk minimizes the number of edits to P, we can assume only substitutions are made.*

Proof. Any substring of P consisting of only $\#$ can be matched without edit cost from any vertex v_i, $1 \leq i \leq |X|$. From vertices u_i, $1 \leq i \leq |X|$, an edge with $\#$ needs to be traversed regardless so it would be suboptimal to delete any $\#$ in P that could be matched on an edge (u_i, v_{i+1}). Combining these, an optimal solution never deletes a substring of P of the form $\#^x$, $x \geq 1$. This leaves only substrings that contain some symbol $Y[i]$. However, the cost for deleting any such substring is at least the cost of substituting $Y[i]$ to a $\#$-symbol. We conclude that no deletions need to be made to P in an optimal solution.

For insertions, a similar argument holds. Any insertion of a substring of the form $\#^x$, $x \geq 1$, is clearly suboptimal since there exist enough $\#$-symbols to traverse from any two vertices v_i and v_j. An insertion that includes a non-$\#$-symbol is also unnecessary, since the edge matched against that symbol could have been not traversed for the same cost.

2.1 Hardness of PMLG over Binary Alphabet

Let $\Sigma' = \Sigma \cup \{\#\}$, $\sigma = |\Sigma'| \geq 3$, and $\ell = 2\lceil \log \sigma \rceil$. We will create our own constant weight binary code (i.e., one where all codewords have the same number of 1's) for Σ'. We first take $t = \lceil \log \sigma \rceil$. This makes $\binom{\ell}{t} \geq \sigma$ and allows us to assign to every symbol in Σ' a distinct binary string of length ℓ containing exactly t 1's. Let $\text{enc}(\alpha)$ denote this encoding for $\alpha \in \Sigma'$. Controlling the number of 1's allows us to compute the cost of an optimal solution, as described next. We modify the earlier reduction by replacing every edge (u, v) (allowing for $v = u$) having label $\alpha \in \Sigma'$ with:

- A directed path from u to v that matches $(10^{t-2+\ell})^t 0^{t-1} \text{enc}(\alpha)$. These paths are called *symbol paths*;
- A parallel directed path starting and ending at the same vertices (or vertex) that matches the string $(10^{t-2+\ell})^t 1^{t-1} 0^\ell$. These are called *escape paths*.

See Fig. 2.

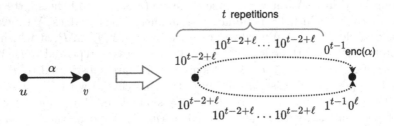

Fig. 2. Conversion of an edge in G with label α from u to v to two paths in G'.

We denote the resulting graph as G'. The pattern P' is created by replacing every symbol $P[i]$ with $(10^{t-2+\ell})^t 1^{t-1} \text{enc}(P[i])$ for $1 \leq i \leq n$. Let

$$\delta' = t(n-k) + (t-1)(|P| - (n-k))$$
$$= t(n-k) + (t-1)(n(\lceil \log n \rceil + 2) - (n-k)).$$

Note that $|V'|$, $|E'|$, and $|P'|$ are $O(n \log n \cdot \log^2 \sigma)$.

Lemma 5. *Any walk in G' matching P' with at most δ' mismatches must start at some vertex corresponding to an original vertex in V.*

Proof. First, consider a walk starting at some vertex w internal to a subdivided path (i.e., one not corresponding to an original vertex in G) and where w does not have an edge with 1 leaving it. By construction, this causes substrings of the form 10^{t-2+l} in P to not be synchronised and creates at least t mismatches for every complete sub-divided path traversed. To see this, for each substring $(10^{t-2+\ell})^t 1^{t-1} \text{enc}(P[i])$ in at most one of the 1's in the prefix $(10^{t-1+\ell})^t$ can be matched to an edge and at least one of the 0's is mismatched to a 1 edge as well.

If the walk starts at a non-original vertex with an edge leaving it labeled 1, but not a edge internal to subpath labeled 1^t in any escape path. Then for each substring $(10^{t-2+\ell})^t 1^{t-1} \text{enc}(P[i])$ in P', the substring $\text{enc}(\alpha)$ is forced to traverse a subpath labeled 0^t, causing at least t mismatches once again. Finally, suppose the walk starts at a non-original vertex with an edge leaving it labeled 1, but an internal to subpath labeled 1^t in some escape path. Then for each substring $(10^{t-2+\ell})^t 1^{t-1} \text{enc}(P[i])$ in P', the prefix $(10^{t-2+\ell})^t$ once again causes t mismatches.

In all cases, at least $t|P| > \delta'$ mismatches are causes in total. $\blacksquare$

Lemma 6. *There exists an LCS of length at least k for strings X and Y iff there exists a walk in G' that matches P' after at most δ' substitutions to P'.*

Proof. First, suppose there exists an LCS of length at least k. Follow the walk in G' corresponding to the walk in G that requires at most δ substitutions to P. When doing so, take the symbol path when the symbol in P matched the corresponding edge in G, and the escape path otherwise. This incurs $t - 1$ mismatches per subdivided edge corresponding to a match and t mismatches per subdivided edge corresponding to a mismatch. Hence the total number of mismatches is at most $t(n - k) + (t - 1)(|P| - (n - k)) = \delta'$.

Next, suppose there exists a walk in G' matching P' with at most δ' mismatches. By Lemma 5, substrings of the form $(10^{t-2+\ell})^t 1^{t-1} \text{enc}(P[i])$ are matched (after substitutions) to sub-paths in the walk that start at the beginning of symbol or escape paths. If $\text{enc}(P[i]) \neq \text{enc}(\alpha)$, then the number of mismatches for that substring is t since the number of mismatches for matching the escape path is t (Hamming distance of $1^{t-1} \text{enc}(P[i])$ and $1^{t-1} 0^\ell$), versus the symbol path, which is at least t (Hamming distance of $1^{t-1} \text{enc}(P[i])$ and $0^{t-1} \text{enc}(\alpha)$). If $\text{enc}(P[i]) = \text{enc}(\alpha)$, by matching the symbol path, the number of mismatches is $t - 1$ (Hamming distance of $1^{t-1} \text{enc}(P[i])$ and $0^{t-1} \text{enc}(\alpha)$). We conclude that in an optimal solution the total number of mismatches is t times the number of mismatched symbols between P and the corresponding walk in G, plus $t - 1$ times the number of matched symbols between P and the corresponding walk in G. Hence, if the number mismatches for G' is at most $\delta' = t(n - k) + (t - 1)(|P| - (n - k))$, the number mismatched symbols in G is at most $n - k = \delta$. By Lemma 3, this implies the LCS of X and Y is at least k. $\blacksquare$

This completes the proof of Theorem 1.

3 Quantum Algorithm for PMLG

We will use Durr et al.'s [11] single-source shortest path algorithm as a black box. Their algorithm is a modification of Dijkstra's algorithm that utilizes a minimum finding version of Grover's search to obtain quantum time/query complexity $\tilde{O}(\sqrt{|V||E|})$ on a graph $G = (V, E)$. It solves the st-shortest path problem correctly with constant probability greater than $\frac{1}{2}$. The version of this algorithm that we are using assumes the graph is represented using adjacency lists (in the form of an oracle). Given that the outdegree v_i is $d^+(v_i)$, the oracles f_i : $[d^+(v_i)] \mapsto \{1, ..., |V|\} \times \mathbb{N}$ have

$$f_i(j) = (\text{the } j^{th} \text{ vertex adjacent to vertex } v_i, \text{ the weight on the corresponding edge}).$$

We assume that $|E| = \Omega(|V|)$ and assign an arbitrary ordering to V.

A reduction from PMLG to the Single Source Shortest Path problem on an *alignment graph* was shown by Amir et al. in [5]. We will be using the oracles for G to implicitly construct an alignment graph for $G = (V, E)$ and $P[1, m]$, denoted G'. The number of vertices in G' is $\Theta(|V|m)$ and the number of edges is $\Theta(|E|m)$. Because of this, if we explicitly constructed the alignment graph, the $\Theta(|E|m)$ edges would result in no speed up over the classical algorithm. The key insight into efficiently using Durr et al.'s algorithm is that G' need not be explicitly constructed to simulate the oracles used by the shortest path algorithm. We show how the output of the oracles for G' can be computed in constant time given the oracles for G and P.

Our algorithm allows for insertions and deletions in addition to substitutions. Assume we have substitution cost S, deletion cost D, and insertion cost I. The alignment graph $G' = (V', E')$ is as follows: The vertex set is

$$V' = \{v_i^j \mid 1 \le i \le |V|, 1 \le j \le m + 1\} \cup \{s, t\}.$$

Indicating edges with the triple (start vertex, end vertex, weight), the edge set is

$$
\begin{aligned}
E' = &\{(s, v_i^1, 0) \mid 1 \le i \le |V|\} \ \cup \\
&\{(v_i^j, v_h^j, I) \mid 1 \le j \le m, (v_i, v_h) \in E\} \ \cup \\
&\{(v_i^j, v_h^{j+1}, 0) \mid 1 \le j \le m, (v_i, v_h) \in E \text{ and } label((v_i, v_h)) = P[j]\} \ \cup \\
&\{(v_i^j, v_h^{j+1}, S) \mid 1 \le j \le m, (v_i, v_h) \in E \text{ and } label((v_i, v_h)) \ne P[j]\} \ \cup \\
&\{(v_i^j, v_i^{j+1}, D) \mid 1 \le i \le |V|, 1 \le j \le m\} \ \cup \\
&\{(v_i^{m+1}, t, 0) \mid 1 \le i \le |V|\}.
\end{aligned}
$$

The linearized index for s is 0, for t it is $(m+1)|V|+1$, and for v_i^j, $1 \le i \le |V|$, $1 \le j \le m + 1$, it is $(j-1)|V| + i$.

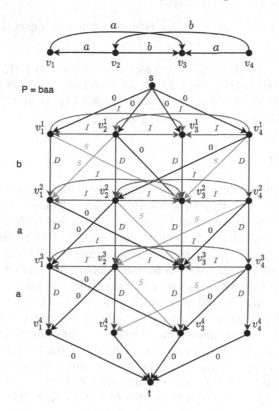

Fig. 3. An alignment graph G' (bottom) that is constructed from the starting graph G (top) and pattern $P = baa$. The edges labeled I correspond to insertion and have weight I; the edges labeled S correspond to substitution and have weight S; the edges labeled D correspond to deletion and have weight D; the black edges correspond to an exact match and have weight 0.

For $1 \le i \le |V|$, $1 \le j \le m$, $d^+(v_i^j) = 2d^+(v_i) + 1$ and the oracle is $f_i^j : [d^+(v_i^j)] \mapsto \{0, ..., |V'| - 1\} \times \mathbb{N}$, where

$$
f_i^j(k) = \begin{cases}
((j-1)|V| + f_i(k), I) & 1 \le k \le d^+(v_i) \\
(j|V| + f_i(k - d^+(v_i)), 0) & d^+(v_i) + 1 \le k \le 2d^+(v_i), \ label((v_i, v_{f_i(k - d^+(v_i))})) = P[j] \\
(j|V| + f_i(k - d^+(v_i)), S) & d^+(v_i) + 1 \le k \le 2d^+(v_i), \ label((v_i, v_{f_i(k - d^+(v_i))})) \ne P[j] \\
(j|V| + i, D) & k = 2d^+(v_i) + 1
\end{cases}
$$

For $1 \le i \le |V|$ and $j = m + 1$, $d^+(v_i^j) = 1$ and $f_i^j(1) = ((m + 1)|V| + 1, 0)$. For vertex s and $1 \le k \le |V|$, $f_0(k) = (k, 0)$. See Fig. 3 for an example alignment graph.

Lemma 7 ([5]). *There exists an st-path in the alignment graph G' with total weight δ iff there exists a walk in G that P matches after δ edits.*

Applying the algorithm of Durr et al. and utilizing the oracles above gives an algorithm running in quantum time $\tilde{O}(\sqrt{|V'||E'|})$ for PMLG. Using that

$|V'| = (m + 1)|V| + 2$ and $|E'| = \Theta(m|E|)$ this has query/time complexity $\tilde{O}(m\sqrt{|V||E|})$.

Theorem 2. *There exists a quantum algorithm that solves PMLG (exact matching, matching with substitutions to P, or matching with substitutions, insertions, and deletions to P) with constant probability greater than $\frac{1}{2}$ and has $\tilde{O}(m\sqrt{|V||E|})$ quantum time and query complexity.*

4   Discussion

We leave open the problem of establishing the same reduction from LCS to PMLG when edits (substitutions, insertion, and deletions) are allowed to the pattern and the PMLG alphabet is binary. Lemma 4 establishes this result for polynomial sized alphabets. Note that the hardness of LCS under NC-QSETH (Lemma 1) holds for constant-sized alphabets, thus Corollary 2 can be extended to PMLG with edits to the pattern for constant-sized alphabets.

Our reduction from LCS creates a sparse graph. A subquadratic time reduction to a dense graph would give an improved quantum algorithm for LCS, suggesting the challenge of finding such a reduction. Moreover, the graph in our reduction is cyclic. This is interesting in light of improvements in the query complexity of quantum algorithms for recognizing if a string is in a regular language when the monoid representation of the regular language is acyclic [2]. If these results for monoids can be efficiently transferred to acyclic NFAs, it suggests the challenge of finding a reduction from LCS to PMLG on DAGs when $\delta = 0$.

Acknowledgement. This research is supported in part by the U.S. National Science Foundation (NSF) grants CCF-2146003 and CCF-2112643.

References

1. Pangaia, November 2020. https://www.pangenome.eu/
2. Aaronson, S., Grier, D., Schaeffer, L.: A quantum query complexity trichotomy for regular languages. In: 2019 IEEE 60th Annual Symposium on Foundations of Computer Science (FOCS), pp. 942–965. IEEE (2019)
3. Akmal, S., Jin, C.: Near-optimal quantum algorithms for string problems. In: Naor, J.S., Buchbinder, N. (eds.) Proceedings of the 2022 ACM-SIAM Symposium on Discrete Algorithms, SODA 2022, Virtual Conference/Alexandria, VA, USA, 9–12 January 2022, pp. 2791–2832. SIAM (2022). https://doi.org/10.1137/1.9781611977073.109
4. Alanko, J., D'Agostino, G., Policriti, A., Prezza, N.: Regular languages meet prefix sorting. In: Chawla, S. (ed.) Proceedings of the 2020 ACM-SIAM Symposium on Discrete Algorithms, SODA 2020, Salt Lake City, UT, USA, 5–8 January 2020, pp. 911–930. SIAM (2020). https://doi.org/10.1137/1.9781611975994.55
5. Amir, A., Lewenstein, M., Lewenstein, N.: Pattern matching in hypertext. J. Algorithms **35**(1), 82–99 (2000). https://doi.org/10.1006/jagm.1999.1063
6. Barenco, A., et al.: Elementary gates for quantum computation. Phys. Rev. A **52**(5), 3457 (1995)

7. Buhrman, H., Loff, B., Patro, S., Speelman, F.: Memory compression with quantum random-access gates. CoRR abs/2203.05599 (2022). https://doi.org/10.48550/arXiv.2203.05599

8. Buhrman, H., Patro, S., Speelman, F.: A framework of quantum strong exponential-time hypotheses. In: Bläser, M., Monmege, B. (eds.) 38th International Symposium on Theoretical Aspects of Computer Science, STACS 2021, Saarbrücken, Germany, 16–19 March 2021 (Virtual Conference). LIPIcs, vol. 187, pp. 19:1–19:19. Schloss Dagstuhl - Leibniz-Zentrum für Informatik (2021). https://doi.org/10.4230/LIPIcs.STACS.2021.19

9. Chen, S., Krusche, P., Dolzhenko, E., Sherman, R.M., Petrovski, R., Schlesinger, F., Kirsche, M., Bentley, D.R., Schatz, M.C., Sedlazeck, F.J., et al.: Paragraph: a graph-based structural variant genotyper for short-read sequence data. Genome Biol. **20**(1), 1–13 (2019). https://doi.org/10.1186/s13059-019-1909-7

10. The Computational Pan-Genomics Consortium: Computational pan-genomics: status, promises and challenges. Briefings Bioinform. **19**(1), 118–135 (2018)

11. Dürr, C., Heiligman, M., Høyer, P., Mhalla, M.: Quantum query complexity of some graph problems. SIAM J. Comput. **35**(6), 1310–1328 (2006). https://doi.org/10.1137/050644719

12. Eggertsson, H.P., et al.: GraphTyper2 enables population-scale genotyping of structural variation using pangenome graphs. Nat. Commun. **10**(1), 1–8 (2019)

13. Eizenga, J.M., et al.: Pangenome graphs. Ann. Rev. Genomics Hum. Genet. **21**, 139–162 (2020)

14. Equi, M., de Griend, A.M., Mäkinen, V.: From bit-parallelism to quantum: breaking the quadratic barrier. CoRR abs/2112.13005 (2021). https://arxiv.org/abs/2112.13005

15. Equi, M., Grossi, R., Mäkinen, V., Tomescu, A., et al.: On the complexity of string matching for graphs. In: 46th International Colloquium on Automata, Languages, and Programming (ICALP 2019). Schloss Dagstuhl-Leibniz-Zentrum fuer Informatik (2019)

16. Gagie, T., Manzini, G., Sirén, J.: Wheeler graphs: a framework for BWT-based data structures. Theor. Comput. Sci. **698**, 67–78 (2017). https://doi.org/10.1016/j.tcs.2017.06.016

17. Garrison, E., et al.: Variation graph toolkit improves read mapping by representing genetic variation in the reference. Nat. Biotechnol. **36**(9), 875–879 (2018)

18. Gibney, D., Hoppenworth, G., Thankachan, S.V.: Simple reductions from formula-SAT to pattern matching on labeled graphs and subtree isomorphism. In: Le, H.V., King, V. (eds.) 4th Symposium on Simplicity in Algorithms, SOSA 2021, Virtual Conference, 11–12 January 2021, pp. 232–242. SIAM (2021). https://doi.org/10.1137/1.9781611976496.26

19. Gibney, D., Thankachan, S.V.: On the hardness and inapproximability of recognizing wheeler graphs. In: Bender, M.A., Svensson, O., Herman, G. (eds.) 27th Annual European Symposium on Algorithms, ESA 2019, Munich/Garching, Germany, 9–11 September 2019. LIPIcs, vol. 144, pp. 51:1–51:16. Schloss Dagstuhl - Leibniz-Zentrum für Informatik (2019). https://doi.org/10.4230/LIPIcs.ESA.2019.51

20. Gibney, D., Thankachan, S.V.: On the complexity of recognizing wheeler graphs. Algorithmica **84**(3), 784–814 (2022). https://doi.org/10.1007/s00453-021-00917-5

21. Gibney, D., Thankachan, S.V., Aluru, S.: The complexity of approximate pattern matching on de Bruijn graphs. In: Pe'er, I. (ed.) RECOMB 2022. LNCS, vol. 13278, pp. 263–278. Springer, Cham (2022). https://doi.org/10.1007/978-3-031-04749-7_16

22. Grover, L.K.: A fast quantum mechanical algorithm for database search. In: Proceedings of the Twenty-Eighth Annual ACM Symposium on Theory of Computing, pp. 212–219 (1996)
23. Gruska, J., et al.: Quantum Computing, vol. 2005. McGraw-Hill, London (1999)
24. Hariharan, R., Vinay, V.: String matching in õ(sqrt(n)+sqrt(m)) quantum time. J. Discrete Algorithms **1**(1), 103–110 (2003). https://doi.org/10.1016/S1570-8667(03)00010-8
25. Hickey, G., et al.: Genotyping structural variants in pangenome graphs using the vg toolkit. Genome Biol. **21**(1), 1–17 (2020). https://doi.org/10.1186/s13059-020-1941-7
26. Jain, C., Zhang, H., Gao, Y., Aluru, S.: On the complexity of sequence-to-graph alignment. J. Comput. Biol. **27**(4), 640–654 (2020)
27. Li, H., Feng, X., Chu, C.: The design and construction of reference pangenome graphs with minigraph. Genome Biol. **21**(1), 1–19 (2020). https://doi.org/10.1186/s13059-020-02168-z
28. Limasset, A., Flot, J.F., Peterlongo, P.: Toward perfect reads: self-correction of short reads via mapping on de Bruijn graphs. Bioinformatics **36**(5), 1374–1381 (2020)
29. Magniez, F., Nayak, A., Roland, J., Santha, M.: Search via quantum walk. SIAM J. Comput. **40**(1), 142–164 (2011)
30. Manber, U., Wu, S.: Approximate string matching with arbitrary costs for text and hypertext. In: Advances in Structural and Syntactic Pattern Recognition, pp. 22–33. World Scientific (1992)
31. Morisse, P., Lecroq, T., Lefebvre, A.: Hybrid correction of highly noisy long reads using a variable-order de Bruijn graph. Bioinformatics **34**(24), 4213–4222 (2018)
32. Navarro, G.: Improved approximate pattern matching on hypertext. Theor. Comput. Sci. **237**(1–2), 455–463 (2000). https://doi.org/10.1016/S0304-3975(99)00333-3
33. Park, K., Kim, D.K.: String matching in hypertext. In: Galil, Z., Ukkonen, E. (eds.) CPM 1995. LNCS, vol. 937, pp. 318–329. Springer, Heidelberg (1995). https://doi.org/10.1007/3-540-60044-2_51
34. Paten, B., Novak, A.M., Eizenga, J.M., Garrison, E.: Genome graphs and the evolution of genome inference. Genome Res. **27**(5), 665–676 (2017)
35. Shor, P.W.: Algorithms for quantum computation: discrete logarithms and factoring. In: 35th Annual Symposium on Foundations of Computer Science, Santa Fe, New Mexico, USA, 20–22 November 1994, pp. 124–134. IEEE Computer Society (1994). https://doi.org/10.1109/SFCS.1994.365700
36. Tzanis, E.: A quantum algorithm for string matching. In: Guimarães, N., Isaías, P.T. (eds.) AC 2005, Proceedings of the IADIS International Conference on Applied Computing, Algarve, Portugal, 22–25 February 2005, vol. 2. pp. 374–377. IADIS (2005)

Pattern Matching Under DTW Distance

Garance Gourdel[1,2]($\boxtimes$), Anne Driemel[3], Pierre Peterlongo[2],
and Tatiana Starikovskaya[1]

[1] DIENS, École normale supérieure de Paris, PSL Research University, Paris, France
`garance.gourdel@gmail.com, starikovskaya@di.ens.fr`
[2] IRISA Inria Rennes, Rennes, France
`pierre.peterlongo@inria.fr`
[3] Hausdorff Center for Mathematics, University of Bonn, Bonn, Germany
`driemel@cs.uni-bonn.de`

Abstract. In this work, we consider the problem of pattern matching under the dynamic time warping (DTW) distance motivated by potential applications in the analysis of biological data produced by the third generation sequencing. To measure the DTW distance between two strings, one must "warp" them, that is, double some letters in the strings to obtain two equal-lengths strings, and then sum the distances between the letters in the corresponding positions. When the distances between letters are integers, we show that for a pattern P with m runs and a text T with n runs:

1. There is an $\mathcal{O}(m + n)$-time algorithm that computes all locations where the DTW distance from P to T is at most 1;
2. There is an $\mathcal{O}(kmn)$-time algorithm that computes all locations where the DTW distance from P to T is at most k.

As a corollary of the second result, we also derive an approximation algorithm for general metrics on the alphabet.

Keywords: Dynamic time warping distance · Pattern matching · Small-distance regime · Approximation algorithms

1 Introduction

Introduced more than forty years ago [27], the dynamic time warping (DTW) distance has become an essential tool in the time series analysis and its applications due to its ability to preserve the signal despite speed variation in compared sequences. To measure the DTW distance between two discrete temporal sequences, one must "warp" them, that is, replace some data items in the sequences with multiple copies of themselves to obtain two equal-lengths sequences, and then sum the distances between the data items in the corresponding positions.

The DTW distance has been extensively studied for parameterized curves—sequences where the data items are points in a multidimensional space—specifically, in the context of locality sensitive hashing and nearest neighbor

This work was partially funded by the grants ANR-20-CE48-0001, ANR-19-CE45-0008 SeqDigger and ANR-19-CE48-0016 from the French National Research Agency.

search [7,9]. In this work, we focus on a somewhat simpler, but surprisingly much less studied setting when the data items are elements of a finite set, the alphabet. Following traditions, we call such sequences *strings*.

The classical textbook dynamic programming algorithm computes the DTW distance between two N-length strings in $\mathcal{O}(N^2)$ time and space. Unfortunately, unless the Strong Exponential Time Hypothesis is false, there is no algorithm with strongly subquadratical time even for ternary alphabets [1,5,17]. On the other hand, very recently Gold and Sharir [12] showed the first weakly sub-quadratic time algorithm (to be more precise, the time complexity of the algorithm is $\mathcal{O}(N^2 \log \log \log N / \log \log N)$). Kuszmaul [17] gave a $\mathcal{O}(kN)$-time algorithm that computes the value of the distance between the strings if it is bounded by k, assuming that the distance between any two distinct letters of the alphabet is at least one, and used it to derive a subquadratic-time approximation algorithm for the general case. Finally, it is known that binary strings admit much faster algorithms: Abboud, Backurs, and Vassilevska Williams [1] showed an $O(N^{1.87})$-time algorithm followed by a linear-time algorithm by Kuszmaul [19].

The problem of computing the DTW distance has also been studied in the sparse and run-length compressed settings, as well as in the low distance regime. In the sparse setting, we assume that most letters of the string are zeros. Hwang and Gelfand [15] gave an $\mathcal{O}((s+t)N)$-time algorithm, where s and t denote the number of non-zero letters in each of the two strings. On sparse binary strings, the distance can be computed in $\mathcal{O}(s+t)$ time [16,24]. Froese et al. [11] suggested an algorithm with running time $\mathcal{O}(mN+nM)$, where M, N are the length of the strings, and m, n are the sizes of their run length encodings. If $n \in \mathcal{O}(\sqrt{N})$ and $m \in \mathcal{O}(\sqrt{M})$, their algorithm runs in time $\mathcal{O}(nm \cdot (n+m))$. For binary strings, the DTW distance can be computed in $\mathcal{O}(nm)$ time [8].

Nishi et al. [25] considered the question of computing the DTW distance in the dynamic setting when the stings can be edited, and Sakai and Inenaga [26] showed a reduction from the problem of computing the DTW distance to the problem of computing the longest increasing subsequence, which allowed them to give polynomial-time algorithms for a series of DTW-related problems.

In this work, we focus on the pattern matching variant of the problem: Given a pattern P and a text T, one must output the smallest DTW distance between P and a suffix of $T[1 \mathinner{.\,.} r]$ for every position r of the text.

Our interest to this problem sparks from its potential applications in Third Generation Sequencing (TGS) data comparisons. TGS has changed the genomic landscape as it allows to sequence reads of few dozens of thousand of letters where previous sequencing techniques were limited to few hundred letters [2]. However, TGS suffers from a high error rate (from ≈ 1 to 10% depending on the used techniques) mainly due to the fact that the DNA sequences are read and thus sequenced at an uneven speed. The uneven sequencing speed has a major impact in the sequencing quality of DNA regions composed of two or more equal consecutive letters. Those regions, called *homopolymers*, are hardly correctly sequenced as, due to the uneven sequencing speed, their size cannot be precisely determined [14]. In particular, a common post-sequencing task consists in aligning the obtained reads to a reference genome. This enables for instance to predict alternative splicing and gene expression [13] or to detect structural variations [23]. All known aligners use the edit distance, most likely, due to the availability of software tools for the latter (see [22] and references therein). However, we find that the nature of TGS errors is much better described by the DTW distance, which we confirm experimentally in Sect. 5.

Our Contribution. As a baseline, the problem of pattern matching under the DTW distance can be solved using dynamic programming in time $\mathcal{O}(MN)$, where M is the length of the pattern and N of the text (Eq. 1).

In this work, we aim to show more efficient algorithms for the low-distance regime on run-length compressible data, which is arguably the most interesting setting for the TGS data processing. Formally, in the k-DTW *problem* we are given an integer $k > 0$, a pattern P and a text T, and must find all positions r of the text such that the smallest DTW distance between the pattern P and a suffix of $T[1 \mathinner{.\,.} r]$ does not exceed k. One might hope that the DTW distance is close enough to the edit distance and thus is amenable to the techniques developed for the latter, such as [20,21]. In the full version, we show that this is indeed the case for $k = 1$:

Lemma 1. *Given run-length encodings of a pattern P and of a text T over an alphabet Σ and a distance $d : \Sigma \times \Sigma \to \mathbb{Z}^+$, the 1-DTW problem can be solved in $\mathcal{O}(m + n)$ time, where m is the number of runs in P and n is the number of runs in T. The output is given in a compressed form, with a possibility to retrieve each position in constant time.*

Unfortunately, extending the approach of [20,21] to higher values of k seems to be impossible as it is heavily based on the fact that in the edit distance dynamic programming matrix the distances are non-decreasing on every diagonal, which is not the case for the DTW distance (see Fig. 1).

In Sect. 3 we develop a different approach. Interestingly, we show that the value of any cell of the bottom row and the right column of a block of the dynamic programming table (i.e. a subtable formed by a run in the pattern and a run in the text) can be computed in constant time given a constant-time oracle access to the left column and the top row. Combining this with a compact representation of the k-bounded values, we obtain the following result:

Theorem 1. *Given run-length encodings of a pattern P and of a text T over an alphabet Σ and a distance $d : \Sigma \times \Sigma \to \mathbb{Z}^+$, the k-DTW problem can be solved in $\mathcal{O}(kmn)$ time, where m is the number of runs in P and n is the number of runs in T. The output is given in a compressed form, with a possibility to retrieve each position in constant time.*

We note that while our algorithm can be significantly faster than the baseline, its worst-case time complexity is cubic. We leave it as an open question whether there exists an $\mathcal{O}(k \cdot (m+n))$-time algorithm. Finally, in Sect. 4 we use Theorem 1 to derive an approximation algorithm for the general variant of pattern matching under the DTW distance.

	G	G	T	T	T	T	C	T	T	A	T	T	T	T	G	G	T	G	A	T	A
0	0	0	0	0	0	0	0	0	0	0	0	0	0	0	0	0	0	0	0	0	0
A ∞	1	1	1	1	1	1	1	1	1	0	1	1	1	1	1	1	1	1	0	1	0
A ∞	2	2	2	2	2	2	2	2	2	0	1	2	2	2	2	2	2	2	0	1	0
T ∞	3	3	2	2	2	2	3	2	2	1	0	0	0	0	1	2	2	3	1	0	1
T ∞	4	4	2	2	2	2	3	2	2	2	0	0	0	0	1	2	2	3	2	0	1
A ∞	5	5	3	3	3	3	3	3	3	2	1	1	1	1	1	2	3	3	2	1	0
T ∞	6	6	3	3	3	3	4	3	3	3	1	1	1	1	2	2	2	3	3	1	1

Fig. 1. Consider $P = AATTAT$ and $T = GGTTTTCTTATTTTGGTGATA$. A cell (i, j) contains the smallest DTW distance between $P[1..i]$ and $T[1..j]$, where the distance between two letters equals one if they are distinct and zero otherwise. A non-monotone diagonal of the table is shown in red. (Color figure online)

2   Preliminaries

We assume a polynomial-size alphabet Σ with σ *letters*. A *string* X is a sequence of letters. If the sequence has length zero, it is called the *empty string*. Otherwise, we assume that the letters in X are numbered from 1 to $n =: |X|$ and denote the i-th letter by $X[i]$. We define $X[i..j]$ to be equal to $X[i] \ldots X[j]$ which we call a *substring* of X if $i \leq j$ and to the empty string otherwise. If $j = n$, we call a substring $X[i..j]$ a *suffix* of X.

Definition 1 (Run, Run-length encoding). *A* run *of a string X is a maximal substring $X[i..j]$ such that $X[i] = X[i + 1] = \ldots = X[j]$. The* run-length encoding *of a string X,* RLE(X) *is a sequence obtained from X by replacing each run with a tuple consisting of the letter forming the run and the length of the run. For example,* RLE$(aabbbc) = (a, 2)(b, 3)(c, 1)$.

Let $d : \Sigma \times \Sigma \to \mathbb{R}^+$ be a distance function such that for any letters $a, b \in \Sigma$, $a \neq b$, we have $d(a, a) = 0$ and $d(a, b) > 0$. The dynamic time warping distance DTW$_d(X, Y)$ between strings $X, Y \in \Sigma^*$ is defined as follows. If both strings are empty, DTW$_d(X, Y) = 0$. If one of the strings is empty, and the other is not, then DTW$_d(X, Y) = \infty$. Otherwise, let $X = X[1]X[2] \ldots X[r]$ and $Y = Y[1]Y[2] \ldots Y[q]$. Consider an $r \times q$ grid graph such that each vertex (i, j) has (at most) three outgoing edges: one going to $(i + 1, j)$ (if it exists), one to $(i + 1, j + 1)$ (if it exists), and one to $(i, j + 1)$ (if it exists). A path π in the graph starting at $(1, 1)$ and ending at (r, q) is called a *warping path*, and its *cost* is defined to be $\sum_{(i,j) \in \pi} d(X[i], Y[j])$. Finally, DTW$_d(X, Y)$ is defined to be the minimum cost of a warping path for X, Y. Below we omit d if it is clear from the context.

Let $M = |P|$, $N = |T|$, and D be an $(M + 1) \times (N + 1)$ table where the rows are indexed from 0 to M, and the columns from 0 to N such that:

1. For all $j \in [0, N]$, $D[0, j] = 0$;
2. For all $i \in [1, M]$, $D[i, 0] = +\infty$;

3. For all $i \in [1, M]$ and $j \in [1, N]$, $D[i, j]$ equals the smallest DTW distance between $P[1 .. i]$ and a suffix of $T[1 .. j]$.

(See Fig. 1). To solve the pattern matching problem under the DTW distance, it suffices to compute the table D, which can be done in $\mathcal{O}(MN)$ time via a dynamic programming algorithm, using the following recursion for all $1 \leq i \leq M, 1 \leq j \leq N$:

$$D[i, j] = \min\{D[i - 1, j - 1], D[i - 1, j], D[i, j - 1]\} + d(P[i], T[j]) \qquad (1)$$

In the subsequent sections, we develop more efficient solutions for the low-distance regime on run-length compressible data. We will be processing the table D by blocks, defined as follows: A subtable $D[i_p .. j_p, i_t .. j_t]$ is called a *block* if $P[i_p .. j_p]$ is a run in P or $i_p = j_p = 0$, and $T[i_t .. j_t]$ is a run in T or $i_t = j_t = 0$. For $i_p, i_t > 0$, a block $D[i_p .. j_p, i_t .. j_t]$ is called *homogeneous* if $P[i_p] = T[i_t]$. (For example, a block $D[3 .. 4][3 .. 6]$ in Fig. 1 is homogeneous.) A block such that all cells in it contain a value q, for some fixed integer q, is called a *q-block*. (For example, a block $D[5 .. 5][11 .. 14]$ in Fig. 1 is a 1-block.) The *border* of a block is the set of the cells contained in its top and bottom rows, as well as first and last columns. Consider a cell (a, b) in B. We say that a block B' is the *top neighbor* of B if it contains $(a - 1, b)$, the *left neighbor* if it contains $(a, b - 1)$, and the *diagonal neighbor* if it contains $(a - 1, b - 1)$.

The following lemma is shown by induction in Appendix A:

Lemma 2. *Consider a block* $B = D[i_p .. j_p, i_t .. j_t]$ *and cell* (a, b) *in it. If* $i_p \leq a < j_p$, *then* $D[a, b] \leq D[a + 1, b]$ *and if* $i_t \leq b < j_t$, *then* $D[a, b] \leq D[a, b + 1]$.

By Eq. 1, inside a homogeneous block each value is equal to the minimum of its neighbors. Therefore, the values in a row or in a column cannot increase and we have the following corollary:

Corollary 1. *Each homogeneous block is a q-block for some value q.*

3   Main Result: $\mathcal{O}(kmn)$-Time Algorithm

In this section, we show Theorem 1 that for a pattern P with m runs and a text T with n runs gives an $\mathcal{O}(kmn)$-time algorithm. We start with the following lemma which is a keystone to our result:

Lemma 3. *For a block* $D[i_p .. j_p, i_t .. j_t]$ *let* $h = j_p - i_p$, $w = j_t - i_t$, *and* $d = d(P[i_p], T[i_t])$. *We have for every* $i_p < x \leq j_p$:

$$D[x, j_t] = \begin{cases} D[i_p, j_t - (x - i_p)] + (x - i_p) \cdot d \text{ if } x - i_p \leq w; \\ D[x - w, i_t] + w \cdot d \text{ otherwise.} \end{cases} \qquad (2)$$

For every $i_t < y \leq j_t$:

$$D[j_p, y] = \begin{cases} D[j_p - (y - i_t), i_t] + (y - i_t) \cdot d \text{ if } y - i_t \leq h; \\ D[i_p, y - h] + h \cdot d \text{ otherwise.} \end{cases} \qquad (3)$$

Proof. For a homogeneous block, we have $d = 0$, and by Corollary 1 all the values in such a block are equal, hence the claim of the lemma is trivially true.

Assume now $d > 0$. Consider x, $i_p < x \leq j_p$, and let us show Eq. 2, Eq. 3 can be shown analogously. Let π be a warping path realizing $D[x, j_t]$. Let (a, b) be the first node of π belonging to the block. We have $a \in [i_p, j_p]$ and $b \in [i_t, j_t]$ and either $a = i_p$ or $b = i_t$. The number of edges of π in the block from (a, b) to (x, j_t) must be minimal, else there would be a shorter path, thus it is equal to $\max\{x - a, j_t - b\}$ and $D[x, j_t] = D[a, b] + \max\{x - a, j_t - b\} \cdot d$ (Fig. 2).

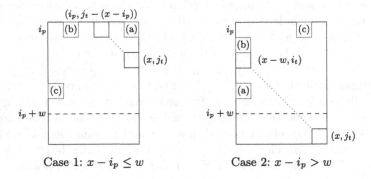

Case 1: $x - i_p \leq w$ Case 2: $x - i_p > w$

Fig. 2. Cases of Lemma 3. Possible locations of the cell (a, b) are shown in blue. (Color figure online)

Case 1: $x - i_p \leq w$. Consider a cell $(i_p, j_t - (x - i_p))$. There is a path from $(i_p, j_t - (x - i_p))$ to (x, j_t) that takes $x - i_p$ diagonal steps inside the block, and therefore $D[x, j_t] \leq D[i_p, j_t - (x - i_p)] + (x - i_p) \cdot d$. We now show that $D[x, j_t] \geq D[i_p, j_t - (x - i_p)] + (x - i_p) \cdot d$, which implies the claim of the lemma.

(a) If $a = i_p$ and $b \geq j_t - (x - i_p)$, then $\max\{x - i_p, j_t - b\} = x - i_p$. We have
$D[x, j_t] = D[i_p, b] + (x - i_p) \cdot d \geq D[i_p, j_t - (x - i_p)] + (x - i_p) \cdot d$ (Lemma 2).
(b) If $a = i_p$ and $b < j_t - (x - i_p)$, then $\max\{x - i_p, j_t - b\} = j_t - b$. As there is a path from $(a, b) = (i_p, b)$ to $(i_p, j_t - (x - i_p))$ of length $(j_t - (x - i_p) - b)$, we have $D[i_p, j_t - (x - i_p)] \leq D[i_p, b] + (j_t - (x - i_p) - b) \cdot d$. Consequently,

$$D[x, j_t] = D[i_p, b] + (j_t - b) \cdot d$$
$$\geq D[i_p, j_t - (x - i_p)] - (j_t - (x - i_p) - b) \cdot d + (j_t - b) \cdot d \text{ (Eq. 1)}$$
$$= D[i_p, j_t - (x - i_p)] + (x - i_p) \cdot d$$

(c) If $b = i_t$, then $i_p \leq a$ and $\max\{x - a, j_t - b\} \leq \max\{x - i_p, w\} = w$. As there is a path from (i_p, i_t) to $(i_p, j_t - (x - i_p))$ of length $(j_t - (x - i_p) - i_t)$, we have $D[i_p, j_t - (x - i_p)] \leq D[i_p, i_t] + (j_t - (x - i_p) - i_t) \cdot d$. Therefore,

$$D[x, j_t] = D[a, i_t] + w \cdot d \geq D[i_p, i_t] + w \cdot d \text{ (Lemma 2)}$$
$$\geq D[i_p, j_t - (x - i_p)] - (j_t - (x - i_p) - i_t) \cdot d + w \cdot d$$
$$= D[i_p, j_t - (x - i_p)] + (x - i_p) \cdot d$$

Case 2: $x - i_p > w$. Consider a cell $(x - w, i_t)$. There is a path from $(x - w, i_t)$ to (x, j_t) that takes w diagonal steps inside the block, and therefore $D[x, j_t] \leq D[x - w, i_t] + w \cdot d$. We now show that $D[x, j_t] \geq D[x - w, i_t] + w \cdot d$, which implies the claim of the lemma.

(a) If $b = i_t$ and $a \geq x - w$, then $\max\{x - a, j_t - b\} = \max\{x - a, w\} = w$ and we have $D[x, j_t] = D[a, i_t] + w \cdot d \geq D[x - w, i_t] + w \cdot d$ (Lemma 2).

(b) If $b = i_t$ and $a < x - w$, then $\max\{x - a, j_t - b\} = \max\{x - a, w\} = x - a$. As there is a path from (a, i_t) to $(x - w, i_t)$ of length $(x - w - a)$, we have $D[x - w, i_t] \leq D[a, i_t] + (x - w - a) \cdot d$ by definition. Therefore,

$$
\begin{aligned}
D[x, j_t] &= D[a, i_t] + (x - a) \cdot d \\
&\geq D[x - w, i_t] - (x - w - a) \cdot d + (x - a) \cdot d \\
&= D[x - w, i_t] + w \cdot d
\end{aligned}
$$

(c) If $a = i_p$, $b \geq i_t$ and thus $\max\{x - a, j_t - b\} \leq \max\{x - i_p, w\} = x - i_p$. Additionally, as there is a path from (i_p, i_t) to $(x - w, i_t)$ of length $(x - w - i_p)$ we have $D[x - w, i_t] \leq D[i_p, i_t] + (x - w - i_p) \cdot d$. Consequently,

$$
\begin{aligned}
D[x, j_t] &= D[i_p, b] + (x - i_p) \cdot d \geq D[i_p, i_t] + (x - i_p) \cdot d \text{ (Lemma 2)} \\
&\geq D[x - w, i_t] - (x - w - i_p) \cdot d + (x - i_p) \cdot d \\
&= D[x - w, i_t] + w \cdot d
\end{aligned}
$$

$\square$

We say that a cell in a border of a block is *interesting* if its value is at most k. To solve the k-DTW problem it suffices to compute the values of all interesting cells in the last row of D. Consider a block $B = D[i_p .. j_p, i_t .. j_t]$ and recall that the values in it are non-decreasing top to down and left to right (Lemma 2). We can consider the following compact representation of its interesting cells. For an integer ℓ, define $q_{\text{top}}^\ell \in [i_t, j_t]$ to be the last position such that $D[i_p, q_{\text{top}}^\ell] \leq \ell$, and $q_{\text{bot}}^\ell \in [i_t, j_t]$ the last position such that $D[j_p, q_{\text{bot}}^\ell] \leq \ell$. If a value is not defined, we set it equal to $i_t - 1$. Analogously, define $q_{\text{left}}^\ell \in [i_p, j_p]$ to be the last position such that $D[q_{\text{left}}^\ell, i_t] \leq \ell$, and $q_{\text{right}}^\ell \in [i_p, j_p]$ the last position such that $D[q_{\text{right}}^\ell, j_t] \leq \ell$. If a value is not defined, we set it equal to $i_p - 1$. Positions $q_{\text{top}}^0, \ldots, q_{\text{top}}^k$ uniquely describe the interesting border cells in the top row of B, $q_{\text{bot}}^0, \ldots, q_{\text{bot}}^k$ in the bottom row, $q_{\text{left}}^0, \ldots, q_{\text{left}}^k$ in the leftmost column, $q_{\text{right}}^0, \ldots, q_{\text{right}}^k$ in the rightmost column.

Lemma 4. *The compact representations of the interesting border cells in the top row and the leftmost column of a block B can be computed in $\mathcal{O}(k)$ time given the compact representation of the interesting border cells in its neighbors.*

Proof. We explain how to compute the representation for the leftmost column of B, the representation for the top row is computed analogously. Let $d = d(P[i_p], T[i_t])$. If $d = 0$ (the block is homogeneous), by Corollary 1 the block is a q-block for some value q which can be computed in $\mathcal{O}(1)$ time by Eq. 1 if it is interesting (and otherwise we have a certificate that the value is not interesting). We can then derive the values q_{left}^{ℓ}, $\ell = 0, 1, \ldots, k$ in $\mathcal{O}(k)$ time.

Assume now $d > 0$. We start by computing $D[i_p, i_t]$ using Eq. 1. We note that if $D[i_p, i_t] \leq k$, then we know the values of its neighbors realizing it and therefore can compute it, otherwise we can certify that $D[i_p, i_t] > k$. Assume $D[i_p, i_t] = v$, which implies that $q_{\text{left}}^0, \ldots, q_{\text{left}}^{\min\{k,v\}-1}$ equal $i_p - 1$. We must now compute $q_{\text{left}}^{\min\{k,v\}}, \ldots, q_{\text{left}}^k$. Consider a cell (q, i_t) of the block with $q > i_p$. The second to the last cell in the warping path that realizes $D[q, i_t] = \ell$ is one of the cells $(q-1, i_t)$, $(q-1, i_t - 1)$ or $(q, i_t - 1)$, and the value of the path up to there must be $\ell - d$. Note that all the three cells belong either to the leftmost column of B, or the rightmost column of its left neighbor. Consequently, for all $\min\{k, v\} < \ell \leq k$, we have $q_{\text{left}}^{\ell} = \min\{\max\{q_{\text{left}}^{\ell-d}, r_{\text{right}}^{\ell-d}\} + 1\}, j_t\}$, and the positions $q_{\text{left}}^0, \ldots, q_{\text{left}}^k$ can be computed in $\mathcal{O}(k)$ time.   □

Lemma 5. *The compact representations of the interesting border cells in the bottom row and the rightmost column of a block B can be computed in $\mathcal{O}(k)$ time given the compact representation of the interesting border cells in its leftmost column and the top row.*

Proof. We explain how to compute the representation for the bottom row, the representation for the rightmost column is computed analogously.

Equation 3 and the compact representations of the leftmost column and the top row of B partition the bottom row of B into $\mathcal{O}(k)$ intervals (some intervals can be empty), and in each interval the values are described either as a constant or as a linear function. (See Fig. 3.) Formally, let $h = j_p - i_p$. By Eq. 3, for $y \in [i_t, j_p + i_t - q_{\text{left}}^k - 1] \cap [i_t, j_t]$ we have $D[j_p][y] > k$. For $y \in [j_p + i_t - q_{\text{left}}^{\ell-1} - 1] \cap [i_t, j_t]$, $\ell = k, k-1, \ldots, 1$, we have $D[j_p][y] = \ell + (y - i_t) \cdot d$. For $y \in [j_p + i_t - q_{\text{left}}^0, j_p + i_t - i_p] \cap [i_t, j_t]$ we have $D[j_p][y] = (y - i_t) \cdot d$. For $y \in [i_t + h, q_{\text{top}}^0 + h - 1] \cap [i_t, j_t]$ we have $D[j_p][y] = h \cdot d$. For $y \in [q_{\text{top}}^{\ell} + h, q_{\text{top}}^{\ell+1} + h - 1] \cap [i_t, j_t]$, $\ell = 0, 1, \ldots, k-1$, we have $D[j_p][y] = \ell + h \cdot d$. Finally, for $y \in [q_{\text{top}}^k + h, j_t]$, there is $D[j_p][y] > k$ again.

By Lemma 2, the values in the bottom row are non-decreasing. We scan the intervals from left to right to compute the values $q_{\text{bot}}^0, \ldots, q_{\text{bot}}^k$ in $\mathcal{O}(k)$ time. In more detail, let q_{bot}^{ℓ} be the last computed value, and $[i, j]$ be the next interval. We set $q_{\text{bot}}^{\ell+1} = q_{\text{bot}}^{\ell}$. If the values in the interval are constant and larger than $\ell + 1$, we continue to computing $q_{\text{bot}}^{\ell+2}$. If the values are increasing linearly, we find the position of the last value smaller or equal to $\ell + 1$, set $q_{\text{bot}}^{\ell+1}$ equal to this position, and continue to computing $q_{\text{bot}}^{\ell+2}$. Finally, if the values in the interval are constant and equal to $\ell + 1$, we update $q_{\text{bot}}^{\ell+1} = j$ and continue to the next interval. As soon as q_{bot}^k is computed, we stop the computation.   □

Since there are $\mathcal{O}(mn)$ blocks in total, Lemmas 4 and 5 immediately imply Theorem 1.

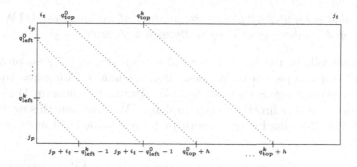

Fig. 3. Compressed representation of interesting border cells.

4 Approximation Algorithm

In this section, we show an approximation algorithm for computing the smallest DTW distance between a pattern P and a substring of a text T. We assume that the DTW distance is defined over a metric on the alphabet Σ. Kuszmaul [17] showed that the problem of computing the smallest DTW distance over an arbitrary metric can be reduced to the problem of computing the smallest distance over a so-called well-separated tree metric:

Definition 2 (Well-separated tree metric). *Consider a rooted tree τ with positive weights on the edges whose leaves form an alphabet Σ. The tree τ specifies a metric μ_τ on Σ: The distance between two leaves $a, b \in \Sigma$ is defined as the maximum weight of an edge in the shortest path from a to b. The metric μ_τ is a well-separated tree metric if the weights of the edges are not increasing in every root-to-leaf path. The depth of μ_τ is defined to be the depth of τ.*

Below we show that Theorem 1 implies the following result for well-separated tree metrics:

Lemma 6. *Given run-length encodings of a pattern P with m runs and a text T with n runs over an alphabet Σ. Assume that the DTW distance is specified by a well-separated tree metric μ_τ on Σ with depth h, and suppose that the ratio between the largest and the smallest non-zero distances between the letters of Σ is at most exponential in $L = \max\{|P|, |T|\}$. For any $0 < \epsilon < 1$, there is an $\mathcal{O}(L^{1-\epsilon} \cdot hmn \log L)$-time algorithm that computes $\mathcal{O}(L^\epsilon)$-approximation of the smallest DTW distance between P and a substring of T.*

By plugging the lemma into the framework of [17], we obtain:

Theorem 2. *Given run-length encodings of a pattern P with m runs and of a text T with n runs over an alphabet Σ. Assume that the DTW distance is specified by a metric μ on Σ, and suppose that the ratio between the largest and the smallest non-zero distances between the letters of Σ is at most exponential in $L = \max\{|P|, |T|\}$. For any $0 < \epsilon < 1$, there is a $\mathcal{O}(L^{1-\epsilon} \cdot mn \log^3 L)$-time* *algorithm that computes $\mathcal{O}(L^\varepsilon)$-approximation of the smallest* DTW *distance between P and a substring of T correctly with high probability*[1].

The proof follows the lines of the full version [18] of [17], we provide it in Appendix B for completeness. We now show Lemma 6. Compared to [17], the main technical challenge is that our k-DTW algorithm (Theorem 1) assumes an integer-valued distance function on the alphabet. We overcome this by developing an intermediary 2-approximation algorithm for real-valued distances (see the two claims below).

Proof of Lemma 6. For brevity, let δ be the smallest DTW_{μ_τ} distance between P and a substring of T.

Claim. Let $0 < \varepsilon < 1$. Assume that for all $a, b \in \Sigma$, $a \neq b$, there is $\mu_\tau(a,b) \geq \gamma$ and that the value of $\mu_\tau(a,b)$ can be evaluated in $\mathcal{O}(t)$ time. There is an $\mathcal{O}(L^{1-\varepsilon}tmn)$-time algorithm which either computes a 2-approximation of δ or concludes that it is larger than $\gamma \cdot L^{1-\varepsilon}$.

Proof. Define a new distance function $\mu'_\tau(a,b) = \lceil \mu_\tau(a,b)/\gamma \rceil$. For all $a, b \in \Sigma$, $a \neq b$, we have $\mu_\tau(a,b) \leq \gamma \cdot \mu'_\tau(a,b) \leq \mu_\tau(a,b) + \gamma \leq 2\mu_\tau(a,b)$. Consequently, for all strings X, Y we have $\mathrm{DTW}_{\mu_\tau}(X,Y) \leq \gamma \cdot \mathrm{DTW}_{\mu'_\tau}(X,Y) \leq 2\mathrm{DTW}_{\mu_\tau}(X,Y)$. Let $\delta' = \min_{S- \text{ substring of } T} \min\{2k+1, \mathrm{DTW}_{\mu'_\tau}(P,S)\}$ for $k = L^{1-\varepsilon}$. By Theorem 1, it can be computed in $\mathcal{O}(L^{1-\varepsilon}tmn)$ time. If $\delta' = 2L^{1-\varepsilon}+1$, we conclude that $\delta \geq \gamma \cdot L^{1-\varepsilon}$, and otherwise, output $\gamma\delta'$.   □

W.l.o.g., the minimum non-zero distance between two distinct letters of Σ is 1 and the largest distance is some value M, which is at most exponential in L. We run the algorithm above for $\gamma = 1$, which either computes a 2-approximation of δ which we can output immediately, or concludes that $\delta \geq L^{1-\varepsilon}$. Below we assume that $\delta \geq L^{1-\varepsilon}$.

Definition 3 (r-simplification). *For a string $X \in \Sigma^*$ and $r \geq 1$, the r-simplification $s_r(X)$ is constructed by replacing each letter a of X with its highest ancestor a' in τ that can be reached from a using only edges of weight $\leq r/4$.*

Fact 3 (Corollary of [17, Lemma 4.6], see also [4]). *For all $X, Y \in \Sigma^{\leq L}$, the following properties hold:*

1. $\mathrm{DTW}_{\mu_\tau}(s_r(X), s_r(Y)) \leq \mathrm{DTW}_{\mu_\tau}(X,Y)$.
2. *If* $\mathrm{DTW}_{\mu_\tau}(X,Y) > Lr$, *then* $\mathrm{DTW}_{\mu_\tau}(s_r(X), s_r(Y)) > Lr/2$.

Fix $r \geq 1$ and $0 < \varepsilon < 1$. In the (L^ε, r)-DTW *gap pattern matching problem*, we must output 0 if the smallest DTW distance between P and a substring of T is at most $L^{1-\varepsilon}r/4$ and 1 if it is at least Lr, otherwise we can output either 0 or 1.

---

[1] The preprocessing time $\mathcal{O}(|\Sigma|^2 \log L)$ that is required to embed μ into a well-separated metric is not accounted for in the runtime of the algorithm.

Claim. The (L^ε, r)-DTW gap pattern matching problem can be solved in $\mathcal{O}(L^{1-\varepsilon} \cdot hmn)$ time.

Proof. Let δ_r be the smallest DTW_{μ_r} distance between $s_r(P)$ and a substring of $s_r(T)$. If $L^{1-\varepsilon} > L/2$, then $L = \mathcal{O}(1)$ and we can compute δ exactly in $\mathcal{O}(1)$ time by Eq. 1. Otherwise, we run the 2-approximation algorithm for $\gamma = r/4$, which takes $O(L^{1-\varepsilon} \cdot hmn)$ time (we can evaluate the distance between two letters in $\mathcal{O}(h)$ time). If the algorithm concludes that $\delta_r > L^{1-\varepsilon}r/4$, then $\delta > L^{1-\varepsilon}r/4$ by Fact 3, and we can output 1. Otherwise, the algorithm outputs a 2-approximation δ'_r of δ_r, i.e. $\delta_r \leq \delta'_r \leq 2\delta_r$. If $\delta'_r \leq L^{1-\varepsilon}r \leq Lr/2$, then we have $\delta_r \leq Lr/2$. Therefore, $\delta \leq Lr$ by Fact 3 and we can output 0. Otherwise, $\delta \geq \delta_r \geq \delta'_r/2 > L^{1-\varepsilon}r/2 > L^{1-\varepsilon}r/4$, and we can output 1.   □

Consider the $(L^\varepsilon/2, 2^i)$-DTW gap pattern matching problem for $0 \leq i \leq \lceil \log ML \rceil$. If the $(L^\varepsilon/2, 2^0)$-DTW gap pattern matching problem returns 0, then we know that $\delta \leq L$, and can return $L^{1-\varepsilon}$ as a L^ε-approximation for δ. Therefore, it suffices to consider the case where the $(L^\varepsilon/2, 2^0)$-DTW gap pattern matching problem returns 1. We can assume, without computing it, that the $(L^\varepsilon/2, 2^{\lceil \log ML \rceil})$-DTW gap pattern matching returns 0 as $\delta \leq ML$. Consequently, there must exist i^* such that $(L^\varepsilon/2, 2^{i^*-1})$-DTW gap pattern matching returns 1 and $(L^\varepsilon/2, 2^{i^*-1})$-DTW returns 0. We can find i^* by a binary search which takes $\mathcal{O}(L^{1-\varepsilon}hmn \log \log ML) = \mathcal{O}(L^{1-\varepsilon}hmn \log L)$ time. We have $\delta \geq 2^{i^*-1}L^{1-\varepsilon}/4$ and $\delta \leq 2^{i^*}L$, and therefore can return $2^{i^*-1}L^{1-\varepsilon}/4$ as a $\mathcal{O}(L^\varepsilon)$-approximation of δ.   □

5   Experiments

This section provides evidence of the advantage of the DTW distance over the edit distance when processing the third generation sequencing (TGS) data. Our experiment compares how the two distances are affected by biological mutation as opposed to sequencing errors, including homopolymer length errors.

We first simulate two genomes, G and G', which can be considered as strings on the alphabet $\Sigma = \{A, C, G, T\}$. The genome G is a substring of the E.coli genome (strain SQ110, NCBI Reference Sequence: NZ_CP011322.1) of length 10000 (positions 100000 to 110000, excluded). The genome G' is obtained from G by simulating biological mutations, where the probabilities are chosen according to [6]. The algorithm initializes G' as the empty string, and pos $= 1$. While pos $\leq |G|$ it executes the following:

1. With probability 0.01, simulate a substitution: chose uniformly at random $a \in \Sigma$, $a \neq G[pos]$. Set $G' = G'a$ and pos $=$ pos $+ 1$.
2. Else, with probability 0.0005 simulate an insertion or a deletion of a substring of length x, where x is chosen uniformly at random from an interval $[1, \mathtt{max\_len\_ID}]$, where $\mathtt{max\_len\_ID}$ is fixed to 10 in the experiments:
   (a) With probability 0.5, set pos $=$ pos $+ x + 1$ (deletion);

   (b) With probability 0.5, choose a string $X \in \Sigma^x$ uniformly at random, set
       $G' = G'X$ and pos = pos + 1 (insertion).
3. Else, set $G' = G'G[\text{pos}]$ and pos = pos + 1.

To simulate reads, we extract substrings of G' and add sequencing errors:

1. For each read, extract a substring R of length 500 at a random position of G'.
   As G' originates from G, we know the theoretical distance from R to G, which
   we call the *"biological diversity"*. The biological diversity is computed as the
   sum of the number of letter substitutions, letter insertions, and letter deletions
   that were applied to the original substring from G to obtain R.
2. Add sequencing errors by executing the following for each position i of R:
   (a) With probability 0.001, substitute $R[i]$ with a letter $a \in \Sigma$, $a \neq R[i]$. The
       letter a is chosen uniformly at random.
   (b) If $R[i] = R[i-1]$, insert with a probability p_{hom} a third occurrence of the
       same letter to simulate a homopolymer error.

   Figure 4 shows the difference between the biological diversity and the smallest edit and DTW distances between a generated read and a substring of G
depending on p_{hom}. It can be seen that the DTW distance gives a good estimation of the biological diversity, whereas, as expected, the edit distance is heavily
affected by homopolymer errors. To ensure reproducibility of our results, our
complete experimental setup is available at https://github.com/fnareoh/DTW.

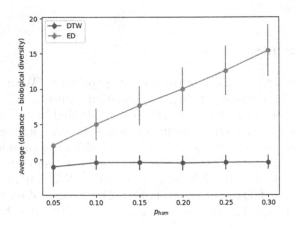

Fig. 4. Edit and DTW distances offset by the biological diversity as a function of p_{hom}.
Each point is averaged over 600 reads ($\times 30$ coverage).

Appendix A

Lemma 2. *Consider a block $B = D[i_p .. j_p, i_t .. j_t]$ and cell (a,b) in it. If $i_p \leq a < j_p$, then $D[a,b] \leq D[a+1,b]$ and if $i_t \leq b < j_t$, then $D[a,b] \leq D[a,b+1]$.*

Proof. Let us first give an equivalent statement of the lemma: if (a, b) and $(a + 1, b)$ are in the same block, then $D[a, b] \leq D[a + 1, b]$, and if (a, b) and $(a, b + 1)$ are in the same block, then $D[a, b] \leq D[a, b + 1]$.

We show the lemma by induction on $a + b$. The base of the induction are the cells such that $a = 0$ or $b = 0$, and for them the statement holds by the definition of D. Consider now a cell (a, b), where $a, b \geq 1$. Assume that the induction assumption holds for all cells (x, y) such that $x + y < a + b$. By Eq. 1, we have:

$$D[a, b] = \min\{D[a - 1, b - 1], D[a - 1, b], D[a, b - 1]\} + d$$
$$D[a + 1, b] = \min\{D[a, b - 1], D[a, b], D[a + 1, b - 1]\} + d$$
$$D[a, b + 1] = \min\{D[a - 1, b], D[a - 1, b + 1], D[a, b]\} + d$$

Assume that (a, b) and $(a + 1, b)$ are in the same block. We have $D[a, b] \leq D[a, b - 1] + d$ and trivially $D[a, b] \leq D[a, b] + d$. By the induction assumption, $D[a, b - 1] \leq D[a + 1, b - 1]$ (the cells $(a, b - 1)$ and $(a + 1, b - 1)$ must belong to the same block). Therefore,

$$D[a + 1, b] = \min\{D[a, b - 1], D[a, b], D[a + 1, b - 1]\} + d$$
$$= \min\{D[a, b - 1] + d, D[a, b] + d, D[a + 1, b - 1] + d\}$$
$$\geq \min\{D[a, b], D[a, b], D[a, b - 1] + d\}$$
$$\geq \min\{D[a, b], D[a, b], D[a, b]\} = D[a, b].$$

Assume now that (a, b) and $(a, b + 1)$ are in the same block. We have $D[a, b] \leq D[a - 1, b] + d$. Furthermore, as $(a - 1, b)$ and $(a - 1, b + 1)$ are in the same block, we have $D[a - 1, b] \leq D[a - 1, b + 1]$ by the induction assumption. Therefore,

$$D[a, b + 1] = \min\{D[a - 1, b], D[a - 1, b + 1], D[a, b]\} + d$$
$$= \min\{D[a - 1, b] + d, D[a - 1, b + 1] + d, D[a, b] + d\}$$
$$\geq \min\{D[a - 1, b] + d, D[a - 1, b] + d, D[a, b]\}$$
$$\geq \min\{D[a, b], D[a, b], D[a, b]\} = D[a, b].$$

This concludes the proof of the lemma. □

Appendix B

Theorem 2. *Given run-length encodings of a pattern P with m runs and of a text T with n runs over an alphabet Σ. Assume that the DTW distance is specified by a metric μ on Σ, and suppose that the ratio between the largest and the smallest non-zero distances between the letters of Σ is at most exponential in $L = \max\{|P|, |T|\}$. For any $0 < \epsilon < 1$, there is a $\mathcal{O}(L^{1-\varepsilon} \cdot mn \log^3 L)$-time algorithm that computes $\mathcal{O}(L^\varepsilon)$-approximation of the smallest DTW distance between P and a substring of T correctly with high probability (See Footnote 1).*

Proof. Any metric μ can be embedded in $\mathcal{O}(\sigma^2)$ time into a well-separated tree metric μ_τ of depth $\mathcal{O}(\log \sigma)$ with expected distortion $\mathcal{O}(\log \sigma)$ (see [10] and [3, Theorem 2.4]). Furthermore, the ratio between the smallest distance and the largest distance grows at most polynomially. Formally, for any two letters a, b we have $\mu(a,b) \leq \mu_\tau(a,b)$ and $\mathbb{E}(\mu_\tau(a,b)) \leq \mathcal{O}(\log \sigma) \cdot d(a,b)$. Therefore, we have:

$$\mathrm{DTW}_\mu(X,Y) \leq \mathrm{DTW}_{\mu_\tau}(X,Y) \tag{4}$$

$$\mathbb{E}(\mathrm{DTW}_{\mu_\tau}(X,Y)) \leq \mathcal{O}(\log \sigma) \cdot \mathrm{DTW}_\mu(X,Y) \tag{5}$$

Let $\delta = \min_{S-\text{ substr. of } T} \mathrm{DTW}_\mu(P,S)$ and $\delta_\tau = \min_{S-\text{ substr. of } T} \mathrm{DTW}_{\mu_\tau}(P,S)$. Assume that δ is realised on a substring X, and δ_τ on a substring X_τ. By Eq. 4, we then obtain:

$$\delta = \mathrm{DTW}_\mu(P,X) \leq \mathrm{DTW}_\mu(P,X_\tau) \leq \delta_\tau$$

And Eq. 5 gives the following:

$$\mathbb{E}(\delta_\tau) \leq \mathbb{E}(\mathrm{DTW}_{\mu_\tau}(P,X)) \leq \mathcal{O}(\log \sigma) \cdot \mathrm{DTW}_\mu(P,X) = \mathcal{O}(\log \sigma) \cdot \delta$$

We apply the embedding $\log L$ times independently to obtain well-separated tree metrics μ_τ^i, $i = 1, 2, \ldots, \log L$. From above and by Chernoff bounds,

$$\min_i \min_{S-\text{ substring of } T} \mathrm{DTW}_{\mu_\tau}^i(P,S)$$

gives an $\mathcal{O}(\log \sigma) = \mathcal{O}(\log L)$ approximation of δ with high probability and can be computed in time $\mathcal{O}(L^{1-\varepsilon} \cdot mn \log^3 L)$ by Lemma 6, concluding the proof of the theorem. $\square$

References

1. Abboud, A., Backurs, A., Williams, V.V.: Tight hardness results for LCS and other sequence similarity measures. In: FOCS 2015, pp. 59–78. IEEE Computer Society (2015). https://doi.org/10.1109/FOCS.2015.14
2. Amarasinghe, S.L., Su, S., Dong, X., Zappia, L., Ritchie, M.E., Gouil, Q.: Opportunities and challenges in long-read sequencing data analysis. Genome Biol. **21**(1), 1–16 (2020)
3. Bansal, N., Buchbinder, N., Madry, A., Naor, J.: A polylogarithmic-competitive algorithm for the k-server problem. In: FOCS 2011, pp. 267–276 (2011). https://doi.org/10.1109/FOCS.2011.63
4. Braverman, V., Charikar, M., Kuszmaul, W., Woodruff, D.P., Yang, L.F.: The one-way communication complexity of dynamic time warping distance. In: SoCG 2019. LIPIcs, vol. 129, pp. 16:1–16:15 (2019). https://doi.org/10.4230/LIPIcs.SoCG.2019.16
5. Bringmann, K., Künnemann, M.: Quadratic conditional lower bounds for string problems and dynamic time warping. In: FOCS 2015, pp. 79–97 (2015). https://doi.org/10.1109/FOCS.2015.15

6. Chen, J.Q., Wu, Y., Yang, H., Bergelson, J., Kreitman, M., Tian, D.: Variation in the ratio of nucleotide substitution and indel rates across genomes in mammals and bacteria. Mol. Biol. Evol. **26**(7), 1523–1531 (2009). https://doi.org/10.1093/molbev/msp063

7. Driemel, A., Silvestri, F.: Locality-sensitive hashing of curves. In: SoCG 2017. LIPIcs, vol. 77, pp. 37:1–37:16 (2017). https://doi.org/10.4230/LIPIcs.SoCG.2017.37

8. Dupont, M., Marteau, P.-F.: Coarse-DTW for sparse time series alignment. In: Douzal-Chouakria, A., Vilar, J.A., Marteau, P.-F. (eds.) AALTD 2015. LNCS (LNAI), vol. 9785, pp. 157–172. Springer, Cham (2016). https://doi.org/10.1007/978-3-319-44412-3_11

9. Emiris, I.Z., Psarros, I.: Products of euclidean metrics and applications to proximity questions among curves. In: SoCG 2018. LIPIcs, vol. 99, pp. 37:1–37:13 (2018). https://doi.org/10.4230/LIPIcs.SoCG.2018.37

10. Fakcharoenphol, J., Rao, S., Talwar, K.: A tight bound on approximating arbitrary metrics by tree metrics. In: STOC 2003, pp. 448–455 (2003). https://doi.org/10.1145/780542.780608

11. Froese, V., Jain, B.J., Rymar, M., Weller, M.: Fast exact dynamic time warping on run-length encoded time series. CoRR abs/1903.03003 (2019)

12. Gold, O., Sharir, M.: Dynamic time warping and geometric edit distance: breaking the quadratic barrier. ACM Trans. Algorithms **14**(4), 50:1–50:17 (2018). https://doi.org/10.1145/3230734

13. Gonzalez-Garay, M.L.: Introduction to isoform sequencing using pacific biosciences technology (Iso-Seq). In: Wu, J. (ed.) Transcriptomics and Gene Regulation. TRBIO, vol. 9, pp. 141–160. Springer, Dordrecht (2016). https://doi.org/10.1007/978-94-017-7450-5_6

14. Huang, Y.T., Liu, P.Y., Shih, P.W.: Homopolish: a method for the removal of systematic errors in nanopore sequencing by homologous polishing. Genome Biol. **22**(1), 95 (2021). https://doi.org/10.1186/s13059-021-02282-6

15. Hwang, Y., Gelfand, S.B.: Sparse dynamic time warping. In: Perner, P. (ed.) MLDM 2017. LNCS (LNAI), vol. 10358, pp. 163–175. Springer, Cham (2017). https://doi.org/10.1007/978-3-319-62416-7_12

16. Hwang, Y., Gelfand, S.B.: Binary sparse dynamic time warping. In: MLDM 2019, pp. 748–759. ibai Publishing (2019)

17. Kuszmaul, W.: Dynamic time warping in strongly subquadratic time: algorithms for the low-distance regime and approximate evaluation. In: ICALP 2019. LIPIcs, vol. 132, pp. 80:1–80:15 (2019). https://doi.org/10.4230/LIPIcs.ICALP.2019.80

18. Kuszmaul, W.: Dynamic time warping in strongly subquadratic time: algorithms for the low-distance regime and approximate evaluation. CoRR abs/1904.09690 (2019). https://doi.org/10.48550/ARXIV.1904.09690

19. Kuszmaul, W.: Binary dynamic time warping in linear time. CoRR abs/2101.01108 (2021)

20. Landau, G.M., Myers, E.W., Schmidt, J.P.: Incremental string comparison. SIAM J. Comput. **27**(2), 557–582 (1998). https://doi.org/10.1137/S0097539794264810

21. Landau, G.M., Vishkin, U.: Fast string matching with k differences. J. Comput. Syst. Sci. **37**(1), 63–78 (1988). https://doi.org/10.1016/0022-0000(88)90045-1

22. Li, H.: Minimap2: pairwise alignment for nucleotide sequences. Bioinformatics **34**(18), 3094–3100 (2018). https://doi.org/10.1093/bioinformatics/bty191

23. Mahmoud, M., Gobet, N., Cruz-Dávalos, D.I., Mounier, N., Dessimoz, C., Sedlazeck, F.J.: Structural variant calling: the long and the short of it. Genome Biol. **20**(1), 1–14 (2019). https://doi.org/10.1186/s13059-019-1828-7

24. Mueen, A., Chavoshi, N., Abu-El-Rub, N., Hamooni, H., Minnich, A.: AWarp: fast warping distance for sparse time series. In: ICDM 2016, pp. 350–359. IEEE (2016)
25. Nishi, A., Nakashima, Y., Inenaga, S., Bannai, H., Takeda, M.: Towards efficient interactive computation of dynamic time warping distance. In: Boucher, C., Thankachan, S.V. (eds.) SPIRE 2020. LNCS, vol. 12303, pp. 27–41. Springer, Cham (2020). https://doi.org/10.1007/978-3-030-59212-7_3
26. Sakai, Y., Inenaga, S.: A reduction of the dynamic time warping distance to the longest increasing subsequence length. In: ISAAC 2020. LIPIcs, vol. 181, pp. 6:1–6:16 (2020). https://doi.org/10.4230/LIPIcs.ISAAC.2020.6
27. Sakoe, H., Chiba, S.: Dynamic programming algorithm optimization for spoken word recognition. IEEE Trans. Acoust. Speech Sig. Process. **26**(1), 43–49 (1978)

Author Index

Printed in the United States
by Baker & Taylor Publisher Services